The Probability Map of the Universe

The Probability Map of the Universe

Essays on David Albert's *Time and Chance*

Edited by
- BARRY LOEWER
- BRAD WESLAKE
- ERIC WINSBERG

HARVARD UNIVERSITY PRESS
Cambridge, Massachusetts & London, England
2023

Copyright © 2023 by the President and Fellows of
Harvard College
All rights reserved
Printed in the United States of America

First printing

Cataloging-in-Publication Data is available from the Library of Congress

ISBN: 978-0-674-96787-8 (alk. paper)

Contents

Introduction   *Barry Loewer, Brad Weslake, and Eric Winsberg*   1

I. OVERVIEW OF *TIME AND CHANCE*

1. The Mentaculus: A Probability Map of the Universe   *Barry Loewer*   13

II. PHILOSOPHICAL FOUNDATIONS

2. The Metaphysical Foundations of Statistical Mechanics: On the Status of PROB and PH   *Eric Winsberg*   57
3. The Logic of the Past Hypothesis   *David Wallace*   76
4. In What Sense Is the Early Universe Fine-Tuned?   *Sean M. Carroll*   110
5. The Meta-Reversibility Objection   *Christopher J. G. Meacham*   142
6. Typicality versus Humean Probabilities as the Foundation of Statistical Mechanics   *Dustin Lazarovici*   178
7. The Past Hypothesis and the Nature of Physical Laws   *Eddy Keming Chen*   204
8. On the Albertian Demon   *Tim Maudlin*   249

III. UNDERWRITING THE ASYMMETRIES OF KNOWLEDGE AND INTERVENTION

9. Reading the Past in the Present   *Nick Huggett*   271
10. Causes, Randomness, and the Past Hypothesis   *Mathias Frisch*   294
11. Time, Flies, and Why We Can't Control the Past   *Alison Fernandes*   312
12. The Concept of Intervention in *Time and Chance*   *Sidney Felder*   335

Conclusion   *David Z Albert*   351

Contributors   *377*
Index   *379*

The Probability Map of the Universe

Introduction

▸ BARRY LOEWER, BRAD WESLAKE,
and ERIC WINSBERG

David Albert's *Time and Chance* (Harvard University Press, 2000) is a tersely brilliant book. It is at once precise and colloquial, at once deep and broad. It addresses issues at the traditional core of the foundations of statistical mechanics, such as the reversibility objections and Maxwell's demon, and also more general philosophical issues concerning the nature of time, the temporal asymmetries of knowledge and influence, the metaphysics of laws, chances and time, and the relationship between fundamental physics and the special sciences. In 2009 Barry Loewer organized a conference on the book at the Rutgers University Center for Philosophy and the Sciences that featured many of the papers included in this volume, and where the idea for this volume was first conceived. We believe that the contributions reflect the galvanizing influence David has had on work on these topics, and hope that this volume will help to stimulate more work along the same lines.

Background

For those who are unfamiliar with statistical mechanics and its philosophical foundations, here is a brief background. For a much more thorough discussion, one cannot do better than to read David Albert's *Time and Chance*.

The science of thermodynamics developed during the nineteenth century with the aim of improving the efficiency of steam engines. Steam engines operate by converting the potential chemical energy of coal into heat, which increases the pressure in a cylinder containing a piston, which in turn does the work of driving the piston that causes a train's wheels to turn. Thermodynamics concerns regularities among the macroscopic quantities heat, energy, work, temperature, pressure, mass density, and so on. A simple example is that when a hot body and a cold body are placed next to each other, heat flows from the hotter to the colder until their temperatures are equal.

Heat is a form of energy. In all thermodynamic processes energy is conserved, but sometimes energy is available to do work, such as move a piston in a steam engine, and sometimes energy is useless for performing work. For example, the heat produced by the friction of a train's wheels on the track dissipates and is no longer useful. A measure of the energy in a system that is no longer available for work in that system is called the system's "entropy." It was discovered that in any thermodynamic interaction in a closed system, entropy never decreases and typically increases. This is the "second law" of thermodynamics. It specifies a particular temporal direction in which thermodynamic processes evolve; entropy increases toward the future. This is quite different from the fundamental laws of classical mechanics. In classical mechanics the state of a system is specified by specifying the positions and momenta of its constituent particles. The classical mechanical laws are deterministic and time reversal invariant. Determinism means that, for a closed system, the mechanical laws and the state of that system at a time t entail the state of that system at any other time. Temporal invariance means that for any sequence of the positions of particles of a mechanical system that are compatible with the laws, the temporally reversed sequence is also compatible with the laws. For example, from watching a video of, say, billiard balls moving on a frictionless table in accordance with classical mechanical laws, one couldn't tell whether the video was being run forward or backward. Classical mechanics (and the same is true for quantum mechanics) has no preferred direction of time.

In the latter half of the nineteenth century, it became clear that matter is composed of particles and that the thermodynamic quantities of a system could be identified with the motions of the system's constituent particles (molecules). For example, heat is identified with the total kinetic energy of the molecules, and temperature with their average kinetic energy. So when, for example, heat "flows" from a hot body to a cold body, it is the result of their constituent particles colliding and exchanging momentum. The entropy of a system was identified with the logarithm of the number (or measure) of the classical mechanical states compatible with the macroscopic quantities of the system. So, when heat "flows" from the hot body to the cold body, entropy increases because there are more (or greater-measure) classical mechanical states compatible with the system at the same temperature than when the bodies are at different temperatures.

But now there is a problem. If macroscopic systems are composed of particles whose motions are governed by temporally invariant laws, how can thermodynamics include a law that specifies a temporal orientation? Statistical mechanics attempts to answer this question. Boltzmann took the first step by recasting the second law as a probabilistic law that says that the entropy of a closed system is very likely to increase. This modification is needed because there are some rare microstates compatible with a system's macrostate whose evolution results in the system's entropy decreasing.

Boltzmann recast the second law by specifying a particular probability distribution, which Albert calls "the statistical postulate" (SP), over the microstates compatible with a system's macrostate and attempted to demonstrate that if a system's entropy is not maximum, it is very likely to increase. But due to the temporal invariance of the classical mechanical laws, it will also follow that a system's entropy was greater in the past. For example, if Boltzmann's account correctly implies that an ice cube in warm water will melt (its entropy will increase), it will also imply that in the past the ice cube was smaller and the water cooler. This is called the reversibility problem.

The response is to assume that when applying statistical mechanics to a system, the system was at lower entropy in the past. Applied to the whole universe, this means that we must assume that the entropy of the universe at the time of the Big Bang was very small. David Albert calls this assumption "the Past Hypothesis" (PH). The project he pursues in *Time and Chance* is to show that the package consisting of the PH, the SP, and the dynamical laws plausibly grounds statistical mechanics and thermodynamics as applied not only to the whole universe but also to its subsystems—ice cubes in warm water, steam engines, and so on.

Our world is full of temporally asymmetric processes—from biological processes like growing old, to chemical processes like combustion, to cosmological processes like the evolution of stars—that involve the increase of entropy and are subsumed by the second law. It is plausible that the PH is also ultimately responsible for these temporal asymmetries. Further, Albert argues that other "arrows of time"—that there are records of the past but not the future, that we can influence the future but not the past, that causes typically are prior to their effects—are also due to the PH. The view of time that emerges from this is that the difference between past and future is not a fundamental feature of time itself but instead is due to the asymmetries of processes in time, and that these are due to the PH.

If this is correct, then the PH doesn't presuppose a distinction between past and future but instead earns its name as "the past" hypothesis by explaining the arrows of time.

This book is divided into three parts: (I) Overview of *Time and Chance*, (II) Philosophical Foundations, and (III) Underwriting the Asymmetries of Knowledge and Intervention. Below are abstracts of the chapters in each part.

Part I: Overview of *Time and Chance*

In Chapter 1, "The Mentaculus: A Probability Map of the Universe," Barry Loewer provides a synopsis and elaboration of one of the central threads of the book. Loewer presents Albert's account of statistical mechanics, explains how in principle it constitutes a complete scientific account of the world (the "Mentaculus"), describes the reversibility paradox, and sketches the way in which the Mentaculus promises to provide explanations for the asymmetries of knowledge, control, counterfactuals, and causation. The Mentaculus contains three components: the fundamental dynamical laws, a probability distribution (PROB), and the Past Hypothesis (PH), a claim about the initial state of the universe. Loewer concludes his chapter by considering three issues: the nature of the probabilities in PROB, the status of PH, and the Boltzmann brains problem. Loewer argues that a best-system theory of probability fits most naturally with the Mentaculus, that PH is a law that does not require (though may ultimately receive) an explanation, and that while the Mentaculus does entail that most brains are Boltzmann brains, it is not likely that *we* are Boltzmann brains.

Part II: Philosophical Foundations

The seven chapters in Part II elaborate or criticize the philosophical foundations of the program of *Time and Chance*. These foundations involve the metaphysics of laws and objective chance, the status of the Past Hypothesis, explanation in statistical mechanics, connections between statistical mechanics and cosmology, and Maxwell's demon.

In Chapter 2, "The Metaphysical Foundations of Statistical Mechanics: On the Status of PROB and PH," Eric Winsberg presents a series of objections to the picture sketched by Loewer. Winsberg's main target is the claim that PH and PROB emerge as laws from a best-system theory of laws. Winsberg distinguishes three ways of formulating the role of informativeness and simplicity, and argues that on every possibility there are fatal objections to the idea that a corresponding best-system theory will confer lawhood on PH and PROB. Either it is implausible that they are laws, or the associated best-system theory is an unviable account of laws. Instead, Winsberg suggests, we should understand the probabilities in PROB as objective degrees of belief. Winsberg goes on to raise a novel line of objection to Loewer's proposal that the lawhood of PH and PROB is required to underwrite the lawhood of the special sciences, connecting this issue with the problem of Boltzmann brains.

In Chapter 3, "The Logic of The Past Hypothesis," David Wallace examines Albert's line of reasoning for positing PH in the first place. Wallace argues that our theory of temporally asymmetric macrophysics should explain, not only the qualitative fact *that* isolated systems increase in entropy, but also the quantitative facts concerning *how long* it takes them to do so. Wallace first provides an overview of a range of mathematical techniques by which macrodynamics can be extracted from microdynamics, for both quantum and classical mechanical systems. He then proposes that the origin of irreversibility can be explained by the postulate that the initial state of the universe is *simple,* in a stipulated sense. As he notes, the crucial difference between this and Albert's PH is that the state is not defined in terms of the macrostates the universe is permitted to initially inhabit. Moreover, Wallace argues that his postulate explains more than PH—and, indeed, that it explains the success of PH itself. Finally, Wallace argues that the postulation of PH and PROB are not required to explain the predictive accuracy of macrophysics: they are stronger than his simplicity condition and hence redundant.

In Chapter 4, "In What Sense Is the Early Universe Fine-Tuned?," Sean Carroll is also concerned with the correct characterization of the initial state of the universe, especially with respect to the sense in which it is correct to characterize it as fine-tuned. Carroll argues that the traditional ways of formulating this, in terms of the horizon problem and the flatness problem, are untenable. Instead, he suggests, we should characterize fine-tuning in terms of the space of cosmological trajectories. Given a standard measure, most such trajectories are inhomogeneous, which is inconsistent with what we take ourselves to know about the early universe. Carroll argues that this helps us to see that the inflationary universe hypothesis does not remove the need for fine-tuning, but instead allows us to see the precise sense in which the initial conditions are finely tuned.

In Chapter 5, "The Meta-Reversibility Objection," Chris Meacham develops an argument for construing statistical mechanical probabilities as objective features of the world rather than as subjective measures of indifference. Meacham presents a detailed reconstruction of the reversibility objections, and considers a number of different ways in which one could reply to them, arguing that the only justifiable response is to postulate PH as a law. Meacham then goes on to argue that the appeal to PH in order to resolve the reversibility objection is available only to the proponent of objective chances. The proponent of indifference accounts, on the other hand, faces a further problem, which Meacham calls the meta-reversibility objection. The objection is that the proponent of indifference accounts is committed to distributing credence over initial conditions in a way that undermines the role that PH must play in order to ground the temporal asymmetry of thermodynamics.

In Chapter 6, "Typicality versus Humean Probabilities as the Foundation of Statistical Mechanics," Dustin Lazarovici discusses two contemporary views about the foundation of statistical mechanics and deterministic probabilities in physics: one that regards a measure on the initial macro-region of the universe as a probability measure that is part of the Humean best system of laws (Mentaculus) and another that relates it to the concept of typicality. The first view is tied to Lewis's Principal Principle, the second to a version of Cournot's principle. He defends the typicality view and addresses open questions about typicality and the status of typicality measures.

In Chapter 7, "The Past Hypothesis and the Nature of Physical Laws," Eddy Chen examines the role of the Past Hypothesis in the Boltzmannian account and defends the view that the Past Hypothesis is a candidate fundamental law of nature. Such a view is known to be compatible with Humeanism about laws, but he argues it is also supported by a minimal non-Humean "governing" conception of laws. Some worries arise from the nondynamical and time-dependent character of the Past Hypothesis as a boundary condition, the intrinsic vagueness in its specification, and the nature of the initial probability distribution. He argues that these worries do not have much force, and that in any case they become less relevant in a new quantum framework for analyzing time's arrows—the Wentaculus. Hence, the view that the Past Hypothesis is a candidate fundamental law should be more widely accepted than it is now.

In Chapter 8, "On the Albertian Demon," Tim Maudlin evaluates Albert's discussion of Maxwell's demon. To begin, Maudlin notes that Maxwell is correct to argue that the possibility of such a demon shows that the second law, as Maxwell understood it, cannot be a logical consequence of the dynamical laws. Maudlin then notes Albert's point that a so-called *Laplacian Demon* could arrange for a closed system to decrease in entropy, by causing it to be in one of the measure-zero regions of phase space from which entropy decreases. But such demons cannot, in Albert's terms, *make us any money*, because we have no reassurance that the total available energy in the world has thereby increased. So the natural question to consider is whether there are any demons that can achieve *that* result. Maudlin points out that it is a consequence of a Boltzmannian definition of entropy plus Liouville's theorem, that "no macropredictable system can macropredictably evolve so that its entropy goes down." But—and this is the loophole that Albert exploits in *Time and Chance*—this leaves open the possibility that a non-macropredictable system could so evolve. This could happen if either the demon were not macropredictable, or the remainder of the system were not macropredictable. Maudlin concludes his chapter by criticizing Albert's proposal to exploit this loophole. In effect, Maudlin argues that Albert has given us no reason to believe that the macrotaxonomy with respect to which such demons are possible is one that is connected with the capacity to do work. Maudlin also argues that even if there were a demon that could do work in this way, it would be practically useless in two senses: first, it could not operate on a cycle; and second, the range of final macrostates such a demon would have to be capable of inhabiting would be prohibitively enormous.

Part III: Underwriting the Asymmetries of Knowledge and Intervention

The four chapters in Part III evaluate Albert's explanations of the temporal asymmetries of knowledge, causation, and control, and the implications of this account for issues in epistemology and free will.

In Chapter 9, "Reading the Past in the Present," Nick Huggett examines the knowledge asymmetry. He begins by reconstructing the argument in *Time and Chance*, emphasizing that it is an account of conditions necessary for knowledge rather than an account of how we actually come to know; and emphasizing the distinction between knowledge in principle available from prediction and retrodiction, on the one hand, and knowledge available from inference from records, on the other. Huggett then argues that reflection on simple information-processing systems should lead us to reject the claim that all knowledge of the past is in principle available to one of these methods.

In Chapter 10, "Causes, Randomness, and the Past Hypothesis," Mathias Frisch considers whether the asymmetry of causation can be explained by the Mentaculus. Frisch begins by arguing that causal inference in physics is widespread and typically depends on an assumption of initial micro-randomness. Frisch then considers the relationship between this assumption and the asymmetry of causation. He begins by considering how the Mentaculus is supposed to explain the asymmetry of records, because it is in terms of this that Albert proposes to explain the asymmetry of causation. The central argument here is what Frisch calls "the constraint argument," which Frisch suggests mistakenly conflates information concerning the phase space volume occupied by a system, on the one hand, and information concerning the macroscopic state of the system, on the other. He also suggests that PH is not sufficient for the reliability of records, since typically there are multiple low-entropy initial conditions from which they (and their modally nearby counterparts) could have evolved. Instead, he suggests that PH is at best a necessary condition for the reliability of records, and that the assumption of initial micro-randomness is also necessary. Because Frisch takes this to be a causal assumption, he doubts that a reduction of the asymmetry of causation can succeed.

In Chapter 11, "Time, Flies, and Why We Can't Control the Past," Alison Fernandes presents an objection to Albert's account of the asymmetry of control, and then suggests a reply to the objection. The source of the problem concerns cases where intentions to act can serve as records of past events, in Albert's sense. If we have control over our intentions, then Albert's account entails that we have control over whatever they serve as records for. The solution Fernandes proposes is a condition on agency: agents must not believe that their intentions are correlated with anything the agent believes at the time of action. This, in conjunction with a condition Fernandes calls the "seriality" of agency, promises to explain at least why agents *believe* that the past is not under their control.

In Chapter 12, "The Concept of Intervention in *Time and Chance*," Sidney Felder also considers Albert's account of the asymmetry of control. Like Fernandes in Chapter 11, Felder is concerned to develop an account of this asymmetry that elaborates on Albert's account. In particular, Felder defends a conception of counterfactual dependence, grounded in the Mentaculus, that permits the dependence of both the past and the future on present states of affairs. However, he then argues that the Mentaculus also provides the resources for an account of why only future-directed dependencies are in general available for our exploitation (and the conditions under which past-directed dependencies would be, as in certain Newcomb cases).

Conclusion

The volume concludes with a contribution by David Albert, who responds to the preceding chapters. Albert's responses are clear and to the point; he states which of the worries articulated in the previous chapters he takes seriously, and what might be done about them, and which ones he thinks are misguided. Readers of *Time and Chance* who want to know how Albert's views on the topics covered in that book have evolved in recent years will find this a very worthwhile read.

PART I

OVERVIEW *of TIME AND CHANCE*

Chapter One

The Mentaculus

A Probability Map of the Universe

▸ BARRY LOEWER

I. What Is the Mentaculus?

In *Time and Chance* David Albert presents a vision of a framework for the complete scientific theory of the world. His vision contains the ingredients for an account of how the laws of thermodynamics, other special science laws, objective probabilities, counterfactuals, causation, and time's arrows connect to fundamental physics. It has consequences for a host of issues in the metaphysics of science, including the natures of laws, objective probability, physicalism, the direction of time, and free will. It is an ambitious and breathtaking proposal. Albert's account has come to be known as "the Mentaculus."[1] The name is taken (with permission) from the Coen Brothers film *A Serious Man,* in which one of the characters uses it as the name of a book he calls "a probability map of the universe."

---

[1] The Mentaculus and some of the same issues addressed in this paper are also discussed in Loewer (2007). A similar view about the arrows of time and related matters is developed in Carroll (2010) with somewhat less emphasis on philosophical issues and more on the cosmology associated with the view.

In what follows I will review the Mentaculus, show that it deserves to be taken seriously as a probability map, develop some of its consequences for metaphysics, and then discuss a number of philosophical issues that arise for those who take it seriously.

The Mentaculus consists of three ingredients:

1. The universe's fundamental space-time arena, ontology, and the dynamical laws that describe the evolution of the universe and its isolated subsystems.[2]
2. A law that characterizes the thermodynamic state of the universe around the time of the big bang, M(0), or shortly after; aka "the past hypothesis" (PH). In agreement with contemporary cosmology Albert posits that M(0) is a state whose entropy is very small.
3. A law (the statistical postulate SP) that specifies a uniform probability distribution (specified by the standard Liouville measure) over the physically possible microstates that realize M(0).

"The Mentaculus" is an appropriate name for this theory because the package of the three ingredients above determines a probability density over the set of physically possible trajectories of microstates emanating from M(0) and thereby conditional probabilities P(B / A) over all macroscopic physical propositions B, A for which P(A) > 0; i.e., a probability map of the world. By saying that it is the framework for "a complete scientific theory," I mean that these conditional probabilities are the basis for accounts of thermodynamics, all physical probabilities, special science laws, counterfactuals, causal relations, and the temporal asymmetries exemplified by physical processes. It thus promises to realize Oppenheim and Putnam's (1958) dream of unifying all of science.

A pervasive feature of special science regularities and, indeed, of our ordinary experiences of the world is that they are temporally asymmetric. The granddaddy of temporal asymmetries is the second law of thermodynamics, which in its original formulations says that the entropy of closed systems never decreases but only can increase over time until it reaches a maximum called "equilibrium." Lawful regularities and processes in geology, chemistry, biology, psychology, social sciences, and so on also typically exhibit temporal asymmetries aligned with the

---

[2] While physics and the philosophical foundations of physics have come a long way there, the task of finding a theory that covers all of physical reality is not complete and the interpretations of those theories that have been proposed as partial accounts—quantum field theory, general relativity, string theory, etc.—are controversial. But this incompleteness will not undermine the main points of this paper.

second law. Further temporal asymmetries include the fact that our epistemic access to the past is very different from our epistemic access to the future—we have ability to control the future but never have control over the past—and the temporal asymmetries of counterfactuals and causation.

It has long been known that these temporal asymmetries cannot be due to the fundamental dynamical laws alone, since these laws are temporally symmetric. By this is meant that whenever a sequence of states in one temporal direction is allowed by the laws, another related sequence of states in the reverse temporal direction is also allowed by the laws. In classical mechanics this implies that if a sequence of particle positions is compatible with the laws, then so is the temporally reversed sequence. For example, because Newtonian laws allow for a sequence of the states of particles that realize a diver jumping off a board, leading to her splashing into a pool of water, it also allows for a sequence of states obtained by reversing the velocities of the particles that realize the diver being ejected from the water and landing feet-first back on the board.[3] It follows that knowing the dynamical laws doesn't suffice to tell whether a movie of the diver is being played forward or backward. So, if the fundamental laws don't themselves underwrite a temporal direction, what does? One answer that appeals to some philosophers is that temporal asymmetries result from the nature of time itself. According to this view, time, although it is a dimension like the three spatial dimensions, is special in that it, unlike the spatial dimensions, passes in one direction and thus orients laws to operate and causal processes to proceed in that direction.[4] On some metaphysical views the present in some sense "moves" along the time dimension, fixing the past as it moves, while the future remains open.[5] The Mentaculus offers a different account of time's arrow that doesn't involve time passing or imputing it with a direction. Instead, it proposes that the temporal asymmetries are based on the probability map contained in the Mentaculus. In this way the Mentaculus unifies the arrows of time along with thermodynamics and the special sciences.

The Mentaculus is an implementation of "physicalism." Physicalism is the metaphysical view that all spatiotemporal phenomena, laws, causation, and explanations are ultimately grounded in the laws and ontology of fundamental

---

[3] The velocities of the particles would be in the reverse direction.
[4] An example is Tim Maudlin's view that space-time possesses an intrinsic directionality that distinguishes past and future and that laws act so as to evolve states from past to future but not the reverse. His account of the direction of time is closely connected to his view of laws. See Maudlin (2007) and Loewer (2012).
[5] Presentism and the growing block accounts of time are metaphysical views that can be understood in this way. See Zimmerman (2011) and Tooley (1997). Attributing an intrinsic direction to time doesn't require either of these metaphysical views.

physics. It is not a new view and, except for its consequences for certain mental phenomena, not terribly controversial.[6] What is novel and controversial about the Mentaculus is the role played by the past hypothesis (PH) and the statistical postulate (SP) in characterizing and defending physicalism. Without the PH and SP, it is difficult to see how macroscopic phenomena emerge from fundamental physics. I will argue that once these are added to the dynamical laws, major obstacles to explaining how the macroscopic world emerges from fundamental physics can be overcome.

Before explaining how the Mentaculus works and why it should be taken seriously, I need to set aside one concern. Most of Albert's and my discussion of the Mentaculus assumes that its ontology and dynamics are those of classical physics and that M(0) is characterized along the lines of a classical version of current cosmology—i.e., that shortly after the moment of the big bang the universe consisted of a tiny, rapidly expanding space-time that contained a very dense soup of fields and elementary particles and was enormously hot with almost uniform density and temperature, and whose associated entropy was very low.[7] Classical mechanics has been superseded by quantum field theory, and contemporary cosmology is formulated in terms of general relativity, for which there is no agreed-upon account of entropy and which awaits a theory of quantum gravity for a completely adequate formulation.[8] The worry is that basing the Mentaculus on classical physics may make our discussion moot. But this is not so. Physics has developed to a point where it is plausible that there is a fundamental theory, yet to

---

[6] There are dissenters—for example, Nancy Cartwright (2012), who, though she may hold that all phenomena are physical, argues that they are not unified by physics.

[7] The claim that the entropy of the early universe was very low may seem surprising, since an ordinary gas at a high temperature spread out uniformly in a container is a system whose entropy is high. But while the effects of gravity are minuscule for the gas in the container, in the very dense early universe the contribution of gravity to entropy is significant. It is widely thought that if gravity is taken into account, the state of the early universe is very low entropy. The reason for this is that in the presence of gravity a very dense uniform distribution of matter / energy is very special (i.e., low entropy) and will likely evolve to higher entropy states as matter / energy clump to form stars etc.

[8] John Earman (2006) points out that there is no agreed-upon account of entropy for general relativistic models of the early universe, and so there is a worry that the PH is "not even false." For the beginning of a response, see Callender (2011). And it hasn't prevented Penrose from estimating the entropy S of the very early universe to be about $1088 k_B$ and the entropy of the universe today to be about a quadrillion times as large: $S = 10103 k_B$. Both of these numbers seem large, but the former number is most definitely low-entropy compared to the latter: it's only 0.0000000000001 percent as large! See Penrose (2005), chap. 27, and Carroll (2010) for discussion.

be discovered, that subsumes quantum field theory and general relativity in their respective domains as these two subsume classical mechanics and electromagnetic theory—and that, as Sean Carroll (2021) argues, whatever this theory turns out to be, it is very plausible that it will preserve classical thermodynamics and the main features of statistical mechanics.[9] Further, although the full characterization of the thermodynamic state of the early universe is not completely known and a theory of its microstates is not available, since that involves quantum gravity, enough is known for it to be plausible that the entropy of the early universe is very low. Indeed, it must have been very low for the second law to have held from the earliest times through to the present. This assumption is central to Albert's accounts of how the Mentaculus grounds thermodynamics and temporal asymmetries. So, it is plausible that arguments that the Mentaculus can account for the thermodynamic laws, the other temporal asymmetries, and objective probabilities for macro propositions will carry over to a version containing future proposals for the correct fundamental ontology and dynamical laws and the correct specification of the macrostate of the early universe needed to formulate the PH.

II. Why the Mentaculus Should Be Taken Seriously as a Probability Map of the World

The claim that the Mentaculus is a faithful probability map of the world that grounds thermodynamics, special science regularities, causation, and time's arrows is outrageously ambitious. Why is it to be believed? The line of reasoning supporting it originates in Boltzmann's statistical mechanical account of how the second law of thermodynamics emerges from the fact that physical systems are composed of particles and the temporally symmetric deterministic dynamics that govern them. This account has been known for over a century, but its importance to philosophy has become increasingly apparent recently. The Mentaculus is new only in that it puts Boltzmann's account into a package, draws out its philosophical consequences, and clarifies its philosophical presuppositions.

---

[9] According to Arthur Eddington, "The law that entropy always increases, holds, I think, the supreme position among the laws of Nature. If someone points out to you that your pet theory of the universe is in disagreement with Maxwell's equations—then so much the worse for Maxwell's equations. If it is found to be contradicted by observation—well, these experimentalists do bungle things sometimes. But if your theory is found to be against the second law of thermodynamics, I can give you no hope; there is nothing for it but to collapse in deepest humiliation" (Eddington 1935, p. 53).

Here is a quick review of the Boltzmannian account.[10] Thermodynamics is the science that concerns regularities involving certain macroscopic quantities, including pressure, temperature, work, density, total energy, average frequency of radiation in regions of space, entropy, and so on.[11] Thermodynamic quantities are ubiquitous and can be characterized for all forms of matter and fields. The second law of thermodynamics says that an isolated system evolves over time until it reaches a state of thermodynamic equilibrium.[12] For example, an ice cube placed in a pail of warm water will dissolve until the system is one in which the water is at a uniform temperature. An isolated system in its equilibrium state is in a state of its greatest entropy. The entropy of a system is a function of its thermodynamic quantities that measures how much of its energy can be used to perform work.[13] The original formulations of the second law of thermodynamics say that the entropy of an isolated system never decreases and typically increases over time until the system reaches its equilibrium state, in which its energy can no longer be transformed into work as long as the system remains isolated.[14] The melting of ice, the diffusion of smoke, the formation of stars, and the aging of our bodies are some of the myriad manifestations of the second law. In each instance, as energy is exchanged among parts of the system, its entropy increases until equilibrium is attained.

A question confronting Maxwell and Boltzmann was how fundamental physical processes give rise to the second law. They assumed that material systems—gasses, liquids, solids—are composed of an enormous number of material particles moving "randomly" according to classical mechanical laws. Boltzmann then identified the entropy of a macrostate M (and the entropy of microstates that realize M) with the logarithm of the volume on the standard Lebesgue measure of the set of microstates that realizes M. On this account, the inverse of the entropy of a system's thermodynamic state specifies how orderly the system is and, on a certain understanding of "information," how much information its thermodynamic macrostate contains about the microstate that realizes it. States with greater entropy

---

[10] The best and most accessible accounts of thermodynamics and statistical mechanics for philosophers are Albert's *Time and Chance* (2000) and Sklar's *Physics and Chance* (1993).

[11] An example of a thermodynamics regularity is the ideal gas law $PV = kT$.

[12] The first law says that the total energy of an isolated system, potential plus kinetic energy, remains constant.

[13] A system performs work on another system when it transfers energy to that system.

[14] The entropy of a macro condition M is given by $SB(M(X)) = k \log |\Gamma M|$, where $|\Gamma M|$ is the volume (on the measure) in Γ associated with the macrostate M, and k is Boltzmann's constant. SB provides a relative measure of the amount of Γ corresponding to each M. Given a partition into macrostates, the entropy of a microstate relative to this partition is the entropy of the macrostate that it realizes.

are less informative than states with less entropy. A system's equilibrium state is realized by the largest number by far of microstates. Or rather, because infinitely many microstates realize a macrostate, by the measure of the volume of the phase space of microstates that realize it, and so contains the least information about its microstate.[15] Boltzmann then interpreted the measure also as specifying a uniform probability measure (the Liouville measure) over the space of all microstates. This captures the idea that the particles are randomly distributed and moving randomly. He and others identified macroscopic thermodynamic quantities with probabilistic functions of fundamental quantities and then argued that probabilistic versions of thermodynamics laws, including the second law, followed.[16] The underlying idea of their arguments, in a nutshell, is that overwhelmingly most (on the Liouville measure) of the microstates that realize the macrostate M of a system not at equilibrium are sitting on trajectories that evolve according to the dynamical laws to realize macrostates of greater entropy. It follows that it is overwhelmingly likely that entropy will increase as the system evolves. The qualification "overwhelmingly likely" is required because there will be some microstates that don't evolve to states of higher entropy, but these anti-thermodynamic states will be very rare and scattered randomly among the set of states that realize M. That is, in every very small (but of a certain finite size) convex region of phase space, almost all states evolve to higher entropy. Thus, Boltzmann modified the second law to say that the entropy of an isolated system doesn't invariably increase but is *very likely* to increase. More generally, work in thermodynamics has shown that it is plausible that if it is a thermodynamic law that system S at time t in macrostate M will evolve to be in macrostate M* at t*, then Boltzmann's probability distribution will recover a probabilistic version of this regularity. Both the assumptions that matter is composed of an enormous number of particles in random motion and Boltzmann's probability hypothesis were thus spectacularly vindicated by their success in accounting for the second law and other thermodynamic laws.

However, a problem was soon noticed. As a consequence of the temporal symmetry of the fundamental dynamical laws, the statistical mechanical arguments entail that the uniform probability distribution applied to a system not at equilibrium at time t entails that the probability that the entropy of the system was greater at times prior to t is also approximately 1. For every microstate realizing

---

[15] A system's equilibrium state is the macrostate that contains the least information about which microstate realizes it.
[16] The question arises of what "probability" means in this context, especially because the fundamental dynamical laws are assumed to be deterministic. I discuss this issue later in the paper.

M that evolves forward in time to a macrostate of higher entropy in one temporal direction, there is a microstate that realizes M, the temporal reverse microstate that evolves to higher entropy in the reverse temporal direction. For example, Boltzmann's probability assumption entails that it is very likely that an ice cube in an isolated glass was smaller an hour ago and even earlier was very likely entirely melted (assuming that the glass has been isolated during that time). More generally, it implies that if M is the macrostate of an isolated system (or the universe) at time t, then it is likely that M is in a state that is an entropy minimum from which it is likely that entropy increases in both temporal directions. The reason is that for each state in which a system's particles are at certain positions, there is another thermodynamically identical state with its particles at the same positions but with the opposite velocities. Of course, this is absurd.[17] If we come upon a partially melted ice cube in a glass that we know has been sitting isolated for an hour, we can be sure that the ice cube was larger, not smaller, in the past. So, while, on the one hand, Boltzmann's probability posit apparently accounts for entropy increasing toward the future, on the other hand it entails the absurdity that entropy was greater in the past. This is the "reversibility paradox."

There are a number of responses to the paradox. One is to construe statistical mechanical probabilities only as recipes for making *predictions* for thermodynamic systems and to refrain from using them for retrodictions. This approach is usually combined with the views that statistical mechanical reasoning should be applied only piecemeal, and only to more or less isolated thermodynamic systems, not to the whole universe, and that its probabilities should be understood epistemically as a measure of an experimenter's knowledge (or lack of knowledge) of a system. The package of these responses amounts to an instrumentalist understanding of statistical mechanics. It avoids the paradox, but it leaves us completely in the dark as to why the recipe works for predictions and in what sense it provides explanations of physical phenomena.[18] In contrast, the Mentaculus is a realist account that construes probabilities as objective features of the world and is applicable to all macroscopic systems, including the entire universe from its earliest times.

---

[17] It is not only absurd but, as Albert points out, leads to cognitive instability if the Boltzmann probability posit is applied to the macro condition of the universe at the present time, since it implies that it is likely that this macro condition arose out of higher entropy states and in particular this means that the "records" in books etc. more likely arose as fluctuations out of chaos than as accurate records of previous events. This undermines the claim that those books report evidence that supports the truth of the dynamical laws and so results in an unstable epistemological situation.

[18] Boltzmann's prescription leads to instrumentalism because it couldn't be literally true, as it prescribes incompatible probabilities at different times, given that the uniform distribution over the macrostate at t will differ from the uniform distribution over the macrostate at other times.

The Mentaculus solves the reversibility problem by positing that the macrostate of the very early universe M(0) is one of very low entropy—i.e., the PH and conditionalizing on it.[19] This results in a probability distribution that gives the same *predictions* (inferences from M(t) to times further away from M(0)) as Boltzmann's prescription when applied to the universe as a whole—i.e., the universe after time 0—and, as we will see, also to its energetically isolated (and quasi-isolated) subsystems while avoiding the disastrous retrodictions we found without the PH. It is clear that conditionalizing on a low-entropy macrostate in the past (e.g., that the ice cube was twice as large an hour ago) constrains the probabilities of trajectories so that entropy is likely lower between now and an hour past. The reason for conditionalizing on M(0) and not, for example, on the lower-entropy macrostate yesterday (or a billion years ago) is that nothing short of placing the low-entropy condition in the very early universe ensures that the second law holds throughout its entire history. The reasons for believing that M(0) is very low entropy are not only that this is required for truth of the second law but also that this is in agreement with cosmological evidence, given plausible proposals for entropy applicable to the early universe.

The probabilistic version of the second law says, not only that the entropy of the entire universe likely increases as long as the universe is not yet at equilibrium, but that this also holds for typical energetically isolated or approximately isolated subsystems—e.g., that the entropy of a system consisting of an ice cube placed in a glass of warm water will likely increase. It may seem astonishing at first that conditionalizing on the state of the universe 13.7 billion years in the past has anything to do with the second law holding for subsystems at much later times—that, for example, the explanation for why ice cubes melt, smoke diffuses, and people grow old now ultimately involves the initial conditions of the universe—but an argument that this is the case is straightforward. Suppose that S is a subsystem of the universe that at time t "branches off" from the rest of the universe to become more or less energetically isolated, and that the macrostate of S is m (t). We can think of the degrees of freedom (i.e., positions and momenta) associated with the microstate of S when it branches off as being selected "at random" conditional on m(t) from the degrees of freedom of the macrostate of the universe M(t). Since "almost all" (i.e., measure almost 1) microstates realizing M(t) in any small region of states that realize M(t) are entropy-increasing and those that are not are randomly scattered in the phase space, it is overwhelmingly likely that the branched-off system will be on an entropy-increasing

---

[19] See Sklar (1993) for a discussion of some other proposals for responding to the reversibility paradox.

trajectory. It would take a meticulously selected special arrangement of particle positions and momenta to produce a system whose entropy doesn't increase. Due to the PH it would also take a meticulously selected state to produce a state whose entropy increases in the direction toward M(0).

The conclusion, then, is that the very low entropy of the early universe is ultimately responsible for its current entropy and, by the argument just sketched, also for the likely entropy increase of subsystems like ice cubes in water. Of course, this doesn't mean that the entropy of *every* subsystem of the universe is likely to increase. Some subsystems are not isolated but are interacting with other parts of the universe so as make it likely that their entropy decreases while the entropy of the other systems increases (e.g., a glass of water in a freezer). Also, there may be systems that are specially prepared so that even when they become isolated, their entropy will very likely decrease.[20] In these cases the second law doesn't hold. But that is as it should be. The job is to get the second law (and other thermodynamic laws) from the Mentaculus *in so far as* the second law is correct; and, arguably, Boltzmann's account and the Mentaculus do exactly that.

Since statistical mechanics applies to all macro systems, including the entire universe, it leads to an assignment of a probability density over the set of all physically possible micro histories and so to a probability for macrostates, whether or not they are expressed in terms of the language of thermodynamics. In other words, the Mentaculus assigns a probability to any proposition corresponding to the probability of the set of micro histories at which it is true, as long as that set is measurable.[21] This is why we call the Mentaculus "a probability map of the universe." The result is an account on which, even though the microdynamics are deterministic, the evolution of macro histories is indeterministic, as depicted in Figure 1.1.

The thin straight lines in Figure 1.1 represent micro histories of the world; the cylinders represent macro histories. A macro history is a history of the world that specifies the values of thermodynamic quantities (temperature, total energy, density, average frequency of radiation) and perhaps other quantities in very small regions of space-time throughout the history of the world. The micro histories included in a cylinder implement the corresponding macro history. The diagram depicts the micro histories as evolving deterministically, because the thin lines

---

[20] See *Time and Chance*, chap. 5, for a discussion of how a system may be prepared so that its entropy reliably decreases.

[21] As an example, Albert observes that the Mentaculus entails the probability that a spatula will be found in your bathtub, given what we know about you, spatulas, and other relevant macro facts (*Time and Chance*, p. 96).

THE MENTACULUS

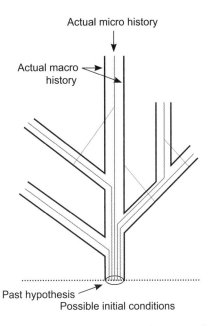

FIGURE 1.1. The universe according to the Mentaculus: Microphysical determinism and macro indeterminism with branching toward the future.

never branch or merge. In contrast, the macro histories evolve non-deterministically as shown by the cylinders constantly branching and occasionally merging.[22] The Mentaculus specifies the probabilities that the macrostate at t evolves into various alternative macrostates at subsequent times.

The illustration is intended to depict the physically possible micro and macro histories of the universe from the time of the PH through some time prior to the time of equilibrium.[23] Of course, there are actually many more branch points at any time, and a typical macrostate evolves into many more than two alternatives. The reason macrostates branch is that small differences in the microstates that implement a macrostate at time t often lead to big differences in the macrostates they implement at later times but it is rarer that different macrostates converge to

---

[22] Merging of micro histories occurs when records of macroscopic events are dissolved into degrees of freedom that don't realize a macroscopic event. If the universe has an equilibrium, then all macro histories will eventually converge to that equilibrium state.

[23] If the universe has an equilibrium, then as possible macro histories get close to equilibrium, they merge, culminating in the equilibrium state.

the same macrostate at later times.[24] As time goes on, the entropy of the universe increases, so many microstates realize its macrostate. This leads to merging of macrostates until equilibrium is reached.[25]

The notion of a macro history has not been fully specified until we have specified exactly what counts as a macroscopic property.[26] Macroscopic properties at least include usual thermodynamic quantities that obtain in small regions of space-time. It is plausible that many other macroscopic properties supervene on the distribution of thermodynamic properties, and so these too are included. Further, the notion of a macro history is vague because "small" is vague, and it is also vague exactly which micro histories implement a macro history since there are borderline cases. So, the Mentaculus's probability map is a bit fuzzy.

The primary support for the Mentaculus comes from its account of thermodynamic laws. Further support comes from manifestations of objective probabilities outside of thermodynamics. Albert points out that since macroscopic systems are composed of an enormous number of particles, the fundamental dynamical laws by themselves put very few restrictions on how they evolve. His example is a rock suddenly changing into a statue of the British royal family. Such bizarre behavior is never encountered because it is enormously unlikely according to the Mentaculus. Further, it is plausible that objective probabilities associated with gambling devices, weather, genetics, Brownian motion, and many other macro phenomena are grounded in the Mentaculus. Here is an example that makes the point. Consider a macro system S consisting of dollar bills dropped from a tower on a windless day. As a dollar bill falls to earth, it will be buffeted by particles of air striking it from various directions with various velocities distributed in the usual manner with a mean determined by the temperature of the air. The bill will describe a kind of random walk through the atmosphere. According to the Mentaculus the dollar bill is as likely to be struck by air particles from one direction as from another, so it is plausible that the Mentaculus yields the statistics describing the distribution of bills once they hit the ground.[27]

---

[24] This is illustrated by the famous "butterfly effect"—that whether or not a butterfly flaps its wings off the coast of Africa can make a difference to the probability of a storm occurring in the Caribbean.

[25] If this were depicted in the diagram, it would show the cylinders growing larger over time until all macro histories merge into the equilibrium state.

[26] I will later discuss exactly what properties count as macroscopic and why these properties are so important.

[27] The example is taken from Nancy Cartwright's *The Dappled World*. She uses it to argue for the incompleteness or falsity of the dynamical laws. To the contrary, when the statistical postulate is added the dynamical laws account for the observed distribution.

A major worry about the Mentaculus is that its probability map is far more detailed than we have any right to expect just on the basis of the fact that it is correct with respect to thermodynamics. It is plausible that there are probability distributions other than the uniform distribution over initial conditions compatible with the PH that are capable of underwriting the laws of thermodynamics but that differ on the probabilities they assign outside of thermodynamics. The thought is that because anti-thermodynamic initial conditions (initial conditions on which entropy decreases) are rare and scattered among the normal thermodynamic initial conditions, and because probability distributions continuous with the Liouville distribution all agree on sets of measure 1 and 0, they will also result in underwriting thermodynamic laws. If all we have an epistemic right to are probability distributions that account for the thermodynamic laws, then it can be argued that we should refrain from assigning a particular probability to propositions on which these probability distributions disagree. Call this the "set Mentaculus." According to it the Mentaculus is a probability map with holes. However, it seems to me that there is reason to conjecture that there is a probability map that assigns conditional probabilities to all pairs of ordinary macroscopic propositions.[28]

Even if there are probability distributions other than the usual statistical mechanical distribution that give results that are the same as or indiscernible from the Boltzmann distribution with respect to thermodynamics, the Lebesgue measure is the *simplest* probability distribution that underwrites the thermodynamic laws; and, as we will discuss later, this plays a part in making it the objectively correct probability distribution. Further, the Mentaculus probability distribution plays a central role in accounts of counterfactuals and causation. So, although I have no argument that the Mentaculus is the uniquely correct probability map of the universe, there is sufficient reason to take it seriously as a conjecture for such a map.

The question naturally arises of what "probability" means in the Mentaculus, given that the fundamental dynamical laws are deterministic and it is often held that fundamental objective probabilities require indeterministic fundamental laws. For this reason, there has been a tendency to construe statistical mechanical probabilities epistemically as characterizing an observer's subjective ignorance of a system's microstate given her knowledge of its macrostate and used by her to make predictions about the system's future behavior. But statistical mechanical probabilities are not merely subjective, since there is something about the world that makes certain probability distributions candidates for being correct. Further,

---

[28] This issue is discussed in more detail in the appendix to this paper, where the Mentaculus is compared with an account of thermodynamics based on "typicality" instead of probability.

an epistemic account is not an appropriate way to understand probabilities in the Mentaculus, where they are involved in causal explanations and not merely predictions of physical phenomena, such as why an ice cube melts in a certain time period. This explanatory role requires that they are objective features of reality whose values make a difference to what happens. An alternative to the epistemic understanding is that statistical mechanical probabilities are actual frequencies of repeatable events. Frequencies are objective, but this account won't work for the Mentaculus because it assigns probabilities or a probability density to entire possible micro histories, and there is only one actual micro history. Later in this paper I will discuss an interpretation of statistical mechanical probabilities according to which they represent objective lawful patterns in the actual micro history that are related to but need not be identical to frequencies, even though the underlying fundamental dynamics is deterministic.

Although I have assumed so far that the fundamental dynamics is deterministic, the Mentaculus can also be developed with indeterministic dynamics. This is what Albert does in the final chapter of *Time and Chance*. The version of quantum mechanics known as the GRW theory (for Ghirardi, Rimini, and Weber) posits an indeterministic dynamical law that specifies how the quantum state of a system at a time evolves. The law says that the state of a system jumps (or, in some versions, moves continuously) to another state with a certain probability, and since almost all states are ones that evolve thermodynamically, the result is a quantum mechanical implementation of statistical mechanics. GRW has almost the same empirical consequences as standard quantum theory, and what differences exist are very difficult to detect. While there are empirical differences between GRW and Newtonian mechanics, these are quantum mechanical and show up macroscopically only in situations in which quantum mechanical effects are relevant. As far as thermodynamics and, more generally, non-quantum-mechanical macroscopic phenomena are concerned, they yield the same results. What is crucial to this working is that the overwhelming majority of states that the system can move to are thermodynamically normal. On this account there is no apparent conflict between events possessing objective probabilities and determinism, but because the dynamics are indeterministic there is still a question of what probability means. That will be addressed later as well.

III. The Mentaculus and Time's Arrows

More, and in my view compelling, reasons for taking the Mentaculus seriously come from the role it plays in accounting for temporal asymmetries ("time's arrows") in addition to the temporal asymmetry of the second law and for its providing the basis for accounts of special science laws, counterfactuals, and causation.

It may appear at first that the Mentaculus explanation of the second law presupposes the past/future distinction rather than explains it, since it posits a low-entropy condition 13.7 billion years or so in the past. But this is misleading. The Mentaculus does not assume a distinction between past and future or a direction to time. Rather, it says that there is a very-low-entropy macro condition M(0) at one temporal boundary (the time of the big bang)—the PH—and posits no similar very-low-entropy condition at any other time. This orients the entropy gradient of the universe from low entropy at the time of the big bang to the time when equilibrium is attained. The past hypothesis earns its name by accounting for the other arrows of time and showing that they are aligned with the entropic arrow. It is because the arrows are so aligned that they are pointing in the direction that we think of as from past to future. Albert's and my claim is that the probabilities entailed by the Mentaculus explain not only the second law but also the asymmetries of knowledge, control, causation, and special science laws, and why these are aligned with one another. This justifies saying that the time between the present and the low-entropy boundary condition *is* the past and the time after the present *is* the future. When evaluated at times further away from the time of the big bang, the number of events that we can control decreases. Further, our experience of these asymmetries underlies our sense that the present marks a sharp difference between past and future that is constantly changing as time, as we say, "passes" or "flows." Saying that time "passes" doesn't literally mean that time or the present itself is moving. That would be absurd. The Mentaculus provides a *scientific* explanation of why time seems to flow in terms of the fundamental laws. Even though the Mentaculus explanations are not complete, Albert's and subsequent work do a lot to advance this project. Here I will give something of an idea along similar lines of how the epistemological, influence/control, and causal arrows are accounted for by the Mentaculus.

The epistemological arrow consists not only in the fact that we can and do know much more about the past than about the future but also in the fact that inferences from the present to the past are grounded differently from inferences about the future. Inferences about the past are based on records and measurements in a way that inferences about the future never are. Why are there records of past events but not of future events? This isn't a matter of the definition of the word "records"; instead it's a question of the probabilistic correlations that can exist between the present and the past and between the present and future events. A feature F of a system S is a macroscopic record of some fact Q if F is a macroscopic property of S that in the circumstances is correlated with Q. One way in which records are produced is by measurements. A measurement M of Q is a record produced by a process that starts in one condition—the ready state R—and interacts with Q, which is temporally between R and M.

The existence of records allows us to have very detailed knowledge of the past; but our knowledge of the future is typically not detailed, and the farther in the future, the less detail we can know and the less reliable our predictions. For example, we now have detailed records of the temperatures and air pressures in New York City over the last 20 years, but even approximately accurate predictions of weather conditions are limited to a few days. Further, once a record of a system S is formed, as long as it is undisturbed it remains a record whatever subsequently happens to S. This is why we can have detailed records of events far in the past. Albert observes that these differences between knowledge of the past and knowledge of the future give the impression that the present marks a sharp and metaphysically fundamental separation between past and future. This is why we think of the past as closed and the future as open. But on the Mentaculus account there are no *metaphysical* differences between past, present, and future. All that is different metaphysically is the time at which one is evaluating these. Physically what is different is that some times are nearer to the time of the PH than other times are and the probabilistic correlations among macroscopic events are different at different times.

Albert argues that what we can know about future weather can be recovered, at least in principle, from the uniform probability distribution and the dynamical laws by conditionalizing on current macro conditions. But, as the reversibility paradox shows, inferring the past from what we know of the present, the dynamical laws, and the uniform distribution yields absurd conclusions about the past. The difference is that what we know about the past depends on also conditionalizing on the PH, and everything we know about past macro conditions can be recovered from the present macro conditions by conditionalizing on the PH.

At first it may appear that the past hypothesis and the probability distribution that accounts for the second law have nothing to do with the epistemological asymmetry. What does the low entropy of the early universe 14 billion years from the present have to do with the reliability of current records? However, if we ask what makes R a record of a situation S, the answer is that R is a record of S in virtue of probabilistic correlations between the condition of R and S. Because the Mentaculus assigns probabilities to every state, it will specify when correlations obtain that make R a record of S. For example, an ice cube half-melted in a pail of warm water at time t that has been isolated for the last hour is a record of the ice cube being less melted an hour earlier at t' because there are correlations between the states of the ice cube at the two times. Conditionalizing on the PH is essential to getting the correct correlations given the statistical mechanical probability distribution. The PH is a necessary part of the explanation of how the state of a system at one time carries information about earlier times for the explanation to apply to all times. Without the entropy being lower at the time the record was formed, and ultimately without the PH, the uniform probability distribution has the same consequences in both temporal directions and so would not support the record.

Measurements are processes that produce records. For example, a thermometer reading 35°C is a record of the temperature of the water in which it is immersed, because there is a correlation between the thermometer reading and the water temperature. Albert points out that the existence of this correlation requires that the thermometer and the rest of the circumstances were in the appropriate state at the time of the measurement. He calls this "the ready condition" for the measurement. For us to know that the thermometer reading is a record of the prior water temperature, we need to know, or for it to be the case, that prior to the measurement the thermometer etc. were in their ready condition. But to know that, we need to have a record that this ready condition obtained, which requires a further ready condition, and so on. Albert argues that the PH plays the role of "the mother of all ready conditions" because it grounds the probabilistic correlations on which records are based and is thus crucial to explaining the epistemological asymmetry. By positing the PH the Mentaculus puts an end to the regress.

The Mentaculus by itself does not entail that any particular system's macrostate is a record of any particular past events. The fact that a particular fossil is a record of an extinct dinosaur depends on facts about how the fossil was produced and its history. However, it is plausible that a universe that conforms to the Mentaculus (i.e., a universe that is in the microstate M(0) at one time) will contain record-bearing systems. Given what we know about M(0), it is likely at that time that as the universe cooled, radiation was released and matter clumped under the force of gravitation, and the universe would evolve to contain atoms, stars, galaxies, planets, and so on.[29] These systems contain systems with states that serve as records of their formation and interactions with other systems. For example, the cosmic microwave background records the temperature and the matter density of the early universe. We don't know the probability of the early universe producing earthlike planets conditional on its prior macrostate, but given the macrostate of the earth 3 billion years ago it is plausible that it was likely it would evolve to contain

---

[29] The formation of stars under the force of gravity is entropy increasing. In Newtonian gravitational theory, as particles clump, their velocities increase, so the system gets more spread out in momentum space, increasing entropy.

animal and plant life and rivers and that some animals would die and become fossilized and so on.

Since the Mentaculus entails a conditional probability distribution (density) over all physically possible micro histories conditional on M(0), if our knowledge of the current macrostate is M(t), then whatever we can know and our credences about the past and the future on the basis of M(t) is given by P(__/ M(0)&M(t)). For example, the credence that we should have that Napoleon was defeated at Waterloo will be given by the Mentaculus conditional probability that Napoleon was defeated at Waterloo given what we know about the current macro conditions. If this were not so, then the Mentaculus would tell us to assign a probability to Napoleon's being defeated at Waterloo (or rather, the disjunction of micro histories that realize the event) conditional on the current macrostate that diverges from our degree of belief. Something would be wrong either with our belief or with the Mentaculus. But the Mentaculus is strongly supported by its success in accounting for statistical mechanical phenomena and our belief supported by historical records. So, we conclude that it is plausible that the two do not diverge and the Mentaculus underlies the reliability of historical records. Of course, I am not claiming that anyone explicitly employs or can employ the Mentaculus to make inferences about Napoleon. Rather, it is that the Mentaculus provides a scientific account of the correlations that ground the existence of records and an account of how the inferences we make from them can be justified by the world's objective probabilities.

The temporal arrow of influence and control is the flip side of the epistemic temporal asymmetry. It consists in the fact that even though we can exert some influence and control over future events (by our decisions), we have absolutely no control, and apparently no influence, over past events. Philosophers who attempt to explain this temporal asymmetry often invoke metaphysical accounts of time; for example, that as time passes from past to future, what becomes past closes while the future remains open. Their idea is that because the past is closed, we cannot influence or control it, whereas the future, being not yet settled, can be settled by our decisions. But without an account of what it is to be open or settled, this isn't so much an explanation as a restatement. In contrast, the Mentaculus explains the asymmetry scientifically as grounded in the probabilistic correlations specified by the Mentaculus.

To evaluate what a person A can influence, I follow Albert and assume that A has *unmediated* control at time *t* only of her decisions at *t* and then ask what that entails about what she can mediately influence. Exactly what it is for a person to have immediate control over her decisions is a difficult question. I assume that decisions are made in a part of the brain I call "the decision-making center" and that this is compatible with microphysical determinism but requires indeterminism at the macroscopic level—that is, the macroscopic state prior to making a decision doesn't determine what decision a person will make, and at the time the decision is made, alternative decisions are probabilistically independent of the macroscopic state, which is external to the agent's brain.

This procedure involves "decision counterfactuals" like "If A were to decide at time t to bring about B at t^*, then the probability that B occurs at t^* would be p." In order for A to influence whether or not B occurs, there must be counterfactual dependence between A's decisions and the probabilities of B occurring and not occurring. It is plausible that the various decisions over which A has immediate control correspond to small differences in the states of the decision-making center of her brain, say the firing of a few neurons.[30] To evaluate "decision counterfactuals" d1 → B and −d1 → B (A choses decision d1, A doesn't chose d1), we find those states of her brain that realize her decision d1 at time t and that are as similar as possible to the actual macroscopic circumstances M(t) external to the decision-making center of A's brain and then determine the probabilities that B and −B occur given d1 and M(t) and given −d1 and M(t). B counterfactually depends on decision d1 iff P(B/d1&M) > P(B/−d1&M)) where M is the macrostate of the universe eternal to her brain's decision-making center at t. That is, in conditions M, whether or not d1 occurs makes a difference to the probability of B. If this is so then I will say that A's decisions have influence over B.

In order for A to have control over B, not only she must have control over decisions that influence B but the influence must be significant (there is a large difference between P(B/d1&M) & P(B/−d1&M)) and she must know or have reason to believe that her decisions have this influence over B. If she didn't have reason to believe this, she would have no reason to make her decision when she desires that B, and consequently she would have no control over the outcome.

According to the Mentaculus alternative, A's decisions made at time t can have significant influence and control over what happens at times after t, but not significant influence or control over what happens prior to t. The reason is that there are macroscopic records external to the decision-making center of the agent's brain of events prior t but there are no records of events subsequent to t. For example,

---

[30] I am assuming that there is a part of the brain devoted to decision making and that various neurophysiological events in this part correspond to different decisions.

suppose that at the moment t there are records of whether A was at the beach yesterday—photographs, A's memories, and such. If there are such records, then the probability that A was at the beach the day before t, given the macrostate of the world external to A's decision-making center, is the same (near 1) whether or not A decides at t to have been at the beach a day prior to t. A's decisions have no influence over the existence of these records, because these are external to her brain. But because there are no records of whether A will be at the beach a day after t, under suitable circumstances it may be that the probability that she will be at the beach is greater if she decides to go than if she doesn't decide to go.

If determinism is true (as we have been assuming), then on the account of influence just described a person's decisions will still have some influence on the past. This is inevitable because alternative decisions will correspond to different micro histories of the world. If my actual decision is d1, then if I had chosen d2 the micro history of the world would have been different. But as long as there are records of a past event E at the time of the decisions, these micro histories must contain E. So, the influence I have is over the exact positions and momentum of particles at past times but not over E or anything I care or can know about. This and the fact that I have no idea what differences there are between these past micro histories this influence doesn't amount to control. That I have this kind of trivial influence over the past micro history may conflict with some philosophers' intuitions and with the view that the past is closed, but so much the worse for these intuitions and that view.

There are a number of objections to this account of the temporal asymmetry of influence and control that have appeared in the literature. One of these, due to Adam Elga (2001), concerns a situation in which there are no records of a past event at the time of the decision. He supposes that Atlantis really existed a few thousand years in the past but was destroyed and left no present records. Then he worried that if I am deciding whether to scratch my ear or not, and I do scratch it, then the counterfactual "If I had scratched my ear, Atlantis would not have existed" is also true. If this were correct, then it would follow that I am able to make it the case that Atlantis never existed. At least this is the result on David Albert's original account of influence. But on the account I sketched above, I have influence over whether Atlantis existed only if my alternative decisions make a difference to the *conditional probability* of Atlantis having existed, given my decision and the macrostate outside my decision—that is, only if $P(\text{Atlantis existed}/d1\&M) =/= P(\text{Atlantis existed}/-d1\&M)$. Since there are no records in M of Atlantis having existed, these two conditional probabilities are very small and equal. I have no influence over whether Atlantis existed.

Mathias Frisch (forthcoming) presented a more challenging example in which A's decision itself is the *only* record of a past macro event E.[31] In this case P(E/d1&M)=/=P)E/d2&M). Since there are no records of E outside of the decision-making center of A's brain, A's decision influences the past event E. I am not sure that there are cases like this in which the only record is a person's decision, but even if there are, I agree with Albert that they are not cases in which A has control over E. My reason is that it is impossible for A to know whether or not there are macro records of E outside of his brain. Further, A couldn't check to see if he succeeded because that would require a further record.[32]

There are other cases in which, given my characterization of control, it may be a consequence of the Mentaculus that a person has control over whether or not a past event occurs.[33] Presented with such a case we can respond either by adding further conditions to control while arguing that these are appropriate or by granting that these really are cases in which an agent has control over a whether a past event occurred. Unless there are cases that fit a characterization of control that enables a person to control a past event in ways that can be tested, such cases do not challenge the Mentaculus explanation of why we cannot control past events. I know of no such cases.

I have argued that the Mentaculus accounts for the temporal asymmetries of knowledge and control, as well as the temporal asymmetry of thermodynamics, and establishes that all these asymmetries are aligned with each other. This goes a long way to showing that it accounts for the "arrow of time." At the bottom of these accounts is the PH, which orients the direction of time and, as previously mentioned, earns its name by its role in accounting for these asymmetries.

A side bonus is that the Mentaculus account of influence and control provides a response to the well-known Consequence Argument that is alleged to show that

---

[31] In Frisch's example, a pianist's decision to play a certain musical passage is correlated with his not having played the passage previously during his playing of the piece, and there are no external records of whether or not he previously played the passage. In this example the decision is supposed to play the roles of both a record that he hadn't played the passage previously (which is not screened off by external records) and a decision to play the passage.

[32] In order to test whether or not A has control over whether or not E occurs (where E is either a past or a future event), we need to find an event E, ask A to decide to make E occur or to make E not occur (say, by flipping a coin to determine which), and then check to see if she succeeded. But this will require having other records of whether or not E occurred that are independent of A's decision after she made it, and it is hard to see how there can be such records without there also being a record of E's occurring at the time of her decision.

[33] Alison Fernandes develops an interesting such case that motivates her to complicate the account of control (Fernandes, forthcoming).

free will is incompatible with determinism. I understand the argument to claim that if determinism is true, then even if one grants that in some sense a person can control her decisions, her decisions do not influence her actions—or to show that she doesn't control her actions. Because acting freely requires that one can control one's actions via one's decisions, the conclusion is that determinism and free will are incompatible. This gives rise to two versions of the argument.

Influence:

(1) Determinism: The past of t and the laws of nature entail the future after t.
(2) One's decision at t does not influence the past.
(3) One's decision at t does not influence the laws of nature.
(4) One's decision at t does not influence the past of t and the laws of nature.

Therefore

(5) One's decision at t does not influence the future of t.

Control:

(1) Determinism: The past of t and the laws of nature entail the future after t.
(2) One's decision at t has no control over the past of t.
(3) One's decision at t has no control over the laws of nature.
(4) One's decision at t has no control over the past of t and the laws of nature.

Therefore

(5) One's decision at t has no control over the future of t.

According to the Mentaculus and our accounts of influence and control, the first argument fails because premise 2 is false. A person's decision at a time does influence the past but only the past micro history in ways that are useless to and unnoticeable by us. The second argument is invalid because it doesn't follow from the fact that we have no control over the laws of nature and no control over the past that we have no control over the future. I may have control over whether or not I raise my left hand, and that control requires that my decision to raise my left hand influences the prior micro history, but for the reasons we discussed that influence does not amount to control.

The Mentaculus account of decision counterfactuals can be extended to accounts of counterfactuals with non-decision antecedents and to an account of causation. Here I will only briefly describe how such accounts can go (Loewer 2007).

Consider Lewis's famous example, "If Nixon had pressed the button, there would have been a nuclear holocaust" (Lewis 1979, p. 467). Lewis proposed an account of the truth conditions of counterfactuals on which A → B is true at a world w iff there is no world in which A&−B is true that is more similar to the actual world than some world in which A&B is true. In evaluating which of two worlds is more similar to the actual world, Lewis proposes that three considerations in order of importance are relevant.

(i) The size of the region and the extent to which events conform to the fundamental laws of the actual world[34]
(ii) The size of the region in which the fundamental facts match perfectly
(iii) The extents to which regions, if not matched, are similar

These three respects determine a family of similarity relations depending on the relative weights attached to each.[35]

Here is how he thinks the account applies to the counterfactual "If Nixon had pressed the button at time t, there would have been a nuclear holocaust." He says that the worlds most similar to the actual world in which Nixon presses the button are worlds just like the actual world up until a time shortly prior to t when there is a small violation of the laws ("small miracle") resulting in Nixon pressing the button and then the laws of the actual world lead to a nuclear holocaust. Lewis thinks that it would take a big violation (a "big miracle") of the laws of the actual world subsequent to Nixon's pressing the button to avoid the nuclear holocaust and lead the counterfactual world to converge to the actual world. More generally, Lewis thinks that for typical counterfactual antecedents only a small violation of laws is needed to lead to a world in which the antecedent is true, whereas a big violation is needed to lead to convergence back to the actual world. He hoped that in this way his account secures the temporal asymmetry of ordinary counterfactuals.

---

[34] It may be that not only the size of the region, but also the extent of the violation, is relevant.
[35] Lewis (1979, p. 472) says that the third of these is "of little or no importance."

There has been much discussion of Lewis's account, but the account most relevant to the Mentaculus is by Adam Elga, who showed that contrary to what Lewis thought, his account fails to obtain the temporal asymmetry of counterfactuals.[36] This is due to the temporal symmetry of the fundamental dynamical laws. Elga showed that if there is a world w that matches the history of the actual world until a time t at which a small miracle occurs that leads by actual law to the antecedent, then there is also a world w* at which the antecedent obtains and matches the entire actual future a short time after the antecedent and that conforms to the actual laws except for a small miracle a short time after the antecedent. On Lewis's account, w* is at least as similar to the actual world as w is. For this reason, contrary to what Lewis thought, his account does not result in the truth of "If Nixon had pressed the button, there would have been a nuclear holocaust" and does not capture the temporal asymmetry of counterfactuals. World w* fails to conform to the PH because entropy increases in both temporal directions from the time of the small violation of law.

The PH plays no role in Lewis's account of similarity, so it is not surprising that his account of counterfactuals doesn't support their temporal asymmetry. It is also clear how to remedy the situation. If it is required that worlds similar to the actual world satisfy the PH, then Elga's counterexample world is eliminated. Thus amended, Lewis's account is not open to Elga's objection; but there are other problems. Contra Lewis, it is plausible that there are worlds that satisfy the past hypothesis, match the actual world exactly until a short time prior to t, where a small miracle leads to Nixon's pressing the button at t, but in which another small miracle shortly after t leads back to match the actual world.[37] If there are worlds like this, then the Nixon counterfactual is false. We can overcome this problem and avoid "miracles" altogether by giving an account of counterfactuals based on the Mentaculus. This account replaces perfect match with respect to the world's micro history with perfect match with respect to its *macroscopic* history and by considering the probabilities of the consequents of counterfactuals. There are worlds that match the actual world perfectly with respect to macroscopic properties until a short time prior to t and then diverge so that Nixon presses the button. No miracles are needed. This is an improvement over Lewis's account, in any case, since it is implausible that if Nixon had pressed the button, the actual laws of nature would have been violated. But there is still a problem, because

---

[36] See Elga (2000) and Loewer (2007).

[37] Lewis says that it takes a big miracle to get the world at which Nixon presses the button to reconverge to the actual world, because this world will contain traces of Nixon's button pressing, which require a big miracle to erase. Most button-pressing worlds are like this, but there are also worlds at which no traces are produced that don't further violate the dynamical laws.

there are worlds conforming to the actual dynamical laws that match the actual world macroscopically until a short time before t, at which time Nixon presses the button and subsequently there is a nuclear holocaust, and there are also worlds a which subsequent to his pressing the button no missiles are launched. However, given the PH and the microstate of the world shortly prior to t, such worlds are rare and so very unlikely on the statistical mechanical probability distribution. So even if Nixon had pressed the button, there might not have been a nuclear holocaust but it is very likely there would have been one.

Once statistical mechanics is taken into account, the counterfactuals of interest have probabilistic consequents even though the dynamical laws are deterministic. For example, "Had the ice cream been left out of the freezer, it would have very likely have melted by now" is true, but "Had the ice cream been left out of the freezer, it *would* have melted by now" is strictly false. In a recent paper Al Hajek also argues that these sorts of counterfactuals are strictly false unless the consequent is probabilistic.[38] Putting this all together, the initial Mentaculus account of counterfactuals is

$A(t) \rightarrow$ probably B is true iff there is a time t' shortly before t at which, given the macrostate of the world at t', $M(t')$ $P(B / A\&M(t'))$ is close to 1.

Although this account involves $M(t')$ where t' is shortly before t, it doesn't assume the temporal asymmetry of counterfactuals but instead derives the temporal asymmetry from the temporal asymmetry of records. "Shortly before" only means between t and the time when the PH holds. Worlds that are macroscopically the same as the actual world until a time t' shortly before t typically contain records of actual events prior to t. So, for example, it will turn out to be true that if Napoleon had won at Waterloo, it still would have been very likely that Caesar would have crossed the Rubicon. Because t' is before t, there can be events occurring after t' but before t for which there are no records at t'. In such cases,

---

[38] Hajek (draft). "For even if our world is deterministic, in the neighborhood of any trajectory of an object (of a billiard ball, or of a jumping human, or what have you) there will typically be some extraordinary trajectories in which things go awry. Here I appeal to statistical mechanics, whose underpinnings are deterministic. The point is familiar from the diffusion of gases, made vivid by Maxwell's demon. (Much as it is fair game for the epistemologist to remind us of the evil demon, it is fair game for me to remind us of Maxwell's demon.) For every set of initial conditions in which the air molecules in my office remain nicely and life-sustainingly spread throughout the room, there is a nearby initial condition in which they deterministically move to a tiny region in one corner— 'nearby' as determined by a natural metric on the relevant phase space. So it is false to say that if I were in my office, I would be breathing normally; the initial conditions *might* be unfortunate ones for me, leading to a phase-space trajectory of the molecules that suffocates me" (p. 20).

P(B / A(t)&N*t') may be close to 1 even if the time of B is prior to t. But this is as it should be. It may be that if Napoleon had won at Waterloo, it would have been likely that reinforcements had left France to join him the previous week.[39] There are further developments of the account that are required to deal with counterfactuals containing temporally complex antecedents and other matters, but this should suffice to see the basics. I hasten to add that this is proposed, not as an account that matches ordinary language speakers' use of counterfactuals—which is a very complex and context-dependent matter—but instead as an account of counterfactuals informed by physics in terms of which notions like reliability and causation can be characterized.

It is plausible that the Mentaculus also explains the temporal asymmetry of causation. It provides the temporally asymmetric counterfactuals involved in accounts like David Lewis's and supplies the probabilistic correlations required by probabilistic accounts like those of Judea Pearl and David Papineau.[40]

The Mentaculus provides the basis for an account of special science laws like the laws that occur in geology, biology, economics, and so on. An example discussed by Callender and Cohen (2010) is the so-called first principle of population dynamics, Malthus's exponential law:

$$P(t) = P_0 e^{rt}$$

where P0 is the initial population (say, of rabbits), r the growth rate, and t the time.

They observe, "This ecological generalization is very powerful. It supports counterfactuals and crucially enters ecological predictions and explanations. It has an undeniably central role in most presentations of the science of ecology." The systems Malthus's law applies to (e.g., rabbits in natural ecologies) are composed of particles moving randomly. Why do these particles behave to implement the law? That they do has seemed to some to be a mysterious conspiracy.[41] It is no

---

[39] Typically there won't be a unique micro history or even unique macro propositions that are likely given the macrostate at t' and A(t).

[40] Lewis's original proposal is in Lewis (1979). It has received much criticism and development, although there is no account that is thought to be fully satisfactory. My point is that the Mentaculus account of counterfactuals can play the role in a satisfactory account. Similarly, while there may be no fully satisfactory probabilistic account of causation, the objective probabilities of accounts like Pearl's (2018) or Papineau's (2022) are supplied by the Mentaculus.

[41] Jerry Fodor says that it is "*moto mysterioso* that Damn near everything we know about the world suggests that unimaginably complicated to-ings and fro-ings of bits and pieces at the extreme micro-level manage somehow to converge on stable macro-level properties" (Fodor, 1998, p. 160).

conspiracy but a consequence of the Mentaculus and background conditions that sustain the special science laws. As the universe evolves, its macrostate may come to be one in which the special science regularity holds according to the Mentaculus. On this account special science laws are not consequences of the Mentaculus but are special in that they follow the Mentaculus probabilities when conditionalized on appropriate background conditions. On this account Malthus's law really has the form $P(P(t) = P^0 e\,(rt)\,/\,C) \approx 1$, where C is the background conditions that enable the law. These conditions follow from the complete microstate of the universe at the time of the PH and the fundamental dynamical laws, but they are matters of chance relative to earlier macrostates, as illustrated in Figure 1.1. Special science laws often seem to be completely autonomous from laws of physics because the physical conditions that enable them are not taken into account. Once such conditions are taken into account, they can be seen as contained within the Mentaculus.

I am not saying that biologists (and other special scientists) need to know the Mentaculus or statistical mechanics in order to discover or confirm laws of biology, but only that if there are such laws, they and the probabilities they posit must be contained within the conditional probabilities entailed by the Mentaculus. Callender and Cohen (2010) disagree. They say:

> There is not a shred of evidence that the chances used in ecology are the ones used in statistical mechanics. A chance is relative to a particular measure over a particular state space.... Are the generalizations that are highly probable in the one space highly probable in the other? We have no idea, and neither does anyone else. The solution in question requires that all of this work out, but we don't see any reason for such confidence. (p. 437)

Instead, they imagine that there may be different probability distributions relative to different languages and the kinds and entities they refer to, and that these probability distributions may disagree. So, on their view it may be likely that, according to the biological probability distribution, a system evolves so as to satisfy Malthus's law while, according to the statistical mechanical distribution, the particles that compose the ecological system are likely to evolve so as to violate the law.[42] But this cannot be right. If the probabilities of statistical mechanics and a special science conflict, they both can't be correct. One of the main roles of probabilities is to advise degrees of belief, and if they disagree they would be giving different advice. Further, there is much more than a shred of evidence

---

[42] Callender and Cohen (2010, p. 437).

that the motions of particles conform to the Mentaculus—primarily the evidence that the laws of thermodynamics hold everywhere, but also the other evidence that was previously mentioned outside of thermodynamics. So, we have good reason to think that if a special scientist discovers that Pspecial(R / S) = x, then Pmentaculus(B / A) = x as long as the pairs R, B and S, A are true in the same micro histories.

This does not mean that every conditional probability entailed by the Mentaculus counts as a law. Callender and Cohen may be right that within a special science and its vocabulary, only those generalizations that form part of a simple and informative system are laws relative to that science. But it does mean that to qualify as a law, the generalization must be backed by a conditional probability entailed by the Mentaculus.

IV. Laws and Probabilities in the Mentaculus

The Mentaculus is a probability map of the universe in the form of laws, including a law specifying probabilities that assigns conditional probabilities to every pair of physically specifiable propositions. But what are laws and what are probabilities in the Mentaculus? These philosophical questions are pressing because the Mentaculus includes elements that would not be counted as fundamental laws or as probabilities on some familiar metaphysical accounts of laws and probabilities. Those who think of all fundamental laws as solely dynamical would not count the PH as a law, because it is a claim about the macrostate of the universe at a time, not about how it evolves over time. The fundamental probability law SP in the Mentaculus is not dynamical since it is an assignment of probabilities to every nomologically possible microstate of the universe at the time of the PH compatible with the PH that together with the fundamental dynamical laws results in probabilities for all nomologically possible histories. What are the probabilities involved in the SP? They are not actual frequencies, because there is just one actual history. Since the fundamental laws are deterministic, these probabilities cannot be understood as non-Humean propensity probabilities. Since they sustain explanations and causation, they are not merely epistemic. What are they?

I think the best account of laws and probabilities in the Mentaculus is provided by a development of David Lewis's Humean Best System Account (BSA) that I describe below. Metaphysical accounts of laws and probabilities divide into the non-Humean and the Humean. Non-Humean views construe the nomological aspects of reality, like laws and chances, as being over and above the non-nomological aspects and as in some way governing, producing, or constraining them. For example, on Maudlin's non-Humean account a law in some way operates on the state of the universe at a moment to produce subsequent states, and propensity accounts of probabilities construe probabilities as a measure of a disposition of one state to produce another.[43] In contrast, Humean views construe the nomological as supervenient on a metaphysically fundamental non-nomological reality. On Humean accounts, laws, including probabilistic laws, describe—they don't produce or constrain, but summarize and organize.

According to Lewis's original BSA, laws are generalizations entailed by an axiom system that best balances informativeness and simplicity in what it says about the entirety of non-nomological reality.

According to Lewis's Humean metaphysics, fundamental reality consists in the distribution of what he calls "perfectly natural properties and entities throughout entire space-time." Perfectly natural properties are fundamental categorical properties instantiated at points in space-time. Lewis calls their distribution throughout all of space-time "the Humean mosaic" (HM), and "Humean supervenience" is his name for the view that all truths—including those about laws, probabilities, counterfactuals, and causation—supervene on the HM. An axiom system organizes the HM by entailing truths about it expressed in a language whose simple predicates express perfectly natural properties and, in addition to logical and mathematical symbols, contains a function $P(t, A) = x$ which specifies the probability at time t of A's holding, where A pertains to a time after t. This enables the axiom system to entail probabilistic laws. Lewis proposes that candidates for law-determining axiom systems are true systems that best balance simplicity, informativeness, and fit. He measures fit of an axiom system to a world in terms of the probability of the world conditional on the axiom system. Lewis is vague about how informativeness, simplicity, and balance are to be understood. It is best to understand his proposal as presupposing that implicit in the practice of science are criteria that an ideal universe aims to satisfy and that informativeness and simplicity represent these criteria. It requires further work to identify these criteria and provide rationales for them.

According to the BSA, probabilities are not fundamental items in the Humean mosaic, as they would be on a propensity account, but earn a place in the law-specifying axiom system by their role in informing and organizing the Humean mosaic. BSA probabilities are not frequencies, since they apply to single events, but frequencies and patterns in the HM play an important role in determining

---

[43] As mentioned in footnote 4, Maudlin's account presupposes that time possess an intrinsic direction. Non-Humean propensity accounts of probabilities seem to share this presupposition.

the best system. A statement $P(t, A) = x$ is true if it is implied by the law-specifying axiom system and it informs by advising what degree of belief to have in A by way of the Principal Principle (discussed below). The idea is that creating an informative description of the mosaic without relying on probability statements is very complicated, but by introducing probability one can describe a complicated sequence of events simply. Consider, for example, the results of a long sequence of coin flips. Describing it in detail will be very complicated. Describing it simply will be very uninformative. But describing it as a sequence that results from many independent flips is both simple and informative.

On the BSA all truths about laws and objective probabilities are entailed by axioms of the BSA. Lewis worried that it might be that a world—and in particular, our world—may have multiple incompatible systems that optimally balance simplicity, informativeness, fit, and whatever other criteria an ideal scientific theory aims to satisfy. Even worse, it might be that no system is good enough to qualify as lawgiving. If the former, Lewis suggested that only those laws and probabilities entailed by all of the best systems obtain. In the case of probabilities, Luke Fenton-Glyn suggested that candidate systems include unsharp probabilities.[44] If our world has no system that is good enough to qualify as lawgiving, then I think it may be right to reject the existence of laws or the BSA. But I don't think there is reason for either worry. In any case, the claim that the Mentaculus is the best system of the actual world is empirical and, as I have argued, sufficiently supported and fruitful to be taken seriously.

There are many virtues of, but also many problems with, Lewis's specific version of the BSA. I won't enter into a detailed discussion of them here, but I do want to make a number of alterations in his account in order to make it more appropriate for the Mentaculus. The first, and most important, is to add to the language whose truths are organized by the law-specifying system. That language contains—in addition to predicates referring to perfectly natural fundamental properties, like "x has mass = m"—also thermodynamic predicates, like "region y has temperature = t", "region y has density = d", "region y has entropy e," etc., where y takes as values small, closed regions of space-time. Thermodynamics language needs to be added in order for the PH to qualify as a law. I will also suppose that candidates for best system may contain sentences connecting thermodynamic properties with perfectly natural properties, like the usual definitions

---

[44] Fenton-Glynn (2019) provides a number of reasons for including imprecise chances into the BSA. His and others' worry that there may be theories that posit probability distributions that differ from the Liouville distribution that are equally good for accounting for statistical mechanics but differ elsewhere are addressed earlier in this paper.

of temperature, pressure, composite mass, and so on. These can be thought of as analytic or as a kind of metaphysical law. They account for how macroscopic facts are grounded in or supervene on microscopic facts.[45] The second alteration is to add to the virtues to be balanced. The most important are explanatory values of the system that are applicable to subsystems. By employing a distinction between initial conditions and the laws, a system may increase both the number and the depth of the explanations it implies and the simplicity of its organization. Third, Lewis restricted his probability axioms to those that specify the probabilities of the possible evolutions of the state of a system at a time, thus making objective probabilities incompatible with determinism. In my view this was a mistake, because the BSA allows for objective probabilities compatible with deterministic dynamics. This is accomplished by allowing axioms that assign a probability distribution or density to initial conditions or trajectories of states. These modifications allow the Mentaculus statistical postulate and the PH to qualify as axioms of a candidate system. Finally, among the axioms may be a specification of an initial condition in a macroscopic language like the PH, which itself counts as a law.

One of the problems with Lewis's original version of the BSA is the way it measures fit. According to Lewis, T fits world w better than T* iff $P(w/T) > P(w/T^*)$. Adam Elga pointed out that the trouble with this is that if w and w* are the micro histories of a world, then for worlds like ours containing infinitely many events $P(w/T) = 0$ or infinitesimal for any candidate T, so T fits each world equally well.[46] Elga suggests that a way of responding to this is to measure fit relative to a finite class of "test" propositions. For each proposition in the class, take the conjunction of it or its negation, depending on which is true. The theory that assigns the greatest likelihood to this conjunction fits the world best relative to the class of propositions. A related, but I think better, way of responding to the problem is to measure fit in terms of the likelihood of a world's macro history. This works as long as the world is spatiotemporally bounded. Even if it's not, we could select finite cutoffs and measure fit relative to those cutoffs. It may be that there is a cutoff such that any theory that best fits relative to that cutoff also fits relative to greater cutoffs.

---

[45] Lewis assumes that naturalness is a metaphysically fundamental feature of some properties. An alternative approach is that candidate systems may include different languages as expressing fundamental properties and that the best system is a package deal that selects the fundamental language and systematization of truths in that language that optimally satisfies criteria. Since the language whose truths are to be systematized is not given on this approach, more needs to be said about how the best package is determined. This is done in my account of laws and properties called "the Package Deal Account" (PDA). (See Loewer, 2008, 2020, 2022.)

[46] Elga (2004).

Modifying the BSA as above allows it to entail probabilistic versions of thermodynamic laws. It also counts as lawful all the conditional probabilities it entails. These provide the basis for special science laws. For "an A is B" to be a law, there must be a true C such that $P(B/A\&C) \approx 1$. In addition, there must be stability so that $P(B/A\&C\&D) \approx 1$ for further true D. Further, we may want to require something along the lines of Callender and Cohen's better best-system account that to qualify as a law the generalization must belong to a system that unifies many conditionals involving similar properties. The modified BSA counts the Mentaculus as a candidate for a law-specifying optimal system. We claim that it is the best system for the actual world.

Although the Mentaculus probabilities are objective, not epistemic, they have consequences for epistemic probabilities. David Lewis's Principal Principle (PP) formulates how objective probabilities and beliefs about objective probabilities should rationally constrain credences. Lewis's original formulation of the PP is

$$(PP) \; Cred(B/Pt(B) = x \; \& E) = x$$

where Pt(B) is the chance of B at time t, and E is a proposition that is "admissible" relative to Pt(B). An admissible proposition is one that provides information about B only by providing information about Pt(B). Lewis thought that objective probabilities are all dynamical and require indeterministic dynamical laws. But statistical mechanical objective probabilities may not be dynamical and are compatible with deterministic dynamics. Exactly how objective probability should be understood in statistical mechanics is a matter we will return to. For now, I just point out that because the Mentaculus probabilities are not time relative but are conditional probabilities, a better formulation appropriate for conditional probabilities is

$$(PPC) \; Cred(B/P(B/A) = x \; \&A) = x$$

There is no need for "admissibility" in this version. It says that one's conditional degree of belief in B, given $P(B/A) = x$ and A, should be x. There is also an "externalist" version of the connection between objective conditional probabilities and conditional credences:

$$\text{External (PPC): Should } (Cred(B/A) = P(B/A))$$

This says that one's credence B given A should be equal to the objective conditional probability (B/A). The sense of "should" is the same as that involved in the prescription that one's beliefs should be true or should aim at the truth. It is a version of Al Hajek's proposal that belief is to truth as credence is to objective probability.[47]

Since objective probability is a guide to credence via something along the lines of Lewis's Principal Principle, the Mentaculus contains within it a guide to what our credences should be, given what we know about the current macrostate.[48] Given the Principal Principle (or a similar principle), the Mentaculus determines an objective rational credence distribution over all physically expressible propositions. It thus lays the foundations for a kind of "objective Bayesianism."[49] But it is quite different from rationalist forms of objective Bayesianism, according to which objectively correct credences are determined *a priori* based on logic, like the principle of indifference or the principle of maximum entropy.[50] In contrast, according to the Mentaculus, objectively correct credences are determined *a posteriori* on the basis of the best theory of the world.

While the Mentaculus provides an objective standard for a person's credences, I am not suggesting that anyone is in a position, or will ever be in a position, to calculate these probabilities from the Mentaculus. Objective conditional probabilities provide the targets that rational credences aim to hit. According to the Mentaculus there is an objective conditional probability (or a range of probabilities) to a woman being elected president of the United States in 2028 given the current macro conditions. We may estimate it on the basis of what we know of politics, history, and so on. To determine it from the Mentaculus would require knowledge of the measure of the set of microstates corresponding to macroscopic conditions and calculational powers akin to the knowledge and abilities of Laplace's demon. Nevertheless, there will be some conditional probabilities, like those involving thermodynamic states, whose values can in principle be extracted from the Mentaculus. And, as mentioned previously, these are borne out by experiment. This and the various other consequences and applications of the

---

[47] Alan Hajek (draft) argues that "truth is to belief as chance is to degree of belief." I am generalizing this to "truth is to belief as conditional objective probability is to conditional credence" (p. 41).

[48] Lewis PP says $P(B\,/\,Ch(B) = x\,\&\,E) = x$, where P is a degree of credence, $Ch(B) = x$ is the chance of B, and E is any proposition admissible. There has been much discussion of exactly how to formulate the PP and the nature of admissibility. See, for example, Lewis (1980), Ismael (2008), Hall (1984). A version of the Principal Principle applicable to the Mentaculus says that a person's credence $C(B\,/\,A)$ ought to be $P(A\,/\,B)$, where P is the Mentaculus distribution.

[49] The central principle of such objective Bayesianism is that a person's rational credences are those obtained from the Mentaculus, conditional on what she knows (explicitly and implicitly) about the present macrostate. Because an inquirer may not know the present macrostate, there is still a need for subjective degrees of belief over various possibilities compatible with what she does know.

[50] See, for example, Rosenkrantz (1989).

Mentaculus we have discussed should be sufficient to establish the Mentaculus as a contender for the title of the probability map of the universe.

Appendix: Probability versus Typicality

There is an alternative to the Mentaculus account of thermodynamics and statistical mechanics that is based not on probability but on the concept "typicality." It is just like the Mentaculus except that the statistical postulate (SP) is replaced by a typicality postulate (TP), which says that typical initial conditions of the universe obey the second law. I will call it the "Typitaculus." Advocates of the Typitaculus explain that a property F is typical among the members of a set S iff S has many members and all or almost all of them are F. Typicality concerns the relative size the set of Fs in S and is not a probability notion. It doesn't involve Fs being randomly selected from S. Because of this the Typitaculus avoids the problem of explaining what statistical mechanical probabilities are. Although typicality is not probability, Fs being typical in S is claimed to justify inferring that x is F on the basis of its being in S, just as learning that Ss are probably F justifies inferring that a randomly selected S is F or justifies having a high degree of belief that it is F. It is also supposed to justify there being no need to explain why x is F once one knows that x is in S.[51]

If S is finite, then "almost all" is understood numerically; but if S is infinite, a measure on S is required to quantify the size of S. In the Typitaculus that measure is the Liouville measure. This measure is chosen among all possible measures on sets of states because it fits well with the dynamics. According to the Liouville measure almost all initial conditions of the universe are entropy increasing, and by an analogue of the argument for subsystems given earlier almost all isolated macroscopic subsystems not at equilibrium evolve toward equilibrium. This is supposed to justify expecting that an isolated or quasi-isolated macroscopic system will evolve thermodynamically and to establish that no further explanation is needed for why a system obeys the second law.

The Mentaculus and the Typitaculus agree with respect to thermodynamic laws and the temporal asymmetry of the second law. One way they differ is that the typicality approach seems to say nothing about propositions whose measure is not near 1 or 0. Although the Liouville measure assigns values to arbitrary propositions unless this value is 0 (or very near 0) or 1 (or very near 1), this merely serves to classify them as typical or atypical. This in almost all cases involves types

---

[51] An excellent recent discussion of typicality is in Wilhelm (2019).

of events that occur many times. For propositions that describe events that occur just once, such as that the president of the United States elected in 2028 is female, it is silent, because the measure of the set of conditions that realize this proposition is, presumably, not near 1 or 0. This contrasts with the Mentaculus, which assigns probabilities to this and many other propositions that refer to events that occur once or only a few times.

Another important difference between the Mentaculus and the Typitaculus is that while the Mentaculus employs the PP, the epistemological principle employed by the Typitaculus is that if it is typical that Ss are F, then on learning that something is S one ought to believe that it is F. This is the typicality version of Cournot's principle that says that one ought to believe to be true propositions whose objective probability is very close to 1.[52]

A third way in which the two approaches differ is that the SP is contingent and is a law because it is an axiom of the best system. In contrast, the TP is closer to being an *a priori* truth or following *a priori* from the dynamics in virtue of expressing the meaning of "most."

While the Typitaculus accounts for thermodynamics in terms of typicality rather than probability, it needn't reject objective probability but supports a frequentist account that is applicable to thermodynamic and other systems. The idea is that there are recurrent situations that typically result in frequencies that exhibit features of randomness in the sense that the frequencies are not describable by a simple rule or are not compressible. For example, tossing a coin under usual conditions one hundred times typically results in random-appearing sequences with 48 to 52 heads. The Typitaculus can explain this by citing the fact that most (on the Liouville measure) of the initial conditions of the situations in which the coin is flipped result in such sequences.

The questions I want to address are whether the Mentaculus and the Typitaculus are in competition or are complementary, and what reasons there may be that favor or disfavor each. There are, I think, two main criticisms that advocates of the Typitaculus make of the Mentaculus.

1. The Mentaculus goes way too far in assigning probabilities to all physical propositions. Even granting that the Mentaculus assigns correct probabilities to thermodynamic propositions, what reason does it give for thinking that there is an objective probability to non-thermodynamic propositions, such as the proposition that a female president will be elected president of the United States in 2028

---

[52] Jacob Bernoulli's formulation of the principle is "Something is morally certain if its probability is so close to certainty that the shortfall is imperceptible." For an extensive discussion of "Cournot's Principle," see Shafer (2006).

conditional on the present political situation and other relevant macro facts? Dustin Lazarovici (2019), Mario Hubert (2021), and others point out that there are infinitely many probability distributions that agree with the Liouville distribution with respect to thermodynamics but, they think, disagree outside of thermodynamics—for instance, regarding the proposition that a female U.S. president will be elected in 2028. The evidence from thermodynamics for the SP equally supports these other distributions. They conclude that the fact that the Mentaculus is successful about thermodynamics provides no reason to believe its probabilities beyond thermodynamics.

2. Lazarovici (2019) and Hubert (2021) seem to think that it doesn't even make sense to assign objective probabilities to single events that haven't been repeated or are not repeatable. At least they claim this for Humean accounts of objective probability like the BSA. Lazarovici puts his objection this way:

> At the end of the day, the best system probability law will be one that informs us about robust regularities and global patterns in the world—"chance making patterns" as Lewis (1994) called them—while the fit to singular events will count little to nothing in the trade-off with simplicity. And if the number that the measure assigns to some singular event has nothing to do with that measure being part of the best systematization (or any systematization at all), it is hard to see why the number should have physical significance, let alone serve an epistemic or behavior-guiding function. (p. 22)

This is closely related to the first objection, because, as Lazarovici points out, there are many measures that agree on the "chance making patterns" while differing on the probabilities they assign to single events like a woman being elected U.S. president in 2028. As he says if two measures that fit the patterns equally well assign different numbers to an event, it is hard to see why they have any physical significance. The conclusion seems to be that it doesn't even make sense to treat the number that the best system assigns to an event as its probability since there is nothing in the Humean mosaic to support one probability distribution over another as long as they agree on thermodynamics. There is no reason to guide one's credences or decisions by the numbers that the Mentaculus calls probabilities that it assigns outside of thermodynamics, and so they are not probabilities.

Arguing along similar lines, Mario Hubert (2021) says:

> It is not immediately clear what the physical meaning of single-case probabilities is in this Humean theory. Let us say that there are two coins, and

the Mentaculus assigns a probability of landing heads of 0.4 to one coin and 0.6 to the other. Each coin is just once tossed and then destroyed. What can these numbers 0.4 and 0.6 mean? These probabilities indeed influence, by the Principal Principle, an agent's attitude and behavior toward the outcome of the coin tosses. For example, an agent will bet differently on an event with probability of 0.4 than on an event with a probability of 0.6. It seems, however, that these single-case probabilities need also to say something about the physical events themselves, whether their occurrence is in some way constrained or not, which is then the basis for an agent to adjust her degree of belief. (p. 5273)

There are a number of misunderstandings in these criticisms. Earlier I discussed the fact that there are many distributions that agree on thermodynamics but may disagree on other macro propositions, so won't return to it again except to reiterate that the uniform distribution is the simplest one that captures thermodynamic laws, and for that reason it counts as a law according to the BSA. The most important point to emphasize, in responding to the objection that it doesn't even make sense to assign probabilities to propositions about non-repeatable events, is that on the Humean best-systems account, probabilities come together with laws as a package. All Mentaculus probabilities are consequences of probabilistic laws. Probability laws are themselves consequences of the axioms that optimally satisfy scientific criteria, including informativeness, simplicity, fit, and other desiderata for an optimal theory that have been developed over the history of science. Because of the intimate connection between laws and probabilities they are involved in counterfactuals. The best system implies conditional probabilities that support counterfactuals. On the BSA it is the role of probability not just to accurately capture regularities, as Lazarovici says, but also to sustain counterfactuals. To accomplish this the best system aims not only for accuracy but for lawful accuracy. If probabilities aimed only for accuracy, then the most accurate distribution would be one that assigned probability 1 to the actual micro history of the universe. This would result in a system that, while assigning probability 1 to all true regularities, doesn't say which are laws and doesn't say what would happen in counterfactual situations. On a Humean best-system account, accuracy is balanced against other lawmaking scientific criteria like simplicity. If the best system of the world assigns a certain probability to a woman being elected U.S. president in 2028 given the current macro conditions, then this means that this is the best estimate of whether or not a woman will be elected given these macro conditions that follows by law. When the Mentaculus assigns a conditional probability to an allegedly unrepeatable event given certain conditions, it says that this is the most accurate *lawful* estimate of whether or not that event will occur given these conditions.

Hubert's worry that there is no physical content to the Mentaculus assignment of probabilities .4 and .6 to two coin tosses is similarly misguided. First, remember that the Mentaculus assigns conditional probabilities—for example, the probability of a coin landing heads given that it has such and such physical structure and so on. Because of this, the probability of a coin landing heads does depend on its physical structure. The probability is the most accurate lawful estimate of its landing heads. Someone who simply guessed that B is true when it is true will be making a more accurate estimate than the Mentaculus probability, but that guess would not be supported by law. Contrary to what Lazarovici and Hubert say, the number assigned as a probability by the Mentaculus has the significance that this is the number specified by a law. The fact that the Mentaculus assigns probabilities that are lawful is a reason for it to guide one's beliefs.

The preceding defense of the Mentaculus depends on a Humean account of laws, and perhaps advocates of the Typitaculus reject Humeanism and for this reason reject the Mentaculus. In fact, it seems that Humeanism and probabilistic accounts of statistical mechanics naturally go together, as do anti-Humeanism and typicality, and I suspect that proponents of the Typitaculus are also anti-Humeans about laws.[53] The dispute between Humean and anti-Humean accounts of laws is large and complicated. I won't attempt to settle it here, but I do think I have rebutted the argument that on the Mentaculus the assignment of probabilities beyond thermodynamics and to non-repeatable propositions do not make sense.

Are the Mentaculus and Typitaculus friends or foes? They are not inconsistent with each other. SP and TP differ in that one says that there is a probability distribution over initial conditions compatible with M(0) whereas the other says that there is a typicality measure over these initial conditions. In both cases it is the Liouville measure but interpreted differently. Due to this they offer different but compatible explanations of thermodynamic behavior. The Mentaculus explains why thermodynamic behavior is very likely whereas the Typitaculus explains why it is typical. As far as I can see, these explanations are compatible and

---

[53] It is possible to combine a non-Humean account of laws with a Humean account of probabilities when the laws are deterministic. This is Carl Hoefer's (2019) position, though his version of Humeanism, unlike the Mentaculus, which is claimed to be the best theory of the universe, is local and closer to an actual frequency account. As Lazarovici (2019) observes, it is also possible to combine a Humean account of laws with a typicality approach to statistical mechanics, but I see little reason for this.

even complementary. The Mentaculus invokes the PP to license the inference from "It is very probable that ice cubes in warm water melt" to a high degree of credence that ice cubes in warm water melt. The Typitaculus invokes CP to license the inference from "It is typical that ice cubes in warm water will melt" to "Ice cubes in warm water will melt." As proponents of typicality emphasize, Fs being typical in S doesn't mean that it is probable that a selected S is F. One cannot conclude just from the fact that crows are typically black that the next crow you see will likely be black. For that one needs a probability assumption, for example that each member of S is as likely to be selected as each other.

Advocates of the Typitaculus find it objectionable that the Mentaculus assigns a probability to the proposition that the president of the U.S. in 2028 is woman given conditions C (describing the macrostate of the world at a time in 2022), but in fact the Typitaculus is committed to something very similar. Suppose that the conditions C occur numerous times in the vast space-time of the universe, just as we can ask what happens when a coin of a certain shape is flipped under such and such conditions. We can ask, What typically happens when conditions C obtain? How often are they followed within a certain time by a woman being elected U.S. president? It may be that it is typical that many repetitions of C result in a female presidency x percent of the time and that the sequences of C have the appearance of randomness. In this case the Typitaculus says that the probability of a female presidency given prior conditions C is .x. This is exactly what the Mentaculus says. I don't mean that for every Mentaculus probability P(B / A) there is a Typitaculus frequency probably P(B / A). It may be that there are not sufficiently many occurrences of A that result in sequences that produce stable frequencies and appear random. But it is not implausible, given the vastness of the universe in both time and space, that there will be many.

Aside from their relations to Humeanism and anti-Humean accounts of laws, the main difference between the Mentaculus and the Typitaculus concerns the epistemological status of the SP and the TP. The SP is an empirical law separate from the dynamical law. The TP is close to an *a priori* principle of rationality derived from the dynamical laws. Even here there is no incompatibility in endorsing both.

So in the spirit of peace, I conclude that the Mentaculus and the Typitaculus can coexist. Where the Mentaculus has an advantage is that its objective probabilities can be used in the accounts of counterfactuals, causation, special science laws, and objective credences outlined in this paper. But I suspect that advocates of either one of them will think they have no need for the other to explain thermodynamics.

Acknowledgments

Thanks to Valia Allori, Craig Callender, Eddy Chen, Heather Demarest, Saakshi Dulani, Martin Glazier, Michael Hicks, Jenann Ismael, Dustin Lazarovici, Brad Weslake, Isaac Wilhelm, and Eric Winsberg for discussions and comments on earlier drafts of this paper. I am especially and enormously grateful to my friend David Albert for having taught me about the main ideas in *Time and Chance* and for the many hours of discussions and debates we have shared on the issues discussed in his book and this paper.

References

Albert, David. (2000). *Time and Chance.* Harvard University Press.
Cartwright, Nancy. (2012). *The Dappled World.* Cambridge University Press.
Callender, Craig. (2011). "The Past Hypothesis Meets Gravity." In *Time, Chance, and Reduction,* ed. Gerhard Ernst and Andreas Hüttemann. Cambridge University Press
Callender, Craig, and Jonathan Cohen. (2010). "Special Sciences, Conspiracy and the Better Best System Account of Lawhood." *Erkenntnis* 73:427–447.
Carroll, Sean. (2010). *From Eternity to Here.* Dutton.
Carroll, Sean. (2021). "The Quantum Field Theory on Which the Everyday World Supervenes." In *Levels of Reality: A Scientific and Metaphysical Investigation* (Jerusalem Studies in Philosophy and History of Science), ed. O. Shenker, M. Hemmo, S. Iannidis, and G. Vishne. https://arxiv.org/abs/2101.07884.
Earman, John. (2006). "The 'Past Hypothesis': Not Even False." *Studies in History and Philosophy of Science Part B: Studies in History and Philosophy of Modern Physics* 37, no. 3: 399–430.
Eddington, Arthur. (1935). *New Pathways in Science.* Cambridge University Press.
Elga, Adam. (2001). "Statistical Mechanics and the Asymmetry of Counterfactual Dependence." *Philosophy of Science* 68, suppl. 1: 313–324.
Elga, Adam. (2004). "Infinitesimal Chances and the Laws of Nature." *Australian Journal of Philosophy* 82, no. 1: 67–76.
Fenton-Glynn, Luke. (2019). "Imprecise Chance and the Best System Analysis." *Philosophers' Imprint.*
Fernandes, Alison. (forthcoming). "Time, Flies, and Why We Can't Control the Pt."
Fodor, Jerry. (1998). "Special Sciences; Still Autonomous after All These Years." In *Philosophical Perspectives,* 11, *Mind, Causation, and World,* 149–163.
Frisch, Matthias. (forthcoming). "Statistical Mechanics and the Asymmetry of Causal Influence." In *David Albert's Time and Chance,* ed. Barry Loewer, Eric Winsberg, and Brad Weslake. Harvard University Press.

Hajek, Alan. (draft). "Most Counterfactuals Are False." https://philosophy.cass.anu.edu.au/people-defaults/alanh/papers/MCF.pdf.

Hall, Ned. (2004). "Two Mistakes about Credence and Chance." *Australasian Journal of Philosophy* 82 (1): 93–111.

Hoefer, Carl. (2019). *Chance in the World: A Skeptic's Guide to Objective Chance*. Oxford University Press.

Hubert, Mario. (2021). "Reviving Frequentism." *Synthese* 199, no. 1–2 (2021): 5255–5284.

Ismael, Jenann. (2008). "Raid! The Big, Bad Bug Dissolved." *Noûs* 42 (2): 292–307.

Lazarovici, Dustin. (2019). "Typicality versus Humean Probabilities as the Foundation of Statistical Mechanics." https://dustinlazarovici.com/wp-content/uploads/Typicality-Probability.pdf.

Lewis, David. (1979). "Counterfactual Dependence and Time's Arrow." *Noûs* 13, no. 4.

Lewis, David. (1980). "A Subjectivist's Guide to Objective Chance." *IFS* 15: 267–297.

Lewis, David. (1986). "A Subjectivist's Guide to Objective Chance." In *Philosophical Papers*, Vol. 2, pp. 83–132. Oxford University Press.

Loewer, Barry. (2001). "Determinism and Chance." *Studies in History and Philosophy of Science Part B: Studies in History and Philosophy of Modern Physics* 32, no. 4: 609–620.

Loewer, Barry. (2004). "David Lewis' Theory of Objective Chance." *Philosophy of Science* 71: 1115–1125.

Loewer, Barry. (2007). "Counterfactuals and the Second Law." In *Causation, Physics, and the Constitution of Reality: Russell's Republic Revisited*, ed. Huw Price and Richard Corry. (Cambridge: Cambridge University Press 2007).

Loewer, Barry. (2012). "Two Account of Laws and Time." *Philosophical Studies* 160, no. 1: 115–137.

Oppenheim, Paul, and Hilary Putnam. (1958). "Unity of Science as a Working Hypothesis." *Minnesota Studies in the Philosophy of Science* 2:3–36.

Papineau, David. (2022). "The Statistical Nature of Causation." *Monist*.

Pearl, Judea. (2018). *The Book of Why*. Allen Lane.

Penrose, Roger. (2005). *The Road to Reality*. Vintage Books.

Rosenkrantz, Roger. (1989). ET Jaynes: Papers on Probability, Statistics and Statistical Physics. Reidel.

Shafer, Glenn. (2006). "Why Did Cournot's Principle Disappear?" http://glennshafer.com/assets/downloads/disappear.pdf.

Sklar, Larry. (1993). *Physics and Chance*. Cambridge University Press.

Tooley, Michael. (1997). *Time, Tense, and Causation*. Oxford University Press.

Wilhelm, I. (2019). "Typical: A Theory of Typicality and Typicality Explanation." *British Journal for the Philosophy of Science*.

Zimmerman, Dean. (2011). "Presentism and the Space-Time Manifold." In *The Oxford Handbook of the Philosophy of Time*, ed. Craig Callender. Oxford University Press.

PART II

PHILOSOPHICAL FOUNDATIONS

Chapter Two

The Metaphysical Foundations of Statistical Mechanics

On the Status of PROB and PH

▶ ERIC WINSBERG

> What if some day or night a demon were to steal after you into your loneliest loneliness and say to you: "This life as you now live it and have lived it, you will have to live once more and innumerable times more"... Would you not throw yourself down and gnash your teeth and curse the demon who spoke thus? Or have you once experienced a tremendous moment when you would have answered him: "You are a god and never have I heard anything more divine."
>
> —NIETZSCHE, *THE GAY SCIENCE*, §341

1. Introduction

One of the central aims of *Time and Chance* (2000) was to bring all of our experience of the macroscopic world under the umbrella of a small package of fundamental physical postulates. Many of the central features of our experience—particularly temporally irreversible ones like the thermodynamic behavior of isolated systems, the asymmetry of knowledge of the past and future, and the temporal asymmetry of causal interventions—look puzzling from the point of view of a time-reversible fundamental physics. The goal of *Time and Chance* was to show how these features of our experience do indeed follow from a package of fundamental laws: a set of time-reversible, deterministic, dynamical microlaws; a hypothesis about the macroscopic state of the early universe (the "past hypothesis," PH); and a statistical postulate about the distribution of microstates compatible with that macrostate (PROB).

Aside from claiming that they are to be understood as laws, however, *Time and Chance* did not have very much to say about the metaphysical underpinnings of PH and PROB. This was despite the fact that PH and PROB appear to be somewhat unusual from the point of view of most ordinary understandings of laws of nature—which ordinarily distinguish between laws and initial conditions. As for the probabilities postulated by PROB, the book contains some brief arguments against construing them as degrees of belief but for the most part is relatively quiet about a potentially puzzling question: How are we to understand the claim that the initial physical state of the universe involves objective chances? There is, after all, only one universe, and it began only once.

In subsequently developing Albert's program, however, Albert and Barry Loewer have given the scientific program of *Time and Chance* a more rigorous metaphysical underpinning. They have done this by embedding Albert's Boltzmannian scientific project into a Humean metaphysical project—one in which PH and PROB have taken on the status of laws of nature of the so-called "best system" variety associated with John S. Mill, Frank Ramsey, and David Lewis, and where the probabilities in PROB are taken to be objective chances of the kind that appear in the extension of that project by Lewis.

In this chapter, I want to attempt to clarify Albert and Loewer's metaphysical project, and subject it to critical scrutiny.[1] The metaphysical project is based, in part, on David Lewis's "Humean" conception of laws and chances; but Lewis's system requires substantial changes in order to try to make it work for statistical mechanics. One of the first things I will do, therefore, is try to clarify exactly what these modifications need to consist in.

Once it is clear what the modifications need to be, and we have a clear view of what Albert and Loewer's version of Humean laws and chances looks like, we can begin to ask how attractive this metaphysical project looks from a variety of points of view. In particular, I will be interested in examining how attractive it looks from the point of view of underwriting what I view as a particularly strong form of "fun-

---

[1] In what follows, I will take for granted the success of what I am calling the scientific project. That is, I take it for granted that PH and PROB, however we are meant to construe them, are sufficient, in conjunction with whatever the dynamical laws are, to underwrite, at least, the thermodynamic behavior of isolated systems. Doubts about this have been expressed in Winsberg (2004), Earman (2006), Callender (2011), and elsewhere; but I will be ignoring those here.

damentalism" that Albert and Loewer endorse and that partially motivates the project in the first place. More will need to be said in due course, but the strong version of fundamentalism I have in mind insists not only that the truths of the special sciences be recoverable from the fundamental laws, but that the lawfulness of their laws must be as well. Ultimately, what I will argue is that the metaphysical picture offered by Albert and Loewer is unsuccessful, and that the version of fundamentalism that motivates it is too strong.

2. The Basic Humean Picture

Let us begin by giving a first pass at the metaphysical picture. How are PROB and PH, which, intuitively, inform us about the initial conditions of the universe, meant to be understood as laws? According to the basic picture of laws offered to us by Lewis (1986, 1994), what is fundamental in the world is the Humean mosaic: a distribution of points in space-time with their natural intrinsic properties. The "laws" of such a world are whatever regularities follow from the axiomatic system that "best" summarizes the facts about the mosaic. The "best system" is the system that maximizes simplicity, informativeness, and fit. Whatever system of axioms, in other words, offers the best compromise between being simple, being informative about the mosaic, and assigning the highest probability to the actual course of events ("fit") is the best system. Its regularities that follow from it are the laws of nature. And any probabilities that are featured in those laws are the objective chances.

Unfortunately, Lewis's conceptions of laws and chances, in his precise original formulation, is ill-suited to understanding PH and PROB. For one thing, PH and PROB are not regularities. More importantly, perhaps, Lewis did not paint a nuanced enough picture of the role of initial conditions in assessing the informativeness and simplicity of a "system" for his account to straightforwardly underwrite the claim that PH and PROB are laws. PH and PROB are, after all, coarse-grained summaries of the initial conditions of the universe. Arguably, the problem is that Lewis expected there to be a relatively simple set of initial conditions, or perhaps boundary conditions, that would help to drive the engine of informativeness of the best system without much complicating its measure of simplicity.

But this now seems unlikely. Albert often illustrates that point by asking us to imagine a scenario in which a physicist is having a conversation with God. The physicist, quite naturally, asks God to tell her what the universe is like. God begins by telling the physicist what the dynamic laws of the universe are. But of course this is not enough. Because she wants to be able to make predictions, the physicist asks in what particular initial condition the universe began. To which God replies: "The universe has 10^{50} particles, how much time do you have?" Because the answer is obviously "not enough," God suggests the following alternative: "Pretend as if there were a uniform probability measure over the set of microconditions that are compatible with whatever macrocondition in which the universe began. From that fact and whatever macroscopic observations you make, you should have enough information to make any practical macroscopic predictions that you like. That should save us some time." The trick to having PH and PROB come out as laws, in other words, is to imagine that a succinct summary of the initial conditions of the universe ought to be regarded as part of the best system. This idea, though, for whatever it is worth, was not on Lewis's radar. Indeed, it is not entirely clear that Lewis thought through the role initial conditions ought to play in his game of best-system analysis carefully enough for us to evaluate the claim that PH and PROB should be part of one.

3. Measuring Informativeness and Simplicity: The Role of Initial Conditions

So let us begin to flesh out the ways in which Albert and Loewer's conception of a best-system analysis might differ from Lewis's by discussing the role that initial conditions might be imagined to play in evaluating the informativeness and simplicity of a system. Informativeness, obviously, involves the comparison of two things: the system, on the one hand, and the Humean mosaic—the distribution of fundamental facts in the world whatever they happen to be—on the other. The question, then, is what role is played in such a comparison by the sorts of things that we—or at least working scientists—ordinarily call initial conditions. There are three ways we can imagine this could go. I describe each of the three below. In each case we can imagine something like the following scenario as a background: Imagine, for example, a world consisting of hard spheres bouncing around in what we would ordinarily think of as in accord with the laws of Newtonian mechanics. Think, then, of two books. In one book would be recorded the distribution of fundamental facts about the world—in this case, presumably, the distribution of the spheres across space-time. The first book, in other words, would record every detail of the Humean mosaic. In the second book would be written a candidate set of laws. The game then is to evaluate the "quality" of the second book: its informativeness, simplicity, and fit with respect to the first book.

1. We could propose that in evaluating the informativeness of a system, what we ordinarily think of as initial conditions can be appealed to, but they are not part of the system for the purposes of measuring the system's simplicity—and importantly, do not count as one of its laws. And so nothing about the positions and velocities of the spheres at any particular time would be recorded in the second book. On this conception of informativeness, then, the informativeness of the second book is measured by how informative it is about what is written in the first book, given some specification of the positions and velocities of the spheres at some instant of time. The initial conditions get plugged into the calculations used when evaluating the informativeness of the second book, but they do not count against its simplicity.
2. As a second possibility, we could imagine that when evaluating the informativeness of a system, all we do is compare the first book with the second book, straight up. In such a conception, the informativeness of the second book in the example above would be rather low—if it has any at all. But by adding some description of the initial state of the world to the second book, we would make it much more informative—while decreasing, obviously, its simplicity.

    Presumably, on this second conception, whether or not the second book contains "initial conditions" is a matter of balance, and whether or not they get included in the version of book two that ends up being the "best system" will depend on whether the reduced simplicity that comes from adding them does or does not balance out the increased informativeness. Of course, it may turn out that a book without any "initial conditions" is never adequately informative, and so it may turn out that any best system, on this conception, will always include what we usually think of as "initial conditions." Or it may turn out that partial description of the initial conditions (something like in the God story above) wins out. It may even turn out that, for the real world with all its complexity, the inclusion of a precise specification of initial conditions never counts as adequately "simple." More on this later.
3. A third possibility is that we think of informativeness in a much more pragmatic way. Rather than thinking of informativeness as a direct, God's-eye comparison between the system and the human mosaic, informativeness is something that is *for us*. So, if I want to compare the informativeness of two competing systems, what I do is try to use

each of those two systems, *in combination with other facts I happen to know* (or think I know), in order to make predictions about things that I can observe. And then I decide which of the two systems is more useful to me in making those sorts of prediction. The one that, in combination with other things I know, enables me to make better predictions, is the more informative system. If system one makes better such predictions than system two without a sacrifice in simplicity, than it is the "better" system.

Here's one thing that should be clear: On the first conception, initial conditions will never be part of the package.[2] On the second and third conceptions, it will be an interesting and open question whether or not they are, and what form they should take (i.e., should they include precise specifications? or coarse-grained descriptions like PH and PROB?). And since it is an open question whether initial conditions will be part of the package and what form they will take, it perhaps makes sense to call the resulting inclusions "laws," despite the fact that they are not regularities. It should also be clear that on the third conception (unlike the other two), how well a system rates for informativeness will depend on the epistemic agent using it. That is because on the third conception, it might very well turn out that a system without initial conditions is very informative to an agent who already knows certain other facts, but much less informative to another sort of agent. How informative a system is, on the third conception, will depend, to a great extent, on whatever other things you happen to know about the world, on what you are hoping to predict, and on other related factors.

Having discussed each of these three ways of thinking about informativeness, we should now ask: When Loewer (2007, p. 305) says, "Adding PH and PROB to the dynamical laws results in a system that is only a little less simple but is vastly more informative than is the system consisting only of the dynamical laws," which conception does he have in mind? More importantly, on which, if any, of the three conceptions above does this claim turn out to be plausible?

---

[2] That is one way of putting the point. A virtually equivalent thing to say is that a precise specification of the initial conditions will always be part of the package. These are virtually equivalent because in either case, no two systems will ever differ in either their informativeness or their simplicity as a result of a difference with respect to this part of the package. And here we see the logic of insisting that the laws be *regularities*. Because however we put the point, there is no point in calling these initial conditions laws. And regardless of how we characterize the point, PROB and PH will not come out as part of the system, or as laws, since they would be redundant given a precise specification of the initial conditions of the universe.

Obviously, the first conception is a nonstarter. On the first conception, you get to use the precise initial conditions for free when evaluating a system that does not contain them. So, on that conception, adding information about the initial conditions to a system will never increase its informativeness. But what about the second and third conceptions? What I want to argue here is that, for various reasons, neither the second nor the third conception supports the idea that PROB should be considered a fundamental law in the way that Albert and Loewer would like it to be.

Let us begin by looking at the second conception.

First, there are reasons to think that the second conception is not a coherent way to think about what a law is. This has been argued by Mathias Frisch. Frisch (2011) gives the example of a simple world consisting of nothing but a smallish collection of objects "governed" by Newtonian gravitation. In such a world, a Lewisian system consisting of Newton's laws would be made vastly more informative, and not much less simple, by including a precise specification of the world's initial condition. But if such a thing is included in the set of laws, then every fact about that world becomes nomically necessary—and the distinction between laws and nomically contingent facts would collapse. This is an unappealing result, and raises doubts that the second conception is coherent.

Of course, the world Frisch imagines is not our world. And it is a premise of Albert and Loewer's proposal (made vivid by the God story) that in our world, a precise specification of the initial conditions is too costly in terms of simplicity to be a candidate for being part of a best system. Perhaps this is so. So perhaps we ought only to conclude that the distinction Frisch is worried about collapses only for other worlds not like ours. So, even though Frisch's example clearly reduces the attractiveness of the second conception of informativeness, we might still want to entertain it as a possibility, and ask instead whether or not it's plausible that, on the second conception, PROB comes out as part the best system. I do not think it does, for two reasons.

The first reason has been articulated by Cohen and Callender (2009). While my primary concern in this paper is with the status of PROB, it should be recalled that Albert and Loewer's argument for the nomic status of PROB also depends on PH being a law. And it is precisely this that Cohen and Callender have called into question. What they point out is that whether or not PH counts as a simple addition to the package depends on what language it is expressed in. The idea is that PH presumably has some relatively simple expression in a thermodynamic language.[3] But if we try to express PH in the language of the microphysics, then

---

[3] Even this might be open to doubt. Notice that Albert formulates PH as the claim that "the universe began in whatever low-entropy macrocondition the normal inferential procedures of

its expression becomes almost as complex as writing down the precise microscopic initial condition of the universe.

A second worry is the following:

Let's call the ordinary dynamical laws that we expect physics to deliver to us at the end of inquiry the P-laws. In most discussions of Boltzmannian statistical mechanics, we assume that the P-laws are ordinary Newtonian mechanics along with some ordinary conservative Hamiltonian. It doesn't really matter, here, what the P-laws turn out to be—but the kind of thing I have in mind is that the P-laws are whatever future development of QFT, or quantum gravity, or whatever, physics has in store for us.[4] By P-laws, in other words, I mean the sort of things that working physicists will expect to find in their best textbooks in the library. What Albert and Loewer seem to be committed to is that, at the end of the day, when we do the best-system analysis of our world, what we will get is the P-laws—the ordinary laws of physics we would find in our textbooks—plus PROB, plus PH.

But is that true? Or is it even plausibly true (on the second conception)? I would argue that it is not, and that it is only a confusion, or a slipping back and forth, between the first conception of informativeness and the second one that makes it seem like it is plausibly true. Here is what I mean:

Let us first of all suppose that one of Lewis's basic intuitions was correct. I argued above that Lewis did not consider the possibility that initial conditions would be complicated enough to play a role in influencing what would turn out the be the simplest system. In effect, therefore, Lewis was working with something like the first conception, above, of informativeness. And so the basic Lewisian intuition was that, at the end of the day, the laws that would come out of a best-system analysis would be the P-laws. If we call the best-system laws the L-laws, then the basic intuition was that (so long as initial conditions don't count too much in terms of simplicity) the L-laws would turn out to be the P-laws. Now, many critics of Lewis's account of laws might very well doubt this intuition. But we will grant the intuition. Unless one believes that this is something Lewis argued for reasonably persuasively, one will not be favorably inclined toward a Humean account of laws in the first place. So we will grant for the sake of argument that, on the first conception of informativeness, or in a world where the precise initial conditions are "cheap," the L-laws are the P-laws.

---

cosmology" tell us it did. In this form, PH is more of a law schema than a candidate law. Its hard to know if PH can be formulated precisely, even in the thermodynamic language.

[4] Unless, of course, the P-laws turn out to be indeterministic, or nonreversible, in just the right sort of way to make PROB unnecessary. I will not be discussing that possibility here. But I certainly do not mean to dismiss it in general.

Let us then also suppose that Loewer's following intuition is correct:

Adding PH and PROB to the dynamical laws results in a system that is only a little less simple but is vastly more informative than is the system consisting only of the dynamical laws (2007, p. 305)

I will assume here that when Loewer talks here of "the dynamical laws," what he means is what I am calling the P-laws.[5] (Let us call the dynamical portion of the package of laws Albert and Loewer want to construct the "ALD-laws"). So, suppose that this claim is true. And continue to suppose, as above, that the L-laws are the P-laws. How does it follow from this that the P-laws, plus PROB, plus PH, will make up the best system? It follows only if we think that there is some antecedent reason to be committed to the proposition that the P-laws necessarily make up a subset of whatever that best system will be, on the second conception. But why should we think that? We are tempted to think that, I would argue, only if we follow some mistaken reasoning.

Here is how the mistaken reasoning goes: First, rely on the first conception of informativeness to decide what the dynamical laws are. Next, choose to believe, as most Humeans will do, that Lewis's intuitions were right, and these "L-laws" (defined in terms of the first conception) will turn out to be the P-laws. Then, switch to the second conception of informativeness, and figure out if any summary of initial conditions, added to those L-laws/P-laws, add to informativeness without overly sacrificing simplicity. If we play this funny sort of game, then—plausibly—we end up with P-laws, PROB, and PH as the best system. And then we end up with the P-laws being the ALD-laws. Indeed, this seems to be the reasoning behind the inference from

P. "Adding PH and PROB to [the P-laws] results in a system that is only a little less simple but is vastly more informative than is the system consisting only of the dynamical laws" (p. 305).

to

C. The P-laws, PROB, and PH make up the best system.

Unless we are following the mistaken reasoning, why would we care what adding something to the P-laws does to informativeness and simplicity?

Of course, this kind of reasoning is not legitimate. No one should think that the laws of nature are the sorts of things that come out of this kind of two-step analysis. But this seems to be the reasoning that Loewer has in mind. If it is not,

---

[5] He can't mean anything like the *de dicto* best-system laws, or the sentence becomes strangely circular.

then the claim that adding PH and PROB to the P-laws results in more informative and only slightly less simple system than just the P-laws is a red herring. Unless we have antecedently assumed that the P-laws will be part of the package, then the claim that adding PH and PROB will be more informative and only a little less simple is irrelevant to determining what is the very best system (as opposed to the best system that is *assumed to contain* the P-laws).

Suppose, then, that we don't antecedently assume that the P-laws will be part of the package. How would we convince ourselves that the P-laws would come out to be part of the best system on the second conception of informativeness. It's hard to see how we would. After all, physicists don't test their putative laws by adding them to PH and PROB. They test them by combining them with detailed microscopic and macroscopic initial conditions. The P-laws are chosen because of their ability, in other words, to make predictions when combined with microscopic initial conditions. They are not chosen (by working physicists) for their ability to work well with PROB and PH.

If we are really trying to guess what will be the best system, where "best" is defined purely in terms of the second conception of informativeness, why not consider the possibility that some much simpler set of dynamical laws than the ones physicists will offer us—added to PH and PROB—will be just as informative, or close, as the desired package. After all, quite plausibly, any set of dynamical laws that is "friendly" to Boltzmannian reasoning (i.e., that takes almost all of the volume of the small regions of state space into large regions, that fibrillates volumes of state space, etc.) will be just as informative, or close, on the second conception, as the genuine P-laws. And surely there are candidate laws that are Boltzmann-friendly that are simpler than the real P-laws, whatever they turn out to be. Or maybe the simplest most informative system will be one without any microscopic dynamical laws at all—i.e., one that gives only thermodynamic laws.

So, the second conception of informativeness is not going to work. It either forces us to rely on an ad hoc, two-step process for determining what is the best system, or else it makes it implausible that the best-system analysis will contain the P-laws as the dynamical portion of the best package.

What, then, of the third conception of informativeness? Recall that the third proposal regarding informativeness is significantly different from the first two in that it does not involve a God's-eye comparison between the world and some fixed system of sentences. Instead, the informativeness of a system is evaluated by an active scientist/participant in the world. This sort of agent has certain pragmatic goals—things that she knows and things that she wants to predict and retrodict. Such an agent will compare the informativeness of two systems by using each of them, in combination with whatever microscopic and macroscopic information she happens to come by in the pursuit of those pragmatic goals, to predict and retrodict whatever interests her. The most informative system, on the third conception, is whichever system is best at doing that.

In one respect, this is an attractive way of thinking about informativeness for grounding the notion of PROB as a fundamental law. It is radically naturalistic and Humean. And it dovetails well with Albert's now-famous line that PROB is "the right probability distribution to use for making inferences about the past and future" (2000, p. 96). And it might be genuinely plausible, I would at least grant here for the sake of argument, that P-laws + PROB + PH will come out on top on this conception of informativeness. That's because an agent like the one we envision here will want to make both microscopic predictions from the bits of microscopic knowledge she can infer to, and will want to make macroscopic predictions about macroscopic objects about which she has only macroknowledge. Doing the former will require that the microlaws in her package be the real P-laws, rather than any old Boltzmann-friendly set, and doing the latter will require that she add PROB and PH to her macroknowledge. These considerations make the third conception attractive.

What makes the third conception much less attractive, in my opinion, is the strong tension that exists between the pragmatic nature of the proposal and a very unpragmatic feature of the rest of what is going on in the overall scheme of things.

Recall, first of all, that what we are after here is a system that will be *constitutive* of the fundamental laws of the universe. Making such a system dependent on the pragmatic goals of a particular agent, or set of agents, strikes me as unattractive. Though I would think someone like Lewis would share such a worry, I admit that perhaps having the worry is not entirely Humean. Perhaps a genuine Humean wants to relativize the fundamental laws to particular kinds of epistemic agents.

If so, however, then the following consideration becomes all the more forceful: There is an ambiguity in the claim that PROB is the "right probability distribution to use." In the sense that is most in harmony with the pragmatic spirit of the third conception of informativeness, PROB is not a probability distribution that can be used *at all*, for making predictions and retrodictions. That's because, pragmatically, we lack the ability to compute predictions using the P-laws at the scale of the entire universe.[6] The only sense in which PROB, along with PH, is the "the right probability distribution to use" is this:

If we encounter some macroscopic system, say a cocktail with ice in it, and we want to predict what it will be like in ten minutes, then here is what we do: We use macroscopic laws that we hold to be correct on the basis of macroscopic

---

[6] And given the physical constraints on computation, we will probably always lack them.

inductive evidence to predict what it will be like in ten minutes. Then, if we are so disposed, we can perhaps convince ourselves that it is plausible, given the correctness of those macroscopic laws, that *had we been able to make the calculation*, it would have come out that that outcome was overwhelmingly likely on the basis of what we knew, the P-laws, PH, and PROB. But that is a long way from the claim that, as a pragmatic matter, we could have predicted that outcome from what we knew, the P-laws, PH, and PROB.

A short example illustrates this point rather well.[7] Famously, in *Time and Chance* Albert makes the claim that the package of the dynamical laws, PH, and PROB not only enables us to predict things about the behavior of traditional thermodynamic systems, it also enables us to predict, from our knowledge of the design of an apartment and the fact that it contains a spatula, that the spatula is very likely to be found in the kitchen. In various debates about the book, the objection has been raised that if we don't know whether or not Martians have visited the apartment, moved around all the household items, and subsequently erased all macro traces of their visit, then we don't know, from that package, what the probability is of the spatula being in the kitchen. Albert's reply has been that the probability of such a Martian visit, given what we know, and the package, is very low—and so, conditional on what we know, the probability of the spatula being in the kitchen is still high. Others are skeptical that the low probability of the Martian hypothesis follows from the package. I take no stand on that issue here. I merely want to point out that all we have here is a clash of intuitions about how much confidence to have in the package. We have nothing even remotely like the ability to even crudely estimate those probabilities. The point, in short, is that from a genuinely *pragmatic* point of view, the package of the P-laws, PH, and PROB is *entirely uninformative*.

I find that these two worries about the pragmatic nature of the third conception of informativeness together combine to make it an unattractive way of grounding the lawfulness of PROB and PH. In sum, then, none of the three conceptions of informativeness provide a clear, coherent, and attractive way of understanding how PROB and PH could be fundamental laws of a Lewisian flavor.

4. What Are PROB and PH After All?

What then, of the arguments that PROB and PH must be laws? Recall that part of the point of this whole enterprise, according to *Time and Chance*, was to under-

---

[7] This example was raised by Stephen Leeds (2003).

stand how the probabilities in statistical mechanics could be objective chances—the kinds of things that are involved "*in bringing it about,* in *making it the case,* that (say) milk dissolves in coffee," rather than being about "mere degrees of belief." So one argument in favor of PROB being a law was that its being a law was the only way PROB could deliver probabilities that were not mere degrees of belief, and that the probabilities in statistical mechanics could not be degrees of belief.

Let us accept, for the moment, that if PROB is not a fundamental law, then it cannot provide objective chances. So what status does it have? And what kind of probabilities does it deliver? I would argue that we ought to understand the probabilities in PROB as objective epistemic degrees of belief, and that we ought to regard PROB itself as a nomically contingent empirical hypothesis.

Elsewhere (Winsberg, 2008) I have given independent arguments for doubting that the probabilities in PROB should be understood as objective chances. I also articulated in more detail the idea that the probabilities in PROB could be understood as representing degrees of belief, and I defended this possibility against the worries, expressed in *Time and Chance,* that degrees of belief cannot explain why the ice melts in my cocktail. Let me just reiterate here that by saying PROB is about epistemic probabilities I do *not* mean to say that they are innocent, or a priori, or reflect uniform ignorance—and I believe it is only against those positions that the arguments in *Time and Chance* apply. Instead, I think the probabilities in PROB are (objectively correct) epistemic probabilities precisely because they are the right probabilities to use given our particular epistemic situation. I also think that it is the job of PROB to *underwrite the legitimacy of our inferences* about what will happen to the ice in my cocktail. It is not the job of the probabilities in PROB to "*make it the case*" that the ice in my cocktail melts. Only the dynamical laws, in conjunction with the specific initial condition of the universe, can do that.

5. Underwriting the Laws of the Special Sciences

Here, I would like to take the space to address the argument, put forward by Loewer, that "PH and PROB *can bestow lawfulness* on [the higher-level laws] only *if they themselves are laws*" (Loewer, 2007, pp. 304–305, my emphasis).

Loewer's argument is rather simple: It appears to be a genuine law, for example, as opposed to an accidental regularity, that ice in a glass of lukewarm water will melt. Therefore, according to Loewer, this fact needs to be underwritten by the lawfulness of PROB and PH. And so the underlying premise is clear: Because thermodynamic laws are not fundamental laws—they are laws of a special science—not only their truth, but also their lawfulness, must follow from fundamental physics.

Loewer's argument, then, depends on a fairly strong variety of what we might call fundamentalism. "Fundamentalism" is word coined for such contexts by Nancy Cartwright, and it is normally understood as the doctrine that there exist fundamental laws of physics whose scope is universal (Cartwright, 1999). According to this version of fundamentalism, then, any truth discovered by a special science ought to be, at least in principle, recoverable from those fundamental laws. But the kind of fundamentalism being appealed to by Loewer here is even stronger. Not only, according to the reasoning above, must the truth of the laws of the special sciences be recoverable from the fundamental laws, but their lawfulness must be as well.

The question, then, is: Does what we know about the relationship between what we take to be the laws of the special sciences and the fundamental laws support confidence in this very strong form of fundamentalism? I believe the answer to this question is no.

To see why this is so, we need only ask ourselves why Loewer insists that both PROB and PH are required in order to underwrite the lawfulness of the special science laws. The reason PH is required, of course, is that the laws of thermodynamics have force in both temporal directions. I can not only predict, using the laws of thermodynamics, that the ice in my cocktail will melt; I can also retrodict that, at some time in the recent past, the ice in my cocktail was less melted than it is now. That this is so will not follow from the dynamical laws and PROB alone—indeed, from just them, the opposite follows. The retrodiction also requires PH. Famously so.

Unfortunately, it does not follow from PROB and PH (plus the dynamical laws) alone either. The reason for this has to do with Poincaré's recurrence theorem and other features we would expect Boltzmannian classical dynamical systems to exhibit. Roughly, Poincaré's recurrence theorem (PR) tells us that any system governed by a conservative Hamiltonian, will, after a sufficiently long period of time, return to a state arbitrarily close to any state it occupied in the past. If the requirements for PR are met, then for any presently observable macrostate of the universe, we should expect a future of maximum entropy followed by endlessly recurring thermal fluctuations back to that macrostate. More generally, we should expect a Boltzmannian universe to fluctuate through every accessible macroscopic condition it can reach, and with predictable frequencies.

This state of affairs leads to what Sean Carroll and others have dubbed the "Boltzmann's Brain" (BB) paradox.[8] To understand the BB, it helps to revisit a sug-

---

[8] See, for example, Dyson et al. (2002), Albrecht and Sorbo (2004), Linde (2007), and Carroll (2010). More precisely, what I have given is the conditions of a "classical" BB paradox. Carroll and others

gestion once made by Boltzmann himself. One of Boltzmann's initial reactions to Poincaré's theorem (when it was put forward as a "reversibility objection" to Boltzmann's attempt to reduce the second law to microphysics) was to postulate that the universe was undergoing a long history of thermal fluctuations, and that we just happened to be on a local fluctuation, with the forward direction of time just being, by definition, whatever direction was locally headed toward equilibrium. I quote the famous passage here:

> There must then be in the Universe, which is in thermal equilibrium as a whole and therefore dead, here and there relatively small regions of the size of our galaxy (which we call worlds), which during the relatively short time of eons deviate significantly from the thermal equilibrium. Among these worlds the state probability [entropy] increases as often as it decreases. For the Universe as a whole the two directions of time are indistinguishable, just as in space there is no up or down. However, just as at a certain place on the Earth's surface we can call "down" the direction toward the center of the Earth, so a living being that finds itself in such a world at a certain period of time can define the time direction as going from less probable to more probable states ... and by virtue of this definition he will find that this small region, isolated from the rest of the universe, is "initially" always in an improbable state. (Boltzmann, 1897, p. 413)

Famously, the problem with this scenario is that it leads to a skeptical paradox. If we suppose that the universe is undergoing thermal fluctuations around equilibrium, then no matter what macrocondition we observe, it becomes overwhelmingly likely that the macrostate we think we are observing is in fact a spontaneous fluctuation out of chaos, rather than a macrostate with a normal macro history (one that runs in accord with the second law.) And so it becomes overwhelmingly likely that any apparent "records" are nothing but spontaneous fluctuations, and hence are not veridical. If we assume that the only real records we have genuine access to are our own mental states, then the assumption that we are in a universe undergoing fluctuations around equilibrium leads to the conclusion that it is overwhelmingly likely that we are nothing but BBs—disembodied brains surrounded by a thermal equilibrium, whose memories and perceptions are not veridical: they

---

are more concerned with a more modern, quantum-gravitational version of the paradox. On this version, we are to take the recent evidence of the presence of dark energy in the universe to be reason to believe that we are headed toward a future universe that is a de Sitter space. In such a universe, vacuum fluctuations would give rise to behavior that was, for all intents and purposes, just like classical Poincaré recurrence. And so in fact the details do not really matter.

are nothing but spontaneous fluctuations out of chaos. And it really does not matter what we take to be the observable macrostate; the conclusion is more general. No matter what we take to be the directly surveyable macrostate, we should conclude that it is nothing but a thermal fluctuation. If you take yourself to be veridically observing that you are in a room with a desk with this book on it, you should conclude that it is overwhelmingly likely that the room with the book is a fluctuation, surrounded—outside of the reach of your eyesight—by thermal chaos, and preceded in time by more of the same.

Strangely, Carroll (2010, p. 314) and others take this result to be "a direct disagreement between theory and observation".[9] Presumably, what they mean is that any set of laws that gives rise to a BB scenario is empirically disconfirmed by what we see. But this is incorrect. In fact, no observation is ever inconsistent with the hypothesis that I am a Boltzmann Brain. In fact, *that* is the problem. Of course, this is no different from the fact that no observation is ever inconsistent with the possibility that the universe is presently in the lowest entropy condition it will be in or ever has been in. That is why, in order to avoid skeptical paradox, we need to postulate PH. PH, we might say, is a condition for the possibility of our having knowledge of the past. As we will see in what follows, something else, something akin to PH, but different, will also be required to block BB from giving rise to similar intractable worries.

What, after all, follows from the fact that we have reason to believe that the universe will eventually reach a state of equilibrium, followed by endless recurrences to macroscopic states arbitrarily close to the one we are in now? It follows that if I reason in the way Albert proposed that I do—that is, if I try to make predictions and retrodictions using a package consisting of only PH, PROB, and DL[10]—then it is overwhelmingly likely that the macrocondition I am presently "observing" is not the one that evolved directly out of the big bang, but is one of the infinitely many fluctuations out of equilibrium that will occur after the universe relaxes. And so it follows that it is overwhelmingly likely that I am a BB and no one will ever read this chapter, which exists only in my mind. More importantly, it is overwhelmingly likely that all the records (including memories) I have of the experiments that made me believe in modern physics in the first place are not veridical. It is also, then, overwhelmingly likely that none of my records are veridical. And it is also the case—and here is the rub—that none of

---

[9] See Winsberg (2012) for a more detailed discussion of this.
[10] Assuming that these dynamical laws are friendly to a Boltzmannian picture. Or, alternatively, that they have the quantum-gravitational structure of the kind Carroll thinks they have, which will give rise to equivalent problems.

the laws of the special sciences will hold, since none of them hold for Boltzmann Brains.

And so rather than being in disagreement with experience, the BB scenario is epistemologically unstable—and denying one of its premises is a precondition for the possibility of veridical records. And so denying one of its premises is a precondition for the possibility of believing in the body of evidence on which we base all of physical theories in the first place, and indeed of reasoning statistical mechanically in the first place.

6. Conclusion

So, just as we need to postulate PH in order to avoid the skeptical paradoxes discussed in *Time and Chance,* we also need to assume one more thing in order to avoid the BB scenario (and a resulting skeptical paradox that is just as bad) as the outcome of our package: we need to assume a postulate, call it the Near Past postulate (NP), which says that our present state lies somewhere between the time of the PH and time when the universe first relaxes. The epistemological status of such a principle would be exactly like the one that Albert attributes to PH in *Time and Chance*—it is a principle that we reason to transcendentally, as it were—it is a condition for the possibility of our having the very knowledge base that forms the empirical evidence base for the physics we were doing in the first place.[11]

The problem for Albert and Loewer's fundamentalism, however, is that, unlike PROB and PH, NP is not even a prima facie contender for being a law. It cannot be a law because it does not describe a fact about the world; it describes a fact about our present location in the world. To put this point most starkly, imagine supposing that it is a law of nature that the universe is 14 billion years old. This is absurd. (Can laws have expiration dates?). But to suppose that NP is a law is akin to this.

So NP cannot be a law. But NP plays exactly the same role as PH (and a role not unlike that of PROB) in "underwriting" the laws of the special sciences. So something must have gone wrong in the inference to the conclusion that everything in the package to which PH and PROB belong must be fundamental laws; it turns out to be impossible to derive the laws of the special sciences from any observable set of conditions plus a package containing only laws.

---

[11] And whether you want to think about that classically (in terms of reaching a state of thermal equilibrium) or in terms of some fancy quantum-gravitation version of entropy (presumably de Sitter space) doesn't seem to me to make any difference at all. And hence I do not really understand the worry, expressed by some, that the BB scenario is a special worry that arises in the context of recent discoveries/proposals in cosmology.

And here I agree wholeheartedly with Albert (and not incidentally, with Boltzmann, Einstein, and Feynman) that this is indeed the correct epistemological status of PH. And hence I think that NP has exactly the same epistemological status. And hence, *pace* Carroll and others, I think it is a mistake to think that a theoretical system that results in a BB scenario, but for which that scenario can be blocked with the addition of NP, is "in disagreement with observation."

So if there is a simple informative package that underwrites the truth of the laws of the special sciences—a package that, when it is added to what we can observe, tells us, for instance, that ice will melt going forward and unmelt going backward—that package must contain NP. And NP is not a law. So it cannot be that the things that are required to be in the package must be laws. So it cannot be that PROB and PH are *required* to be laws so that they can underwrite the lawfulness of the laws of the special sciences. But anyway: the various possible ways in which one could have made sense of the claim that PH and PROB are laws were unattractive. What should we conclude? I would argue that we should conclude that the strong form of fundamentalism that motivated the desire to elevate them to the status of laws in the first place was ill-advised. The lawfulness of the laws of the special sciences must be autonomous features of their respective levels of description.

References

Albert, D. (2000). *Time and Chance*. Harvard University Press.
Albrecht, A., and Sorbo, L. (2004). "Can the Universe Afford Inflation?" Physical Review D 70, 063528.
Boltzmann, L. (1897). "On Zermelo's Paper 'On the Mechanical Explanation of Irreversible Processes'" [Zu Hrn. Zermelo's Abhandlung Ober die mechanische Erklärung irreversibler Vorgange]. *Annalen der Physik* 60:392–398. English trans. ed. Stephen G. Brush, *The Kinetic Theory of Gases: An Anthology of Classic Papers with Historical Commentary Kinetic Theory*, 403–411. Imperial College Press, 2003.
Callender, C. (2011). "The Past Histories of Molecules." In Claus Beisbart and Stephan Hartmann, eds., *Probabilities in Physics*. Oxford University Press.
Carroll, Sean. (2010). *From Eternity to Here: The Quest for the Ultimate Theory of Time*. New York: Dutton.
Cartwright, Nancy. (1999). *The Dappled World: A Study of the Boundaries of Science*. Cambridge University Press.
Cohen, Jonathan, and Callender, Craig. (2009). "A Better Best System Account of Lawhood." *Philosophical Studies* 145:1, 1–34.

Dyson, L., Kleban, M., and Susskind, L. (2002). "Disturbing Implications of a Cosmological Constant." *Journal of High Energy Physics,* 0210 011.

Earman, J. (2006). "The 'Past Hypothesis': Not Even False." *Studies in History and Philosophy of Science* 37:399–430.

Frisch, M. (2011). "From Arbuthnot to Boltzmann: The Past Hypothesis, the Best System, and the Special Sciences." *Philosophy of Science* 78 (5):1001–1011.

Leeds, S. (2003). "Foundations of Statistical Mechanics—Two Approaches." *Philosophy of Science* 70 (1):126–144.

Lewis, D. (1986). A Subjectivist's Guide to Objective Chance. *Philosophical Papers,* vol. 2. Oxford University Press.

Lewis, D. (1994). "Humean Supervenience Debugged." *Mind* 103:473–489.

Linde, A. (2007). "Sinks in the Landscape, Boltzmann Brains, and the Cosmological Constant Problem." *Journal of Cosmology and Astroparticle Physics,* doi:10.1088/1475-7516/2007/01/022.

Loewer, B. (2007). "Counterfactuals and the Second Law." In H. Price & R. Corry, eds., *Causation, Physics, and the Constitution of Reality: Russell's Republic Revisited.* Oxford University Press.

Winsberg, E. (2004). "Can Conditioning on the 'Past Hypothesis' Militate Against the Reversibility Objections?" *Philosophy of Science* 71:489–504.

Winsberg, E. (2008). "Laws and Chances in Statistical Mechanics." *Studies in History and Philosophy of Science Part B: Studies in History and Philosophy of Modern Physics* 39: 872–888.

Winsberg, E. (2012). "Bumps on the Road to Here (from Eternity.)" *Entropy* 14 (3):390–406; https://doi.org/10.3390/e14030390.

Chapter Three

The Logic of the Past Hypothesis

▶ DAVID WALLACE

1. Introduction

There are no consensus positions in philosophy of statistical mechanics, but the position that David Albert eloquently defends in *Time and Chance* (Albert, 2000) is about as close as we can get.[1] It hinges on two views (in *Time and Chance*, the latter gets most of the air time, but both play crucial roles):

1. The tendency of systems' entropy to increase is basically just a consequence of the geometry of phase space. That region of phase space corresponding to a system being at equilibrium is so very large compared to the rest of phase space that unless either the dynamics or the initial state is (as Goldstein, 2001, puts it) "ridiculously special," the system will, in fairly short order, end up in the equilibrium region.
2. The observed asymmetry in statistical mechanics—in particular, the tendency of entropy to increase rather than decrease—can be derived

---

[1] Albert-like claims are also espoused by Goldstein (2001), Lebowitz (2007), Callender (2009), and Penrose (1989, 2004).

from time-symmetric microphysics provided we are willing to postulate that the entropy of the early universe is very low compared to the current entropy of the universe—what Albert has memorably dubbed the "Past Hypothesis."

There is something rather puzzling about both views. Take the first: it seems to suggest that any given system, unless it is "ridiculously special," will quickly end up in equilibrium. But of course, in the real world, we very frequently find systems far from equilibrium—indeed, life itself depends on it. And many of those systems, even when isolated from their surroundings, refuse to evolve into equilibrium. A room filled with a mixture of hydrogen and oxygen, at room temperature, can remain in that state for years or decades, yet one has only to strike a spark in that room to be reminded that it is not an equilibrium state. Indeed, a room filled with hydrogen at room temperature is not really at equilibrium: it is thermodynamically favorable for it to fuse into iron, but you would have to wait a long time for this to happen.

Furthermore, we have a detailed, quantitative understanding of exactly how quickly systems in various non-equilibrium states evolve toward equilibrium. In particular, chemists (whether of the ordinary or nuclear variety) have precise and thoroughly tested dynamical theories which predict, from the microdynamics, just how quickly systems complete their irreversible movement toward equilibrium. It is, at best, very difficult to see how these quantitative theories of the approach to equilibrium fit into the very general argument for equilibration given by Albert, Goldstein, and others.

The Past Hypothesis is puzzling in a different way. It suggests, or seems to suggest, that our knowledge of the low entropy of the early universe is somehow special: we are not supposed to know the Past Hypothesis in the way we usually know information about the past, but rather, we are justified in postulating it because without that postulate, all of our beliefs about the past would be unjustified. But there is something a little odd here: after all, we have (or think we have) rather detailed knowledge of the macroscopic state of the early universe gained from cosmology, and we can calculate its entropy fairly accurately. It is also not clear (see in particular the trenchant criticisms of Earman, 2006) exactly how imposing a low entropy at the beginning of time can lead to irreversible physics here and now.

And yet... for all that, there is clearly something to both views. There does seem to be some important sense in which irreversibility is connected with phase space volume and the behavior of typical—that is, not-ridiculously-special—systems. And it does seem that, absent time asymmetry in microphysics, as a matter of logic there must be some link between the boundary conditions of the universe and the observed time asymmetry in macrophysics.

My purpose in this paper is to try to get as clear as possible on just how the logic of deriving macrophysical irreversibility from microdynamics-plus-Past-Hypothesis is supposed to go. My starting point is the observation above: that we actually have a large body of quantitative theory about irreversible physical processes, and any adequate account of irreversibility needs to explain the quantitative success of these theories and not just the qualitative observation that systems tend to equilibrium. So from §2 to §4, I set aside philosophical concerns and try to get as clear as possible on what the mathematical route is by which we derive empirically reliable irreversible macrodynamics from reversible microdynamics. From §5 to §8, I examine just when this mathematical route is physically justified and conclude that a Past Hypothesis is indeed needed, but of a rather different character from what is usually argued. I conclude by making contact again with the two views mentioned above, and in particular, with Albert's own approach.

I should draw attention to two distinctive features of my approach. First, one frequent theme in the criticism of the two views given above has been on their lack of mathematical rigor and care (see, in particular, Frigg's (2009) criticism of Goldstein and Earman's (2006) objections to any assignment of entropy to the early universe). By contrast, I am perfectly happy to allow their proponents to make whatever plausible-sounding mathematical conjectures they like (and indeed, make several such myself in my own account). My concern, rather, is in understanding just what those conjectures are supposed to achieve and why they can be expected to achieve it. The purpose of the philosopher of physics, it might be argued, is not to prove theorems but to see which theorems are worth proving.

Second, it seems all but universal to conduct discussions of statistical mechanics at the classical level. Sklar's account of the reasons for this appears to be fairly typical:

> [T]he particular conceptual problems on which we focus—the origin and rationale of probability distributions over initial states, the justification of irreversible kinetic equations on the basis of reversible underlying dynamical equations, and so on—appear, for the most part, in similar guise in the development of both the classical and quantum versions of the theory. The hope is that by exploring these issues in the technically simpler classical case, insights will be gained that will carry over to the understanding of the corrected version of the theory. ... This way of doing things is not idiosyncratic, but common in the physics literature devoted to foundational issues. (Sklar, 1993, p. 12)

But I am not convinced that the classical case really is "technically simpler" (at least where study of general features of the theory, rather than rigorous analysis of specific systems, is our goal), nor am I confident that the conceptual problems really do appear "in similar guise." Notably, quantum mechanics contains probability at an essential level; it also includes its own form of irreversibility in the form of decoherence-induced branching. So my approach is in general to study the classical and quantum cases in parallel, and to neglect the classical theory in favor of the quantum one where they differ in important respects. If we are interested in understanding irreversibility in our world, after all, classical systems should be of interest to us only insofar as they are good approximations to quantum systems.

In discussing quantum mechanics, I assume that (i) the quantum state represents the physical features of a system, (ii) it evolves unitarily at all times, and (iii) there are no hidden variables. That is, I basically assume the Everett interpretation (discussed and developed *in extenso* in Wallace, 2012, and Saunders, Barrett, Kent, and Wallace, 2010). In doing so, of course, I part company with Albert. *Time and Chance* is an admirable exception to the usual classical-physics-only trend, but its quantum-mechanical discussions are largely confined to explicitly time-asymmetric dynamical-collapse theories. Much of what I say should, however, carry over to versions of quantum theory with hidden variables of one kind or another, such as modal interpretations or pilot-wave theories.

2. The Macropredictions of Microdynamics

For present purposes, classical and quantum mechanics have, essentially, a similar dynamical form. In both cases, we have

- A state space (phase space or (projective) Hilbert space);
- A deterministic rule determining how a point on that state space evolves over time (generated by the classical Hamiltonian and the symplectic structure, or the quantum Hamiltonian and the Hilbert-space structure, as appropriate);
- Time-reversibility, in the sense that given the state at time t, the dynamics are just as well-suited to determine the state for times before t as for times after t.

I will also assume that, whatever particular version of each theory we are working with, both theories have something which can reasonably be called a "time-reversal" operator. This is a map τ from the state space to itself, such that if the t-second evolution of x is y, then the t-second evolution of τy is τx; or, equivalently, if $x(t)$ solves the dynamical equations then so does $\tau x(-t)$. I'm not going to attempt a formal criterion for when something counts as a time-reversal operator; in classical and quantum mechanics, we know it when we see it. (Though in quantum field theory, it is the transformation called CPT, and not the one usually called T, which deserves the name).

Both theories also have what might be called, neutrally, an "ensemble" or "distributional" variant, though here they differ somewhat. In the classical case, the deterministic dynamics induce a deterministic rule to evolve functions over phase space, and not just points on phase space: if the dynamical law is given schematically by a function φ_t, so that $\varphi_t(x)$ is the t-second evolution of x, then $\varphi_{t*}\rho = \rho \cdot \varphi_t$. In more concrete and familiar terms, this takes us from the Hamiltonian equations of motion for individual systems to the Liouvillian equations for ensembles.

In the quantum case, we instead transfer the dynamics from pure to mixed states. If the t-second evolution takes state $|\psi\rangle$ to $\hat{U}_t|\psi\rangle$, the distributional variant takes density operator ρ to $\hat{U}_t \rho \hat{U}_t^\dagger$.

I stress: the existence of these distributional variants is a purely mathematical claim; no statement of their physical status has yet been made. The space of functions on, or density operators over, the state space can be thought of, mathematically speaking, as a state space in its own right, for the distributional variant of the theory.

In principle, the way we use these theories to make predictions ought to be simple: If we want to know the state of the system we're studying in t seconds' time, we just start with its state now and evolve it forward for t seconds under the microdynamics. And similarly, if we want to know its state t seconds ago, we just time-reverse it, evolve it forward for t seconds, and time-reverse it again. (Or equivalently, we just evolve it forward for $-t$ seconds.)

And sometimes, that's what we do in practice too. When we use classical mechanics to predict the trajectory of a cannonball or the orbit of a planet, or when we apply quantum mechanics to some highly controlled situation (say, a quantum computer), we really are just evolving a known state under known dynamics. But of course, in the great majority of situations this is not the case, and we have to apply approximation methods. Sometimes that's glossed as being because of our lack of knowledge of the initial state, or our inability to solve the dynamical equations exactly, but this is really only half the story. Even if we were able to calculate (say) the expansion of a gas in terms of the motions of all its myriad constituents, we would have missed important generalizations about the gas by peering too myopically at its microscopic state. We would, that is, have missed important, robust higher-level generalizations about the gas. And in quantum mechanics, the emergent behavior is frequently the only one that physically realistic observers can have epistemic access to: decoherence strongly constrains our ability to see genuinely unitary dynamical processes, because it's too difficult to avoid getting entangled with those same processes.

The point is that in general we are not interested in all the microscopic details of the systems we study, but only in the behavior of certain more coarse-grained details. It is possible (if, perhaps, slightly idealized) to give a rather general language in which to talk about this: suppose that $t_1, \ldots t_N$ is an increasing sequence of times, then a set of *macroproperties* for that sequence is an allocation, to each time t_i in the sequence, of either

1. in the classical case, a Boolean algebra of subsets of the system's phase space whose union is the entire phase space; or
2. in the quantum case, a Boolean algebra of subspaces of the system's Hilbert space whose direct sum is the entire Hilbert space.

In both cases, it is normal to specify the macroproperties as being unions or direct sums (as appropriate) of macro*states:* a set of macrostates for a (classical / quantum) system is a set of mutually (disjoint / orthogonal) (subsets / subspaces) whose (union / direct sum) is the entire state space. Throughout this paper, I will assume that any given set of macroproperties is indeed generated from some set of macrostates in this way. (And in most *practical* cases, the choice of macrostate is time-independent.) For the sake of a unified notation, I will use $\oplus$ to denote the union operation for classical sets and the direct sum operation for quantum subspaces, $\subset$ to denote the subset relation for classical sets and the subspace relation for quantum subspaces, and "disjoint" to mean either set-theoretic disjointness or orthogonality, as appropriate.

The idea of this formalism is that knowing that a system has a given macroproperty at time t_i gives us some information about the system's properties at that time, but only of a somewhat coarse-grained kind. We define a *macrohistory* α of a system as a specification, at each time t_i, of a macroproperty $\alpha(t_i)$ for that time; the set of all macrohistories for a given set of macroproperties is the *macrohistory space* for that set. It should be fairly clear that given the macrohistory space of a given set of macroproperties, we can recover that set; hence, I speak interchangeably of a macrohistory space for a theory and a set of macroproperties for the same theory. For simplicity, I usually drop the "macro" qualifier when this is not likely to cause confusion.

A few definitions: by a history of length K (where $K<N$), I mean a history which assigns the whole state space to all times t_i for $i>M$. Given histories α and β of lengths K and K' (with $K<K'$), α is then an *initial segment* of β if $\alpha(t_i)=\beta(t_i)$ for $i \leq M$. Given macrohistories α and β, we can say that α is a *coarsening* of β if $\beta(t_i) \subset \alpha(t_i)$ for each time t_i at which they are defined, and that α and β are *disjoint* if $\beta(t_i)$ and $\alpha(t_i)$ are disjoint at each t_i. A history β is the *sum* of a (countable) set of mutually disjoint histories $\{\alpha_j\}$ (write $\beta = \oplus_j \alpha_j$) if $\beta(t_i) = \oplus_j \alpha_j(t_i)$ for all t_i, and, in particular, a set of disjoint histories is *complete* if their sum is the trivial history $\hat{1}$ whose macroproperty at each time is just the whole state space. And a *probability measure* Pr for a given history space is a real function from histories to [0, 1] such that

1. If $\{\alpha_j\}$ is a countable set of disjoint histories then $\Pr(\oplus_j \alpha_j) = \sum_j \Pr(\alpha_j)$.
and
2. $\Pr(\hat{1}) = 1$.

The point of a probability measure over a history space is that it determines a set of (generally stochastic) dynamics: given two histories α and β where α is an initial segment of β, we can define the *transition probability* from α to β as $\Pr(\beta)/\Pr(\alpha)$. A set of *macrodynamics* for a (classical or quantum) system is then just a history space for that system, combined with a probability measure over that history space. A set of macrodynamics is *branching* if and only if whenever α and β agree after some time t_m but disagree at some earlier time, either $\Pr(\alpha)=0$ or $\Pr(\beta)=0$; it is *deterministic* if whenever α and β agree before some time t_m but disagree at some later time, either $\Pr(\alpha)=0$ or $\Pr(\beta)=0$.

With this formalism in place, we can consider how classical and quantum physics can actually induce macrodynamics: that is, when it will be true, given the known microdynamics, that the system's macroproperties obey a given set of macrodynamics. The simplest case is classical mechanics in its non-distributional form: any given point x in phase space will have a determinate macrostate at any given time, and so induces deterministic macrodynamics: if $U(t) \cdot x$ is the t-second evolution of x under the classical microdynamics, then

$$\Pr_x(\alpha) = 1 (if\ U(t_n - t_1) \cdot x \in \alpha(t_n)\ \text{for all } n)$$
$$\Pr_x(\alpha) = 0\ \text{(otherwise)} \tag{1}$$

To get stochastic dynamics from classical microdynamics, we need to consider the distributional version. Suppose that at time t_1 the probability of the system having state x is $\rho(x)$; then the probability at time t_n of it having state x is given by evolving ρ forward for a time $t_n - t_1$ under the distributional (Liouville) dynamics. Writing $L(t) \cdot \rho$ for the t-second evolution of ρ and $P(M) \cdot \rho$ for the restriction of ρ to the macrostate M, we define the *history super-operator* $H(\alpha)$ by

$$H(\alpha) \cdot \rho = P(\alpha(t_n)) \cdot L(t_n - t_{n-1}) \cdot P(\alpha(t_{n-1})) L(t_{n-1} - t_{n-2}) \ldots \\ L(t_2 - t_1) P(\alpha(t_1)) \cdot \rho \qquad (2)$$

$H(\alpha) \cdot \rho$ is the distribution obtained from ρ by alternately evolving forward and then restricting to the successive terms in α. So we have that the probability of history α given initial distribution ρ is

$$Pr_\rho(\alpha) = \int H(\alpha) \cdot \rho \qquad (3)$$

where the integral is over all of phase space.

A formally similar expression can be written in quantum mechanics. There, we write ρ for the system's density operator at time t_1, $L(t) \cdot \rho$ for the t-second evolution of ρ under the unitary dynamics (so if $\hat{U}(t)$ is the t-second unitary time translation operator, $L(t) \cdot \rho = \hat{U}(t) \rho \hat{U}^\dagger(t)$), and $P(M) \cdot \rho$ for the projection of ρ onto the subspace M (so that if $\hat{\Pi}_M$ is the standard projection onto that subspace, $P(M) \cdot \rho = \hat{\Pi}_M \rho \hat{\Pi}_M$). Then (2) can be understood quantum-mechanically, and (3) becomes

$$Pr_\rho(\alpha) = Tr(H(\alpha) \cdot \rho). \qquad (4)$$

The resemblance is somewhat misleading, however. For one thing, in classical physics the macrodynamics are probabilistic because we put the probabilities in by hand in the initial distribution ρ. But in quantum physics, (4) generates stochastic dynamics even for the pure-state version of quantum theory (relying on one's preferred solution to the Measurement Problem to explain why the weights of histories deserve to be called "probabilities"). And for another, (4) only defines a probability measure in special circumstances. For if we define the *history operator* $C(\alpha)$ by

$$\hat{C}(\alpha) = \hat{\Pi}_{\alpha_n} \hat{U}(t_n - t_{n-1}) \hat{\Pi}_{\alpha_{n-1}} \ldots \hat{U}(t_2 - t_1) \hat{\Pi}(\alpha_1), \qquad (5)$$

we can express $H(\alpha)$ by

$$H(\alpha) \cdot \rho = \hat{C}(\alpha) \rho \hat{C}^\dagger(\alpha) \qquad (6)$$

and rewrite (4) as

$$Pr_\rho(\alpha) = Tr(\hat{C}(\alpha)\rho\hat{C}^\dagger(\alpha)), \qquad (7)$$

in which case

$$Pr_\rho\left(\sum_j \alpha_j\right) = \sum_{j,k} Tr(\hat{C}(\alpha_j)\rho\hat{C}^\dagger(\alpha_k)), \qquad (8)$$

which in general violates the requirement that $Pr_\rho(\sum_j \alpha_j) = \sum_j Pr_\rho(\alpha_j)$. To ensure that this requirement is satisfied, we need to require that the history space satisfies the decoherence condition: that the decoherence function

$$d_\rho(\alpha,\beta) \equiv Tr(\hat{C}(\alpha)\rho\hat{C}^\dagger(\beta)) \qquad (9)$$

vanishes unless α is a coarsening of β. (A weaker requirement—that the real part of the decoherence functional vanishes—would be formally sufficient but seems to lack physical significance.) In general, this is ensured in practical examples by environment-induced decoherence (cf. Wallace, 2012, chapter 3, and references therein for further discussion).

Before moving on, I should stress that the entire concept of a history operator, as defined here, builds in a notion of time asymmetry: by construction, we have used the system's distribution at the initial time t_1 to generate a probability measure over histories defined at that time and all subsequent times. However, we could equally well have defined histories running backward in time—"antihistories," if you like—and used the same formalism to define probabilities over antihistories given a distribution at the final time for those antihistories.

3. Coarse-Grained Dynamics

The discussion so far has dealt entirely with how macroscopic dynamics can be extracted from the microscopic equations, assuming that the latter have been solved exactly. That is, the framework is essentially descriptive: it provides no shortcut to determining what the macrodynamics actually are. In reality, though, it is almost never the case that we have access to the exact micro-level solutions to a theory's dynamical equations; instead, we resort to certain approximation schemes both to make general claims about systems' macrodynamics and to produce closed-form equations for the macrodynamics of specific systems. In this section, I wish to set out what I believe to be *mathematically* going on in these approximation schemes, and what assumptions of a purely technical nature need to be made. For now, I set aside philosophical and conceptual questions, and ask the reader to do likewise.

The procedure being used is intended to allow for the fact that we are often significantly ignorant of, and significantly uninterested in, the microscopic details of the system and instead wish to gain information of a more coarse-grained nature, which seems to go like this. First, we identify a set of macroproperties (defined as above) in whose evolution we are interested. Second, we define a map $\mathcal{C}$—the *coarse-graining map*—which projects from the distribution space onto some subset S_C of the distributions. By "projection" I mean that $\mathcal{C}^2 = \mathcal{C}$, so that the distributions in S_C—the "coarse-grained" distributions—are unchanged by the map. It is essential to the idea of this map that it leaves the macroproperties (approximately) unchanged—or, more precisely, that the probability of any given macroproperty being possessed by the system is approximately unchanged by the coarse-graining map. In mathematical terms, this translates to the requirement that for any macroproperty M,

$$\int_M \mathcal{C}(\rho) = \int_M \rho \tag{10}$$

in the classical case, and

$$Tr(\Pi_M \mathcal{C}(\rho)) = Tr(\Pi_M \rho) \tag{11}$$

in the quantum case. I will also require that $\mathcal{C}$ commutes with the time-reversal operation (so that the coarse-graining of a time-reversed distribution is the time-reverse of the coarse-graining of the distribution).

We then define the *forward dynamics induced by* $\mathcal{C}$—or the $\mathcal{C}$+ dynamics for short—as follows: take any distribution, coarse-grain it, time-evolve it forward (using the microdynamics) by some small time interval Δt, coarse-grain it again, time-evolve it for another Δt, and so on. (Strictly speaking, then, Δt ought to included in the specification of the forward dynamics. However, in practice we are only interested in systems where (within some appropriate range) the induced dynamics are insensitive to the exact value of Δt).

By a *forward dynamical trajectory induced by* $\mathcal{C}$, I mean a map from (t_i, ∞) into the coarse-grained distributions (for some t_i), such that the distribution at t_2 is obtained from the distribution at t_1 by applying the $\mathcal{C}$+ dynamics whenever $t_2 > t_1$. A *section* of this trajectory is just a restriction of this map to some finite interval $[t, t']$.

What is the coarse-graining map? It varies from case to case, but some of the most common examples are:

The coarse-grained exemplar rule: Construct equivalence classes of distributions: two distributions are equivalent if they generate the same probability function over macroproperties. Pick one element in each equivalence class, and let the coarse-graining map take all elements of the equivalence class onto that element. This defines a coarse-graining rule in classical or quantum physics; in practice, however, although it is often used in foundational discussions, rather few actual applications make use of it.

The measurement rule: Replace the distribution with the distribution obtained by a nonselective measurement of the macrostate: that is, apply

$$\rho \to \sum_M \Pi_M \rho \Pi_M \qquad (12)$$

where the sum ranges over macrostates.[2] (This obviously counts as a coarse-graining only in quantum mechanics; the analogous classical version, where ρ is replaced by the sum of its restrictions to the macrostates, would be trivial.)

The correlation-discard rule: Decompose the system's state space into either the Cartesian product (in classical physics) or the tensor product (in quantum physics) of state spaces of subsystems. Replace the distribution with that distribution obtained by discarding the correlations between subsystems (by replacing the distribution with the product of its marginals or the tensor product of its partial traces, as appropriate).

One note of caution: the correlation-discard rule, though very commonly used in physics, will fail to properly define a coarse-graining map if the probability distribution over macroproperties itself contains nontrivial correlations between subsystems. In practice, this only leads to problems if the system does not behave deterministically at the macroscopic level, so that such correlations can develop from initially uncorrelated starting states. Where this occurs, the correlation-discard rule needs generalizing: decompose the distribution into its projections onto macrostates, discard correlations of these macrostates individually, and re-sum. Note, though, that in quantum mechanics this means that two

---

[2] To avoid problems with the quantum Zeno effect (Misra and Sudarshan, 1977; see Home and Whitaker, 1997, for a review) for very small, the measurement rule strictly speaking ought to be slightly unsharpened (for instance, by using some POVM formalism rather than sharp projections onto subspaces); the details of this do not matter for our purposes.

coarse-grainings are being applied: to "decompose the distribution into its projections onto macrostates and then re-sum" is just to perform a non-selective measurement on it—that is, to apply the measurement rule for coarse-graining.

Another example is again often used in foundational discussions of statistical mechanics but turns up rather less often in practical applications.

The smearing rule: Blur the fine structure of the distribution by the map

$$\rho' = \int dq' dp' f(q', p') T(q', p') \cdot \rho \quad (13)$$

where $T(q', p')$ is a translation by (q', p') in phase space, and f is some function satisfying $\int f = 1$, whose macroscopic spread is small. A simple choice, for instance, would be to take f to be a suitably normalized Gaussian function, so that

$$\rho' = \mathcal{N} \int dq' dp' \exp\left[-\frac{(q-q')^2}{(\Delta q^2)}\right] \exp\left[-\frac{(p-p')^2}{(\Delta p^2)}\right] \rho(q, p) \quad (14)$$

where ρ is to be read as either the phase-space probability distribution (classical case) or the Wigner-function representation of the density operator (quantum case).

For a given system of $C+$ dynamics, I will call a distribution *stationary* if its forward time evolution, for all times, is itself. (So, stationary distributions are always coarse-grained.) Classic examples of stationary distributions are the (classical or quantum) canonical and microcanonical ensembles. Distributions involving energy flow (such as those used to describe stars) look stationary, but generally aren't, as the energy eventually runs out.

How do we generate empirical predictions from the coarse-grained dynamics? In many cases, this is straightforward because those dynamics are deterministic at the macroscopic level ("macrodeterministic"): if we begin with a coarse-grained distribution localized in one macrostate, the $C+$ dynamics carry it into a coarse-grained distribution still localized in one (possibly different) macrostate.

More generally, though, what we want to know is: how probable is any given sequence of macrostates? That is, we need to apply the history framework used in §2. All that this requires is for us to replace the (in-practice-impossible-to-calculate) macrodynamics induced by the microdynamics with the coarse-grained dynamics: if $L^{C+}(t) \cdot \rho$ is the t-second evolution of ρ under the $C+$ dynamics, and

$P(M) \cdot \rho$ is again the projection of ρ onto the macroproperty M, then we can construct the coarse-grained history super-operator

$$H^{C+}(\alpha) = P(\alpha(t_n)) \cdot L^{C+}(t_n - t_{n-1}) \cdot P(\alpha(t_{n-1})) \cdot$$
$$L^{C+}(t_{n-1} - t_{n-2}) \ldots L^{C+}(t_2 - t_1) P(\alpha(t_1)). \quad (15)$$

(It should be pointed out for clarity that each $L^{C+}(t_k - t_{k-1})$ typically involves the successive application of many coarse-graining operators, alternating with evolution under the fine-grained dynamics; put another way, typically $t_k - t_{k-1} \gg \Delta t$. Even for the process to be well defined, we have to have $t_k - t_{k-1} \geq \Delta t$; in the limiting case where $t_k - t_{k-1} = \Delta t$, we obtain $H^{C+}(\alpha)$ by alternately applying *three* operations: evolve, coarse-grain, project.)

We can then define the probability of a history by

$$Pr_\rho^{C+}(\alpha) = \int H^{C+}(\alpha) \cdot \rho \quad (16)$$

in the classical case and

$$Pr_\rho^{C+}(\alpha) = Tr(H^{C+}(\alpha) \cdot \rho) \quad (17)$$

in the quantum case.

The classical expression automatically determines a set of (generally stochastic) macrodynamics (that is, a probability measure over histories); the quantum expression does also, provided that all the coarse-grained distributions are diagonalized by projection onto the macrostates: that is, provided that

$$\mathcal{C} \cdot \rho = {}_M\sum_M P(M) \cdot \mathcal{C} \cdot \rho \quad (18)$$

where the sum ranges over macrostates. This condition is satisfied automatically by the measurement and correlation-discard rules (the latter rules, recall, build in the former); it will be satisfied by the coarse-grained exemplar rule provided that the exemplars are chosen appropriately; it will be satisfied approximately by the smearing rule given that the smearing function is small on macroscopic scales.

Examples in physics where this process is used to generate macrodynamics include the following.[3]

---

[3] It is of interest to note that all these examples—and indeed, all the examples of which I am aware—use the correlation-discard coarse-graining rule or the coarse-grained exemplar rule. The other rules, so far as I know, are used in foundational discussions but not in practical applications—though I confess freely that I have made no systematic study to verify this.

Boltzmann's derivation of the H-theorem: Boltzmann's "proof" that a classical gas approached the Maxwell-Boltzmann distribution requires the "Stosszahlansatz" hypothesis—the assumption that the momenta of gas molecules are uncorrelated with their positions. This assumption is in general very unlikely to be true (cf. the discussion in Sklar, 1993, pp. 224–247), but we can reinterpret Boltzmann's derivation as the forward dynamics induced by the coarse-graining process of simply discarding those correlations.

More general attempts to derive the approach to equilibrium: As was already noted, the kind of mathematics generally used to explore the approach of classical systems to equilibrium proceeds by partitioning phase space into cells and applying a smoothing process to each cell. (See Sklar, 1993, pp. 212–214, for a discussion of such methods; I emphasize once again that at this stage of the discussion I make no defense of their conceptual motivation.)

Kinetic theory and the Boltzmann equation: Pretty much all of nonequilibrium kinetic theory operates, much as in the case of the H-theorem, by discarding the correlations between different particles' velocities. Methods of this kind are used in weakly interacting gases, as well as in the study of galactic dynamics (Binney and Tremaine, 2008). The BBGKY hierarchy of successive improvements of the Boltzmann equation (cf. Sklar, 1993, pp. 207–210, and references therein) can be thought of as introducing successively more sophisticated coarse-grainings, which preserve N-body correlations up to some finite N but not beyond.

Environment-induced decoherence and the master equation: Given our goal of understanding the asymmetry of quantum branching, quantitative results for environment-induced decoherence are generally derived by (in effect) alternating unitary (and entangling) interactions of system and environment with a coarse-graining defined by replacing the entangled state of system and environment with the product of their reduced states (derived for each system by tracing over the other system).

Local thermal equilibrium: In pretty much all treatments of heat transport (in, for instance, oceans or stars), we proceed by breaking the system up into regions large enough to contain many particles but also small enough to treat properties such as density or pressure as constant across them. We then take each system to be at instantaneous thermal equilibrium at each time and study their interactions.

In most of the above examples, the coarse-graining process leads to deterministic macrodynamics. Some (rather theoretical) examples where it does not, include the following:

Rolling dice: We don't normally do an explicit simulation of the dynamics that justifies our allocation of probability 1/6 to each possible outcome of rolling a die. But qualitatively speaking, what is going on is that (i) symmetry considerations tell us that the region of phase space corresponding to initial conditions that lead to any given outcome has Liouville volume 1/6 of the total initial-condition volume; (ii) because the dynamics are highly random, any reasonably large and reasonably Liouville-smooth probability distribution over the initial conditions will therefore overlap to degree 1/6 with the region corresponding to each outcome; and (iii) any coarse-graining process that delivers coarse-grained states which are reasonably large and reasonably Liouville-smooth will therefore have probability 1/6 of each outcome.

Local thermal equilibrium for a self-gravitating system: Given a self-gravitating gas, the methods of local thermal equilibrium can be applied, but (at least in theory) we need to allow for the fact that a distribution which initially is fairly sharply peaked on a spatially uniform (and so, non-clumped) state will, in due course, evolve through gravitational clumping into a sum of distributions peaked on very non-uniform states. In this situation, the macrodynamics will be highly nondeterministic, and so, if we want to coarse-grain by discarding long-range correlations, we first need to decompose the distribution into macroscopically definite components.

Decoherence of a system with significant system-environment energy transfer: If we have a quantum system being decohered by its environment, and if there are state-dependent processes that will transfer energy between the system and its environment, then macro-level correlations between, say, the position of the system's center of mass and the temperature of the environment may develop, and tracing these out will be inappropriate. Again, we need to decompose the system into components with fairly definite macroproperties before performing the partial trace.

4. Time-Reversibility in Coarse-Grained Dynamics

The process used to define forward dynamics is—as the name suggests—explicitly time-asymmetric, and this makes it at least possible that the forward dynamics are themselves time-irreversible. In fact, that possibility is in general fully realized, as we shall see in this section.

Given a dynamical trajectory of the microdynamics, we know that we can obtain another dynamical trajectory by applying the time-reversal operator and then running it backward. Following this, we will say that a given segment of a dynamical trajectory of the coarse-grained dynamics is time-reversible if the corresponding statement holds true. That is, if $\rho(t)$ is a segment of a dynamical trajectory (for $t \subset [t_1, t_2]$), then it is reversible if and only if $T_\rho(-t)$ is a segment of a dynamical trajectory (for $t \in [-t_1, t_2]$).[4]

Although the microdynamics are time-reversible, in general the coarse-graining process is not, and this tends to prevent the existence of time-reversible coarse-grained trajectories. It is, in fact, possible to define a function S_G—the *Gibbs entropy*—on distributions, such that S_G is preserved under microdynamical evolution and under time-reversal, but such that for any distribution ρ, $S_G(C\rho) \geq S_G(\rho)$, with equality only if $C\rho = \rho$. (And so, since the forward dynamics consist of alternating microdynamical evolution and coarse-graining, S_G is nondecreasing on any dynamical trajectory of the forward dynamics.) In the classical case, we take

$$S_G(\rho) = -\int \rho \ln \rho \qquad (19)$$

and in the quantum case we use

$$S_G(\rho) = -Tr(\rho \ln \rho). \qquad (20)$$

(At the risk of repetitiveness: I am assuming *absolutely nothing* about the connection or otherwise between this function and thermodynamic entropy; I use the term "entropy" purely to conform to standard usage.) Of the coarse-graining methods described above, the fact that correlation-discarding, measurement, and smearing increase Gibbs entropy is a well-known result of (classical or quantum) information theory; the exemplar rule will increase Gibbs entropy provided that the exemplars are chosen to be maximal-entropy states, which we will require.

The existence of a Gibbs entropy function for C is not itself enough to entail the irreversibility of the $C+$ dynamics. Some coarse-grained distributions might actually be carried by the microdynamics to other coarse-grained distributions, so that no further coarse-graining is actually required.

---

[4] Note that I assume, tacitly, that the dynamics are time-translation invariant, as is in fact the case in both classical and quantum systems in the absence of explicitly time-dependent external forces.

I will call a distribution *Boring* (over a given time period) if evolving its coarse-graining forward under the microdynamics for arbitrary times within that time period leads only to other coarse-grained distributions, and *Interesting* otherwise. The most well-known Boring distributions are stationary distributions—distributions whose forward time evolution under the microdynamics leaves them unchanged—such as the (classical or quantum) canonical and microcanonical distributions; any distribution whose coarse-graining is stationary is also Boring. On reasonably short timescales, generic states of many other systems—planetary motion, for instance—can be treated as Boring or nearly so.[5] However, if the ergodic hypothesis is true for a given system (an assumption which otherwise will play no part in this paper), then on sufficiently long timescales, the only Boring distributions for that system are those whose coarse-grainings are uniform on each energy hypersurface.

If a segment of a dynamical trajectory of the $C+$ dynamics contains any distributions that are Interesting on timescales short compared to the segment's length, that segment is irreversible. For in that case, nontrivial coarse-graining occurs at some point along the trajectory, and so the final Gibbs entropy is strictly greater than the initial Gibbs entropy. Time-reversal leaves the Gibbs entropy invariant, so it follows that for the time-reversed trajectory, the initial Gibbs entropy is higher than the final Gibbs entropy. But we have seen that Gibbs entropy is nondecreasing along any dynamical trajectory of the forward dynamics, so the time-reversed trajectory cannot be allowed by those dynamics.

So: the coarse-graining process C takes a dynamical system (classical or quantum mechanics), which is time-reversal invariant, and generates a new dynamical system ($C+$, the forward dynamics induced by C), which is irreversible. Where did the irreversibility come from? The answer is hopefully obvious: it was put in by hand. We could equally well have defined a *backward* dynamics induced by C ($C-$ for short) by running the process in reverse: starting with a distribution, coarse-graining it, evolving it backward in time by some time interval, and iterating. And of course, the time-reversal of any dynamical trajectory of $C+$ will be a dynamical trajectory of $C-$, and vice versa.

It follows that the forward dynamics and backward dynamics in general make contradictory claims. If we start with a distribution at time t_i, evolve it forward in time to t_f using the $C+$ dynamics, and then evolve it backward in time using the $C-$ dynamics, in general we do *not* get back to where we started.

---

[5] More precisely, in general a system's evolution will be Boring on timescales short relative to its Lyapunov timescale.

This concludes the purely mathematical account of irreversibility. One more physical observation is needed, though: the forward dynamics induced by coarse-graining classical or quantum mechanics has been massively empirically successful. Pretty much all of our quantitative theories of macroscopic dynamics rely on it, and those theories are in general *very* well confirmed by experiment. With a great deal of generality—and never mind the conceptual explanation as to *why* it works—if we want to work out quantitatively what a large physical system is going to do in the future, we do so by constructing coarse-graining-induced forward dynamics.

On the other hand (of course), the *backward* dynamics induced by basically any coarse-graining process is not empirically successful at all: in general, it wildly contradicts our actual records of the past. And this is inevitable given the empirical success of the forward dynamics: on the assumption that the forward dynamics are not only predictively accurate now but also were in the past (a claim supported by very extensive amounts of evidence), then—since they are in conflict with the backward dynamics—it cannot be the case that the backward dynamics provide accurate ways of retrodicting the past. Rather, if we want to retrodict, we do so via the usual methods of scientific inference: we make tentative guesses about the past, and test those guesses by evolving them forward via the forward dynamics and comparing them with observation. (The best-known and best-developed account of this practice is the Bayesian one: we place a credence function on possible past states, deduce how likely a given present state is conditional on each given past state, and then use this information to update the past-state credence function via Bayes' theorem.)

5. Microdynamical Underpinnings of the Coarse-Grained Dynamics

In this section and §6, I turn my attention from the practice of physics to the justification of that practice. That is: given that (we assume) it is really the macrodynamics induced by the microdynamics—and not the coarse-grained dynamics—that describe the actual world, under what circumstances do those two processes give rise to the *same* macrodynamics?

There is a straightforward technical requirement which will ensure this: we need to require that for every history α,

$$CH(\alpha)\rho = H^{C+}(\alpha)\rho. \tag{21}$$

That is, the result of alternately evolving ρ forward under the fine-grained dynamics and restricting it to a given term in a sequence of macroproperties must be the same, up to coarse-graining, as the result of doing the same with the coarse-grained dynamics. If ρ and C jointly satisfy this condition (for a given history space), we say that ρ is *forward predictable* by C on that history space. (Mention of a history space will often be left tacit.) Note that in the quantum case, if ρ is forward predictable by C, it follows that the macrohistories are decoherent with respect to ρ.

I say "forward" because we are using the coarse-grained *forward* dynamics. Pretty clearly, we can construct an equivalent notion of backward predictability, using the backward coarse-grained dynamics and the antihistories mentioned in §2. And equally clearly, ρ is forward predictable by C if and only if its time-reverse is backward predictable by C.

Forward predictability is closely related to the (slightly weaker) notion of *forward compatibility*. A distribution ρ is *forward compatible* with a given coarse-graining map C if evolving ρ forward under the microdynamics and then coarse-graining at the end gives the same result as evolving ρ forward (for the same length of time) under the coarse-grained dynamics. (Note that forward compatibility, unlike forward predictability, is not defined relative to any given history space.) Forward predictability implies forward compatibility (just consider the trivial history, where the macrostate at each time is the whole state space), and the converse is true in systems that are macrodeterministic. More generally, if $H(\alpha)\rho$ is forward compatible with C for all histories α in some history space, then ρ is forward predictable by C on that history space.

Prima facie, one way in which forward compatibility could hold is if the coarse-graining rule is actually physically implemented by the microdynamics: if, for instance, a distribution ρ is taken by the micrograined dynamics to a distribution $C\rho$ on timescales short compared to those on which the macroproperties evolve, then all distributions will be forward compatible with C. And indeed, if we want to explain how one set of coarse-grained dynamics can be compatible with another even coarser-grained set of dynamics, this is very promising. We can plausibly explain the coarse-graining rule for local equilibrium thermodynamics, for instance, if we start from the Boltzmann equation and deduce that systems satisfying that equation really do evolve quickly into distributions which are locally canonical. (Indeed, this is the usual defense given of local thermal equilibrium models in textbooks.)

But clearly, this cannot be the explanation of forward compatibility of the *fine-grained* dynamics with any coarse-graining rule. For by construction, the coarse-graining rules invariably increase Gibbs entropy, whereas the fine-grained dynamics leave it static. One very simple response, of course, would be just to postulate an explicit modification to the dynamics which enacts the coarse-graining. In classical mechanics, Ilya Prigogine has tried to introduce such modifications (see, e.g., Prigogine, 1984, and references therein); in quantum mechanics, of course, the introduction of an explicit, dynamical rule for the collapse of the wave function could be thought of as a coarse-graining, and the final chapter of *Time and Chance* can be seen as developing this idea.

However, at present there remains no direct empirical evidence for any such dynamical coarse-graining. For this reason, I will continue to assume that the unmodified microdynamics (classical or quantum) should be taken as exact.

Nonetheless, it would not be surprising to find that distributions are, in general, forward compatible with coarse-graining. Putting aside exemplar rules for coarse-graining, there are strong heuristic reasons to expect a given distribution generally to be forward compatible with the other three kinds of rules:

- A distribution will be forward compatible with a smearing coarse-graining rule whenever the microscopic details of the distribution do not affect the evolution of its overall spread across phase space. Whilst one can imagine distributions where the microscopic structure is very carefully chosen to evolve in some particular way contrary to the coarse-grained prediction, it seems heuristically reasonable to suppose that generically this will not be the case, and that distributions (especially reasonably widespread distributions) which differ only on very small length scales at one time will tend to differ only on very small length scales at later times. (However, I should note that I find this heuristic only *somewhat* plausible, and in light of the dearth of practical physics examples which use this rule, I'd be relaxed if readers are unpersuaded!)
- A distribution will be forward compatible with a correlation-discard coarse-graining rule whenever the details of the correlation do not affect the evolution of the macroscopic variables. Since macroscopic properties are typically local, and correlative information tends to be highly delocalized, heuristically, one would expect that generally the details of the correlations are mostly irrelevant to the macroscopic properties—only in very special cases will they be arranged in just such a way as to lead to longer-term effects on the macroproperties.
- A distribution will be forward compatible with a measurement coarse-graining rule (which, recall, is nontrivial only for quantum theory)

whenever interference between components of the distribution with different macroproperties does not affect the evolution of those macroproperties. This is to be expected whenever the macroproperties of the system at a given time leave a trace in the microproperties at that time which is not erased at subsequent times: when this is the case, constructive or destructive interference between branches of the wave function cannot occur. Decoherence theory tells us that this will very generically occur for macroscopic systems: particles interacting with the cosmic microwave background radiation or with the atmosphere leave a trace in either; the microscopic degrees of freedom of a nonharmonic vibrating solid record a trace of the macroscopic vibrations, and so forth. These traces generally become extremely delocalized, and are therefore not erasable by local physical processes. In principle, one can imagine that eventually they relocalize and become erased—indeed, this will certainly happen (on absurdly long timescales) for spatially finite systems—but it seems heuristically reasonable to expect that on any realistic timescale (and for spatially infinite systems, perhaps on any timescale at all) the traces persist.

At least in the deterministic case, forward compatibility implies forward predictability; even in probabilistic cases, these kinds of heuristics suggest—again, only heuristically—that forward predictability is generic.

In any case, my purpose in this paper is not to prove detailed dynamical hypotheses but to identify those hypotheses that we need. So—given the above heuristic arguments—we could try postulating a:

Bold Dynamical Conjecture: For any system of interest to studies of irreversibility, all distributions are forward predictable by the appropriate coarse-graining of that system on the appropriate history space for that system.

It is clear that, were the Bold Dynamical Conjecture correct, it would go a long way toward explaining why coarse-graining methods work.

But the line between boldness and stupidity is thin, and—alas—the Bold Dynamical Conjecture strides Boldly across it. For suppose $X = C\rho$ is the initial state of some Interesting segment of a dynamical trajectory of the forward coarse-grained dynamics (Interesting so as to guarantee that Gibbs entropy increases on this trajectory) and that X' is the final state of that trajectory (say, after time t). Then by the Bold Dynamical Conjecture, X' can be obtained by evolving ρ

forward for time t under the fine-grained dynamics (to some state ρ', say) and then coarse-graining.

Now suppose we take the time-reversal TX' of X' and evolve it forward for t seconds under the coarse-grained forward dynamics. By the Bold Dynamical Conjecture, the resultant state could be obtained by evolving $T\rho'$ forward for t seconds under the fine-grained dynamics and then coarse-graining. Since the fine-grained dynamics are time-reversible, this means that the resultant state is the coarse-graining of $T\rho$. And since coarse-graining and time-reversal commute, this means that it is just the time-reverse TX of X.

But this yields a contradiction. For Gibbs entropy is invariant under time-reversal, so $S_G(TX) = S_G(X)$ and $S_G(TX') = S_G(X')$. It is nondecreasing on any trajectory, so $S_G(TX) \geq S_G(TX')$. And it is increasing (since the trajectory is Interesting) between X and X', so $S_G(X') > S_G(X)$. So, the Bold Dynamical Conjecture is false; and, more generally, we have shown that if $C\rho$ is any coarse-grained distribution on a trajectory of the forward coarse-grained dynamics which has higher Gibbs entropy than the initial distribution on that trajectory, then $T\rho$ is *not* forward compatible with C.

So much for the Bold Dynamical Conjecture. But just because not *all* distributions are forward compatible with C, it does not follow that none are; it does not even follow that most aren't. Indeed, the (admittedly heuristic) arguments above certainly seem to suggest that distributions, which are in some sense "generic" or "typical" or "non-conspiratorial" or some such term, will be forward compatible with the coarse-graining rules. In general, the only known way to construct *non*-forward-compatible distributions is to evolve a distribution forward under the fine-grained dynamics and then time-reverse it.

This suggests a more modest proposal:

Simple Dynamical Conjecture: (For a given system with coarse-graining C): Any distribution whose structure is at all simple is forward predictable by C; any distribution *not* so predictable is highly complicated and, as such, is not specifiable in any simple way *except* by stipulating that it is generated via evolving some other distribution in time (for instance, by starting with a simple distribution, evolving it forward in time, and then time-reversing it).

Of course, the notion of "simplicity" is hard to pin down precisely, and I will make no attempt to do so here. (If desired, the Simple Dynamical Conjecture can be taken as a family of conjectures, one for each reasonable precisification of "simple.") But for instance, any distribution specifiable in closed functional form (such as the microcanonical or canonical distributions), or any distribution that's uniform over a given (reasonably simply specified) macroproperty, would count as specifiable in a simple way.'

In fact, it will be helpful to define a *Simple* distribution as any distribution specifiable in a closed form in a simple way, without specifying it via the time evolution of some other distribution. Then the Simple Dynamical Conjecture is just the conjecture that all Simple distributions are forward predictable by the coarse-graining. Fairly clearly, for any precisification of the notion of Simple, a distribution will be Simple if and only if its time-reverse is.

Are individual states (that is, classical single-system states or quantum pure states) Simple? It depends on the state in question. Most classical or quantum states are not Simple at all: they require a great deal of information to specify. But there are exceptions: some product states in quantum mechanics will be easily specifiable, for instance, and so would states of a classical gas where all the particles are at rest at the points of a lattice. This in turn suggests that the Simple Dynamical Conjecture may well fail in certain classical systems (specifically, those whose macrodynamics are in general indeterministic): Simple classical systems will generally have highly unusual symmetry properties and so may behave anomalously. For example, a generic self-gravitating gas will evolve complex and highly asymmetric structure because small density fluctuations get magnified over time, but a gas with no density fluctuations whatsoever has symmetries which cannot be broken by the dynamics, and so will remain smooth at all times.

This appears, however, to be an artifact of classical mechanics that disappears when quantum effects are allowed for. A quantum system with a similar dynamics will evolve into a superposition of the various asymmetric structures; in general, the classical analogue of a localized quantum wave function is a narrow Gaussian distribution, not a phase-space point. So I will continue to assume that the Simple Dynamical Conjecture holds of those systems of physical interest to us.

6. Microdynamical Origins of Irreversibility: The Classical Case

It is high time to begin addressing the question of what all this has to do with the real world. I begin with the classical case, although, of course, the quantum case is ultimately more important. The question at hand is: On the assumption that classical microphysics is true for some given system, what additional assumptions need to be made about that system in order to ensure that its macroscopic behavior is correctly predicted by the irreversible dynamics generated by coarse-graining?

The most tempting answer, of course, would be "none." It would be nice to find that absolutely any system has macroscopic behavior well described by the coarse-grained dynamics. But we know that this cannot be the case: the coarse-grained dynamics are irreversible, whereas the microdynamics are time-reversal-invariant, so it cannot be true that all microstates of a system evolve in accordance with the coarse-grained dynamics. (A worry of a rather different kind is that the coarse-grained dynamics are in general probabilistic, whereas the classical microdynamics are deterministic.)

This suggests that we need to supplement the microdynamics with some restrictions on the actual microstate of the system. At least for the moment, I will assume that such restrictions have a probabilistic character; I remain neutral for now as to how these probabilities should be understood.

A superficially tempting move is just to stipulate that the correct probability distribution over microstates of the system is at all times forward predictable by the coarse-graining. This would be sufficient to ensure the accuracy of the irreversible dynamics, but it is all but empty: to be forward predictable by the coarse-graining *is* to evolve, up to coarse-graining, in accordance with the irreversible dynamics.

Given the Simple Dynamical Conjecture, an obvious alternative presents itself: stipulate that the correct probability distribution over microstates is at all times Simple. This condition has the advantage of being non-empty, but it suffers from two problems: it is excessive, and it is impossible. It is excessive because the probability distribution at one time suffices to fix the probability distribution at all other times, so there is no need to independently impose it at more than one time. And it is impossible because, as we have seen, in general the forward time evolution of a Simple distribution is not Simple. So, if we're going to impose Simplicity as a condition, we'd better do it once at most.

That being the case, it's pretty clear when we have to impose it: at the beginning of the period of evolution in which we're interested. Imposing Simplicity at time t guarantees the accuracy of the forward coarse-grained dynamics at times later than t; but by time-reversibility (since the time-reverse of a Simple distribution is Simple), it also guarantees the accuracy of the *backward* coarse-grained dynamics at times earlier than t, which we need to avoid. So, we have a classical recipe for the applicability of coarse-grained methods to classical systems: they will apply, over a given period, only if at the beginning of that period the probability of the system having a given microstate is specified by a Simple probability function.

So, exactly when should we impose the Simplicity criterion? There are basically two proposals in the literature:

1. We should impose it, on an ad hoc basis, at the beginning of any given process that we feel inclined to study.
2. We should impose it, once and for all, at the beginning of time.

The first proposal is primarily associated with the objective Bayesian approach pioneered by Jaynes (see, e.g., Jaynes, 1957a, 1957b, 1968)—and I have to admit to finding it incomprehensible. In no particular order:

- We seem to be reasonably confident that irreversible thermodynamic processes take place even when we're not interested in them.
- Even if we are uninterested in the fact that our theories predict anti-thermodynamic behavior of systems before some given time, they still do (i.e., the problem that our theories predict anti-thermodynamic behavior doesn't go away just because they make those predictions before the point at which we are "inclined to study" the system in question).
- The direction of time is put in by hand, via an a priori assumption that we impose our probability measure at the beginning, rather than the end, of the period of interest to us. This seems to rule out any prospect of understanding (for instance) humans themselves as irreversible physical systems.

Perhaps the most charitable way to read the first proposal is as a form of strong operationalism, akin to the sort of operationalism proposed in the foundations of quantum mechanics by, e.g., Fuchs and Peres (2000). In this paper, though, I presuppose a more realist approach to science, and from that perspective the second proposal is the only one that seems viable: we must impose Simplicity at the beginning of time. The time asymmetry in irreversible processes is due to the asymmetry involved in imposing the condition at one end of time rather than the other.

(Incidentally, one can imagine a cosmology—classical or quantum—according to which there is no well-defined initial state—for instance, because the state can be specified at arbitrarily short times after the initial singularity but not at the singularity itself, or because the notion of spacetime itself breaks down as one goes farther into the past. If this is the case, some more complicated formulation would presumably be needed, but it seems unlikely that the basic principles would be unchanged. For simplicity and definiteness, I will continue to refer to "the initial state.")

At this point, a technical issue should be noted. My definition of the Simple Dynamical Conjecture was relative to a choice of system and coarse-graining; what is the appropriate system if we want to impose Simplicity at the beginning of time? The answer, presumably, is that the system is the universe as a whole, and the coarse-graining rule is just the union of all the coarse-graining rules we wish to use for the various subsystems that develop at various times. Presumably there ought to exist a (probably imprecisely defined) maximally fine-grained choice of coarse-graining rule such that the Simple Dynamical Conjecture holds for that rule; looking ahead to the quantum-mechanical context, this seems to be what Gell-Mann and Hartle (2007) mean when they talk about a maximal quasi-classical domain.

So: if the probabilities we assign to possible initial states of the universe are given by a Simple probability distribution, and if we accept classical mechanics as correct, we would predict that the coarse-grained forward dynamics are approximately correct predictors of the probability of the later universe having a given state. We are now in a position to state an assumption which suffices to ground the accuracy of the coarse-grained dynamics.

> **Simple Past Hypothesis (classical version):** There is some Simple distribution ρ over the phase space of the universe such that for any point x, $\rho(x)\delta V$ is the objective probability of the initial state of the universe being in some small region δV around x.

(By "objective probability" I mean that the probabilities are not mere expressions of our ignorance, but are in some sense objectively correct.)

To sum up: if (a) the world is classical, (b) the Simple Dynamical Conjecture is true of its dynamics (for given coarse-graining C), and (c) the Simple Past Hypothesis is true, then the initial state of the world is forward predictable by the C dynamics: the macrodynamics defined by the C dynamics are the same as the macrodynamics induced by the microdynamics.

7. Microdynamical Origins of Irreversibility: The Quantum Case

Rather little of the reasoning above actually made use of features peculiar to classical physics. So the obvious strategy to take in the case of quantum mechanics is just to formulate a quantum-mechanical version of the Simple Past Hypothesis involving objective chances of different pure states, determined by some Simple probability distribution.

There are, however, two problems with this: one conceptual, one technical. The technical objection is that quantum distributions are density operators, and the relation between density operators and probability distributions over pure states is one-to-many. The conceptual objection is that quantum mechanics already incorporates objective chances, and it is inelegant, to say the least, to introduce additional such.

However, it may be that no such additional objective chances are in fact necessary, for two reasons.

1. There may be many pure states that are Simple and that are reasonable candidates for the state of the very early universe.
2. It is not obvious that pure, rather than mixed, states are the correct way to represent the states of individual quantum systems.

To begin with the first: as I noted previously, there is no problem in quantum mechanics in regarding certain pure states as Simple, and the (as always, heuristic) motivations for the Simple Dynamical Conjecture are no less true for these states. As for the second, mathematically speaking, mixed states do not seem obviously more alien than pure states as representations of quantum reality. Indeed, if we wish to speak at all of the states of individual systems in the presence of entanglement, the only option available is to represent them by mixed states. And since the universe appears to be open, and the vacuum state of the universe appears to be entangled on all length scales (cf. Redhead, 1995, and references therein), even the entire observable universe cannot be regarded as in a pure state.

This being the case, I tentatively formulate the quantum version of the Simple Past Hypothesis as follows.

Simple Past Hypothesis (quantum version): The initial quantum state of the universe is Simple.

What is the status of the Simple Past Hypothesis? One way to think of it is as a hypothesis about whatever law of physics (fundamental or derived) specifies the state of the very early universe: that that law requires a Simple initial state. Indeed, if one assumes that probabilistic physical laws must be simple (which seems to be part of any reasonable concept of "law"), and that simplicity entails Simplicity, all the Simple Past Hypothesis amounts to is the

Past Law Hypothesis: The initial quantum state of the universe is determined by some law of physics.

Alternatively, we might think of the Simple Past Hypothesis as a (not very specific) conjecture about the contingent facts about the initial state of the universe, unmediated by law. Indeed, it is not clear that there is any very important difference between these two readings of the Hypothesis. In either case, the route by which we come to accept the Hypothesis is the same: because of its power to explain the present-day observed phenomena, and in particular, the success of irreversible macrodynamical laws. And on at least some understandings of "law" (in particular, on a Humean account like that of Lewis (1986) where laws supervene on the actual history of the universe) there is not much metaphysical gap between (i) the claim that the initial state of the universe has particular Simple form X and this cannot be further explained, and (ii) the claim that it is a law that the initial state of the universe is X.

8. A Low-Entropy Past?

The suggestion, espoused by Albert, that the origin of irreversibility lies in constraints on the state of the early universe is hardly new: it dates back to Boltzmann, and has been espoused in recent work by, among others, Penrose (1989, 2004), Goldstein (2001), and Price (1996). But their Past Hypotheses differ from mine in an interesting way. Mine is essentially a constraint on the *microstate* of the early universe but is silent on its macrostate (on the assumption that for any given macroscopic state of the universe, there is a Simple probability distribution concentrated on that macrostate). But the conventional hypothesis about the past is instead a constraint on the macrostate of the early universe.

Low-Entropy Past Hypothesis: The initial macrostate of the universe has very low thermodynamic entropy.

Is such a hypothesis needed in addition to the Simple Past Hypothesis? I think not. For if the Simple Past Hypothesis is true (and if the Simple Dynamical Conjecture is correct), then it follows from that Hypothesis and our best theories of microdynamics that the kind of irreversible dynamical theories we are interested in—in particular, those irreversible theories which entail that thermodynamic entropy reliably increases—entail that the entropy of the early universe was at most no higher than that of the present universe, and was therefore "low" by comparison to the range of entropies of possible states (since there are a great many states with thermodynamic entropy far higher than that of the present-day universe). So the

Low-Entropy Past "Hypothesis" is not a hypothesis at all, but a straightforward prediction of our best macrophysics—and thus, indirectly, of our best microphysics combined with the Simple Past Hypothesis.

It will be helpful to expand on this a bit. On the assumption that the relevant irreversible dynamics (in this case, non-equilibrium thermodynamics) are predictively accurate, predictions about the future can be made just by taking the current state of the universe and evolving it forward under those dynamics. Since the dynamics do not allow retrodiction, our route to obtain information about the past must (as noted earlier) be more indirect: we need to form hypotheses about past states and test those hypotheses by evolving them forward and comparing them with the present state. In particular, the hypothesis that the early universe was, in a certain sharply specified way, very hot, very dense, very uniform, and very much smaller than the current universe—and therefore, much lower in entropy than the current universe[6]—does very well under this method: conditional on that hypothesis, we would expect the current universe to be pretty much the way it in fact is. On the other hand, other hypotheses—notably the hypothesis that the early universe was much higher in entropy than the present-day universe—entail that the present-day universe is fantastically unlikely, and therefore very conventional scientific reasoning tells us that these hypotheses should be rejected.

In turn, we can derive the assumption that our irreversible dynamical theories are predictively accurate by assuming (i) that our microdynamical theories are predictively accurate, and (ii) that the Simple Past Hypothesis and the Simple Dynamical Conjecture are true. So, these hypotheses jointly give us good reason to infer that the early universe had the character we believe it to have had. On the other hand, (i) alone does not give us reason to accept (ii). Rather, we believe (ii) because combined with (i), it explains a great deal of empirical data—specifically, the success of irreversible dynamical theories.

The difference between the Simple Past Hypothesis and the Low-Entropy Past Hypothesis, then, does not lie in the general nature of our reasons for believing them: both are epistemically justified as inferences by virtue of their explanatory power. The difference is that the Simple Past Hypothesis, but not the Low-Entropy Past Hypothesis, is justified by its ability to explain the success of thermodynamics (and other irreversible processes) *in general*. The Low-Entropy Past Hypothesis, by contrast, is justified by its ability to explain *specific features* of our current world. (Although the hypothesis that does this is better understood as a specific

---

[6] It is widely held that (i) such a universe ought to be much *higher* in entropy than the present-day universe, but (ii) this supposed paradox is solved when gravity is taken into account. This is very confused; I attempt to dispel the confusion in Wallace (2009).

cosmological hypothesis about the state of the early universe, rather than the very general hypothesis that its entropy was low.)

Albert himself gives a particularly clear statement of his framework for inducing the (Low-Entropy) Past Hypothesis, which makes for an interesting contrast with my own. He makes three assumptions:

1. That our best theory of microdynamics (which for simplicity he pretends is classical mechanics) is correct.
2. That the Low-Entropy Past Hypothesis is correct.
3. That the correct probability distribution to use over current microstates is the uniform one, conditionalized on whatever information we know (notably, the Low-Entropy Past Hypothesis).

He also makes a tacit mathematical conjecture, which is a special case of the Simple Dynamical Conjecture: in my terminology, he assumes that those distributions that are uniform over some given macrostate and zero elsewhere are forward compatible with coarse-graining.

Now, (2) and (3) together entail that the correct distribution to use over initial states (and Albert is fairly explicit that "correct" means something like "objective-chance-giving") is the uniform distribution over whatever particular low-entropy macrostate is picked out by the Low-Entropy Past Hypothesis. Since these distributions are Simple, Albert's two assumptions entail the Simple Past Hypothesis. But the converse is not true: there are many Simple distributions which are not of the form Albert requires, but which (given the Simple Dynamical Conjecture) are just as capable of grounding the observed accuracy of irreversible macrodynamics.

Put another way: let us make the following abbreviations.

SPH: Simple Past Hypothesis
LEPH: Low-Entropy Past Hypothesis
UPH: Uniform Past Hypothesis: the hypothesis that the initial distribution of the universe was a uniform distribution over some macrostate
SDC: Simple Dynamical Conjecture
PAμ: Predictive Accuracy of Microphysics (our current best theory of microphysics is predictively accurate)
PAM: Predictive Accuracy of Macrophysics (the macrodynamics derived from microphysics by coarse-graining are predictively accurate)

My argument is that

$$SPH + SDC + PA\mu \to PAM. \tag{22}$$

Albert's (on my reading) is that

$$LEPH + UPH + SDC + PA\mu \to PAM. \tag{23}$$

But in fact,

$$UPH \to SPH. \tag{24}$$

So actually, LEPH appears to play no important role in Albert's argument. All that really matters is that the initial distribution was uniform over some macrostate; the fact that this macrostate was lower-entropy than the present macrostate is then a straightforward inference from PAM and the present-day data.

9. Conclusion

There are extremely good reasons to think that, in general and over timescales relevant to the actual universe, the process of evolving a distribution forward under the microdynamics of the universe commutes with various processes of coarse-graining, in which the distribution is replaced by one in which certain fine structures—most notably the small-scale correlations and entanglements between spatially distant subsystems—are erased. The process of alternately coarse-graining in this manner and evolving a distribution forward leads to dynamical processes which are irreversible: for instance, when probabilistic, they will have a branching structure; where a local thermodynamic entropy is definable, that entropy will increase. Since coarse-graining, in general, commutes with the microdynamics, in general we have good grounds to expect distributions to evolve under the microdynamics in a way which gives rise to irreversible macrodynamics, at least over realistic timescales.

Given that the microdynamics are invariant under time-reversal, then if this claim is true, so is its time-reverse, so we have good reason to expect that, in general, the evolution of a distribution both forward and backward in time leads to irreversible macrodynamics on realistic timescales. It follows that the claim can be true only "in general" and not for *all* distributions, since—for instance—the time-evolution of a distribution which does behave this way cannot in general behave this way. However, we have no reason to expect this anomalous behavior except for distributions with extremely carefully chosen fine-scale structure (notably those generated from other distributions by evolving them forward in time). I take this to be a more accurate expression of Goldstein's idea of "typicality": it is not that systems are guaranteed *to achieve equilibrium* unless they or their dynamics are "ridiculously special"; it is that only in "ridiculously special" cases will the microevolution of a distribution not commute with coarse-graining. Whether, and how fast, a system approaches thermal equilibrium is then something that can be determined via these coarse-grained dynamics.

In particular, it seems reasonable to make the Simple Dynamical Conjecture that reasonably simple distributions do not show anomalous behavior. If the correct distribution for the universe at some time t is simple in this way, we would expect that macrophysical processes after t are well described by the macrodynamics generated by coarse-graining (and so exhibit increases in thermodynamic entropy, dispersal of quantum coherence, etc.), in accord with the abundant empirical evidence that these macrodynamics are correct. But we would also expect that macrophysical processes *before t* are not at all described by these macrodynamics—are described, in fact, by the time-reversal of these macrodynamics—in wild conflict with the empirical evidence. But if t is the first instant of time (or, at least, is very early in time), then no such conflict will arise.

It follows that any stipulation of the boundary conditions of the universe according to which the initial distribution of the universe is reasonably simple will (together with our microphysics) entail the correctness of our macrophysics. Since any law of physics specifying the initial distribution will (essentially by the nature of a law) require that initial distribution to be reasonably simple, it follows that any law which specifies the initial distribution suffices to ground irreversible macrodynamics.

It is virtually tautologous that if microscopic physics has no time asymmetry but the emergent macroscopic dynamics do have a time asymmetry, that time asymmetry must be due to an asymmetry in the initial conditions of the universe. The most common proposal for this asymmetry is the proposal that the initial distribution is the uniform distribution over a low-entropy macrostate. From the point of view of explaining irreversibility, all the work in this proposal is being done by the "uniform distribution" part: the low-entropy part alone is neither necessary nor sufficient to establish the correctness of the irreversible macrodynamics, though of course if the initial macrostate is a maximum-entropy state, then its macroevolution will be very dull and contradicted by our observations.

And in fact, the only special thing about the uniformity requirement is that we have good (if heuristic) grounds to expect the microdynamical evolution of uniform distributions to be compatible with coarse-graining. But we have equally good (if equally heuristic) grounds to expect this of any simply specified distribution. So really, the asymmetry of the universe's macroscopic dynamics is not a product of the particular form of the physical principle that specifies the initial conditions of the universe: it is simply a product of some such principle being imposed at one end of the universe rather than at the other.

Acknowledgments

I'd like to take the opportunity to thank David Albert for many stimulating discussions over the last ten years: from no one with whom I disagree even half as much have I learned even half as much.

I'd also like to thank Jeremy Butterfield and Simon Saunders for helpful comments on earlier versions of this paper, and to acknowledge valuable conversations with Harvey Brown, Wayne Myrvold, Roman Frigg, and Jos Uffink.

References

Albert, D. Z. (2000). *Time and Chance*. Cambridge, MA: Harvard University Press.
Binney, J., and S. Tremaine. (2008). *Galactic Dynamics* (2nd ed.). Princeton: Princeton University Press.
Callender, C. (2009). The past hypothesis meets gravity. In G. Ernst and A. Hütteman (Eds.), *Time, Chance and Reduction: Philosophical Aspects of Statistical Mechanics*. Cambridge: Cambridge University Press.
Earman, J. (2006). The "past hypothesis": Not even false. *Studies in History and Philosophy of Science Part B: Studies in History and Philosophy of Modern Physics* 37 (3), 399–430. https://doi.org/10.1016/j.shpsb.2006.03.002.
Frigg, R. (2009). Typicality and the approach to equilibrium in Boltzmannian statistical mechanics. *Philosophy of Science* 76 (5), 997–1008. https://doi.org/10.1086/605800.
Fuchs, C., and A. Peres. (2000). Quantum theory needs no "interpretation." *Physics Today* 53 (3), 70–71. https://doi.org/10.1063/1.883004.
Gell-Mann, M., and J. B. Hartle. (2007). Quasiclassical coarse graining and thermodynamic entropy. *Physical Review A* 76 (2), 022104. https://doi.org/10.1103/PhysRevA.76.022104.
Goldstein, S. (2001). Boltzmann's approach to statistical mechanics. In J. Bricmont, D. Dürr, M. Galavotti, F. Petruccione, and N. Zanghi (Eds.), *Chance in Physics: Foundations and Perspectives*. Berlin: Springer.
Home, D., and M. A. B. Whitaker. (1997). A conceptual analysis of quantum Zeno: Paradox, measurement and experiment. *Annals of Physics* 258 (2), 237–285. https://doi.org/10.1006/aphy.1997.5699.

Jaynes, E. (1957a). Information theory and statistical mechanics. *Physical Review* 106 (4), 620–630. https://doi.org/10.1103/PhysRev.106.620.

Jaynes, E. (1957b). Information theory and statistical mechanics II. *Physical Review* 108 (2), 171–190. https://doi.org/10.1103/PhysRev.108.171.

Jaynes, E. (1968). Prior probabilities. *IEEE Transactions on Systems Science and Cybernetics* 4 (3), 227–241. https://doi.org/10.1109/TSSC.1968.300117.

Lebowitz, J. (2007). From time-symmetric microscopic dynamics to time-asymmetric macroscopic behavior: An overview. Available online at http://arxiv.org/abs/0709.0724.

Lewis, D. (1986). Philosophical Papers, vol. 2. Oxford: Oxford University Press.

Misra, B., and E. C. G. Sudarshan. (1977). The Zeno's paradox in quantum theory. *Journal of Mathematical Physics* 18, 756. https://doi.org/10.1063/1.523304.

Penrose, R. (1989). *The Emperor's New Mind: Concerning Computers, Minds, and the Laws of Physics*. Oxford: Oxford University Press.

Penrose, R. (2004). *The Road to Reality: A Complete Guide to the Laws of the Universe*. London: Jonathan Cape.

Price, H. (1996). *Time's Arrow and Archimedes' Point*. Oxford: Oxford University Press.

Prigogine, I. (1984). *Order out of Chaos: Man's New Dialogue with Nature*. New York: Bantam Books.

Redhead, M. (1995). More ado about nothing. *Foundations of Physics* 25, 123–137. https://doi.org/10.1007/BF02054660.

Saunders, S., J. Barrett, A. Kent, and D. Wallace. (Eds.) (2010). *Many Worlds? Everett, Quantum Theory, and Reality*. Oxford: Oxford University Press.

Sklar, L. (1993). *Physics and Chance*. Cambridge: Cambridge University Press.

Wallace, D. (2009). Gravity, entropy, and cosmology: In search of clarity. *British Journal for the Philosophy of Science* 61 (3), 513–540.

Wallace, D. (2012). *The Emergent Multiverse*. Oxford: Oxford University Press.

Chapter Four

In What Sense Is the Early Universe Fine-Tuned?

▶ SEAN M. CARROLL

I. Introduction

The issue of the initial conditions of the universe—in particular, the degree to which they are "unnatural" or "fine-tuned," and possible explanations thereof—is obviously of central importance to cosmology, as well as to the foundations of statistical mechanics. The early universe was a hot, dense, rapidly expanding plasma, spatially flat and nearly homogeneous along appropriately chosen space-like surfaces.[1] The question is, *why* was it like that? In particular, the thinking goes, these conditions don't seem to be what we would expect a "randomly chosen" universe to look like, to the extent that such a concept makes any sense. In addition to the obvious challenge to physics and cosmology of developing a theory of initial conditions under which these properties might seem natural, it is a useful

---

[1] Our concern here is the state of the universe—its specific configuration of matter and energy, and the evolution of that configuration through time—rather than the coupling constants of our local laws of physics, which may also be fine-tuned. I won't be discussing the value of the cosmological constant, or the ratio of dark matter to ordinary matter, or the matter/antimatter asymmetry.

exercise to specify as carefully as possible the sense in which they don't seem natural from our current point of view.

Philosophers of science (and some physicists: Penrose 1989; Carroll and Chen 2004) typically characterize the kind of fine-tuning exhibited by the early universe as being a state of *low entropy*. This formulation has been succinctly captured in David Albert's *Time and Chance* (2003) as "the Past Hypothesis." A precise statement of the best version of the Past Hypothesis is the subject of an ongoing discussion (see, e.g., Earman 2006; Wallace 2011). But it corresponds roughly to the idea that the early universe—at least the observable part of it, i.e., the aftermath of the Big Bang—was in a low-entropy state with the right microstructure to evolve in a thermodynamically sensible way into the universe we see today.

Cosmologists, following Alan Guth's influential paper on the inflationary-universe scenario (1981), tend to describe the fine-tuning of the early universe in terms of the horizon and flatness problems.[2] The horizon problem, which goes back to Misner (1969), is based on the causal structure of an approximately homogeneous and isotropic (Friedmann-Robertson-Walker) cosmological model in general relativity. If matter and radiation (but not a cosmological constant) were the only sources of energy density in the universe, then regions observed using the cosmic background radiation that are separated by an angle of more than about one degree could never have been in causal contact—their past light-cones, traced back to the Big Bang, could not intersect. It is, therefore, mysterious how these regions could be at the same temperature today, despite the impossibility in a matter/radiation-dominated universe of any signal passing from one such region to another. The flatness problem, meanwhile, was elucidated by Dicke and Peebles (1979). Spatial curvature grows with respect to the energy density of matter and radiation, so the curvature needs to be extremely small at early times so as to not completely dominate today. (For some history of the horizon and flatness problems, see Brawer 1995). Brawer notes that neither the horizon problem nor the flatness problem were considered of central importance to cosmology until inflation suggested a solution to them.)

The horizon and flatness problems are clearly related in some way to the low entropy of the early universe, but they also seem importantly different. In this essay, I will try to clarify the nature of the horizon and flatness problems and argue that they are *not* the best way of thinking about the fine-tuning of the early universe.

---

[2] Guth also discussed the overabundance of magnetic monopoles predicted by certain grand unified theories. This problem was the primary initial motivation for inflation, but it is model-dependent in a way that the horizon and flatness problems don't seem to be.

The horizon problem gestures in the direction of a real puzzle, but the actual puzzle is best characterized as fine-tuning within the space of cosmological *trajectories*, rather than an inability of the early universe to equilibrate over large distances. This reformulation is important for the status of inflationary cosmology, as it makes clear that inflation by itself does not solve the problem (though it may play a crucial role in an ultimate solution). The flatness problem, meanwhile, turns out to simply be a misunderstanding; the correct measure on cosmological trajectories predicts that all but a set of measure zero should be spatially flat. Correctly describing the sense in which the early universe was fine-tuned helps us understand what kind of cosmological models physicists and philosophers should endeavor to construct.

II. What Needs to Be Explained

In order to understand the claim that the state of the universe appears fine-tuned, we should specify what features of the state we are talking about. According to the standard cosmological model, the part of the universe we are able to observe is expanding and emerged from a hot, dense Big Bang about fourteen billion years ago.[3] The distribution of matter and radiation at early times was nearly uniform, and the spatial geometry was very close to flat. We see the aftermath of that period by observing the radiation constituting the cosmic microwave background (CMB) from the time the universe became transparent about 380,000 years after the Big Bang, known as the "surface of last scattering." This radiation is extremely isotropic; its observed temperature is 2.73 K and is smooth to within about one part in 10^5 across the sky (Ade et al. 2014). Our observable region contains about 10^{88} particles, most of which are photons and neutrinos, with about 10^{79} protons, neutrons, and electrons, as well as an unknown number of dark matter particles. (The matter density of dark matter is well-determined, but the mass per particle is extremely uncertain.) The universe has evolved today to a collection of galaxies and clusters of galaxies within a web of dark matter, spread homogeneously on large scales over an expanse of tens of billions of light-years.

Note that our confidence in this picture depends on assuming the Past Hypothesis. The most relevant empirical information concerning the smoothness of

---

[3] In classical general relativity, such a state corresponds to a curvature singularity; in a more realistic but less well-defined quantum theory of gravity, what we call the Big Bang may or may not have been the absolute beginning of the universe, but at the least it is a moment prior to which we have no empirical access.

the early universe comes from the isotropy of the CMB, but that does not strictly imply uniformity at early times. It is a geometric fact that a manifold that is isotropic around every point is also homogeneous; hence, the observation of isotropy and the assumption that we do not live in a special place are often taken together to imply that the matter distribution on spacelike slices is smooth. But we don't observe the surface of last scattering directly; what we observe is the isotropy of the temperature of the radiation field reaching our telescopes here and now in a much older universe. That temperature is determined by a combination of two factors: the intrinsic temperature at emission, and the subsequent redshift of the photons. The intrinsic temperature at the surface of last scattering is mostly set by atomic physics, but not quite; because there are so many more photons than atoms, recombination occurs at a temperature of about 3eV, rather than the 13.6eV characterizing the ionization energy of hydrogen. Taking the photon-to-baryon ratio as fixed for convenience, our observations imply that the cosmological redshift is approximately uniform between us and the surface of last scattering in all directions on the sky. But that is compatible with a wide variety of early conditions. The redshift along any particular path is a single number; it is not possible to decompose it into "Doppler" and "cosmic expansion" terms in a unique way (Bunn and Hogg 2009). This reflects the fact that there are many ways to define spacelike slices in a perturbed cosmological spacetime. For example, we can choose to define spacelike surfaces such that the cosmological fluid is at rest at every point ("synchronous gauge"). In those coordinates there is no Doppler shift, by construction. But the matter distribution at recombination could conceivably look very inhomogeneous on such slices; that would be compatible with our current observations as long as a direction-dependent cosmological redshift conspired to give an isotropic radiation field near the Earth today. Alternatively, we could choose spacelike slices along which the temperature was constant; the velocity of the fluid on such slices could be considerably nonuniform, yet the corresponding Doppler effect could be canceled by the intervening expansion of space along each direction. Such conspiratorial conditions seem unlikely to us, but they are more numerous (in the measure to be discussed below) in the space of all possible initial conditions. Of course, we also know that of all possible past conditions that lead to a half-melted ice cube in a glass of water, most of them look like a glass of liquid water at a uniform temperature rather than a glass of warmer water with an unmelted ice cube that we would generally expect. In both cases, our conventional reasoning assumes the kind of lower-entropy state postulated by the Past Hypothesis.

With the caveat that the Past Hypothesis is necessary, let us assume that the universe we are trying to account for is one that was very nearly uniform (and spatially flat) at early times. What does it mean to say that such a state is fine-tuned?

Any fine-tuning is necessarily a statement about one's expectations about what would seem natural or non-tuned. In the case of the initial state of the universe, one might reasonably suggest that we simply have no right to have any expectations at all, given that we have only observed one universe. But this is a bit defeatist. While we have not observed an ensemble of universes from which we might abstract ideas about what a natural one looks like, we know that the universe is a physical system, and we can ask whether there is a sensible *measure* on the relevant space of states for such a system, and then whether our universe seems generic or highly atypical in that measure. In practice, we typically use the classical Liouville measure, as we will discuss in Section IV. The point of such an exercise is not to say "We have a reliable theory of what universes should look like, and ours doesn't fit." Rather, given that we admit that there is a lot about physics and cosmology that we don't yet understand, it's to look for clues in the state of the universe that might help guide us toward a more comprehensive theory. Fine-tuning arguments of this sort are extremely important in modern cosmology, although the actual measure with respect to which such arguments are made is sometimes insinuated rather than expressly stated, and usually posited rather than carefully derived under a more general set of assumptions.

The fine-tuning argument requires one element in addition to the state and the measure: a way to coarse-grain the space of states. This aspect is often overlooked in discussions of fine-tuning, but it is again implicit. Without a coarse-graining, there is no way to say that any particular state is "natural" or "fine-tuned," even in a particular measure on the entire space of states. Each individual state is just a point, with measure zero. What we really mean to say is that states *like* that state are fine-tuned, in the sense that the corresponding macrostate in some coarse-graining has a small total measure. The coarse-graining typically corresponds to classifying states as equivalent if they display indistinguishable macroscopically observable properties. This is usually not problematic, although, as we will see, it is necessary to be careful when we consider quantities such as the spatial curvature of the universe.

Given a measure and a choice of coarse-graining, it is natural to define the entropy of each microstate, defined by Boltzmann to be the logarithm of the volume of the corresponding macrostate. Penrose (1989) has famously characterized the fine-tuning problem in these terms, emphasizing how small the entropy of the early universe was compared to the current entropy, and how small the current entropy is compared to how large it could be. At early times, when inhomogeneities were small, we can take the entropy to simply be that of a gas of particles in the absence of strong self-gravity, which is approximately given by the number of particles; for our observable universe, that's about 10^{88}. Today the entropy is dominated by supermassive black holes at the centers of galaxies, each of which has a Bekenstein-Hawking entropy $S = 4G\, M^2$ in units where $\hbar = c = k_B = 1$. The best estimates of the current entropy give numbers of order 10^{103} (Egan and Lineweaver 2010). Penrose suggests a greatest lower bound on the allowed entropy of our observable universe by calculating what the entropy would be if all of the matter were put into one gigantic black hole, obtaining the famous number 10^{122}. Since the universe is accelerating, distant galaxies will continue to move away from us rather than collapse together to form a black hole, but we can also calculate the entropy of the de Sitter phase to which the universe will ultimately evolve, obtaining the same number 10^{122} (Banks 2000; Banks and Fischler 2001). (The equality of these last two numbers can be explained by the cosmological coincidence of the density in matter and vacuum energy; they are both of order $1/GH_0^2$, where H_0 is the current Hubble parameter.) By any measure, the entropy of the early universe was fantastically smaller than its largest possible value, which seems to be a result of fine-tuning.[4] Part of my goal in this essay is to relate this formulation to the horizon and flatness problems.

To close this section, note that the early universe was in an extremely *simple* state, in the sense that its "apparent complexity" was low: macroscopic features of the state can be fully characterized by a very brief description (in contrast with the current state, where a macroscopic description would still require specifying every galaxy, if not every star and planet). But apparent complexity is a very different thing than entropy; the universe is simple at the earliest times, and will be simple again at much later times, while entropy increases monotonically (Aaronson et al. 2014). The simplicity of the early universe should not in any way be taken as a sign of its "naturalness." A similar point has been emphasized by Price (1997), who notes that any definition of "natural" that purportedly applies to the early universe should apply to the late universe as well. The early universe appears fine-tuned because the macroscopic features of its state represent a very small region of phase space in the Liouville measure, regardless of how simple it may be.

---

[4] Even though the co-moving volume corresponding to our observable universe was much smaller at early times, it was still the same physical system as the late universe, with the same space of states, governed by presumably reversible dynamical laws. It is, therefore, legitimate to calculate the maximum entropy of the early universe by calculating the maximum entropy of the late universe.

III. The Horizon Problem

a. Defining the Problem

We now turn to the horizon problem as traditionally understood. Consider a homogeneous and isotropic universe with scale factor a(t), energy density ρ(a), and a fixed spatial curvature κ. In general relativity, the evolution of the scale factor is governed by the Friedmann equation,

$$H^2 = \frac{8\pi G}{3}\rho - \frac{\kappa}{a^2}$$

where $H = \dot{a}/a$ is the Hubble parameter and G is Newton's constant of gravitation, and we use units where the speed of light is set equal to unity, $c = 1$. Often, the energy density will evolve as a simple power of the scale factor, $\rho \propto a^{-n}$. For example, "matter" to a cosmologist refers to any species of massive particles with velocities much less than the speed of light, for which $\rho_M \propto a^{-3}$, while "radiation" refers to any species of massless or relativistic particles, for which $\rho_R \propto a^{-4}$. (In both cases, the number density decreases as the volume increases, proportional to a^{-3}; for matter, the energy per particle is constant, while for radiation it diminishes proportional to a^{-1} due to the cosmological redshift.)

When the energy density of the universe is dominated by components with $n > 2$, $\dot{a}$ will be decreasing, and we say that the universe is "decelerating." In a decelerating universe, the horizon size at time t_* (i.e., the distance a photon can travel between the Big Bang and t_*) is approximately

$$d_{\mathrm{hor}}(t_*) \approx t_* \approx H_*^{-1}.$$

We call H^{-1} the "Hubble distance" at any given epoch. Sometimes the Hubble distance is conflated with the horizon size, but they are importantly different; the Hubble distance depends only on the instantaneous expansion rate at any one moment in time, while the horizon size depends on the entire past history of the universe back to the Big Bang. The two are of the same order of magnitude in universes dominated by matter and radiation, with the precise numerical factors set by the abundance of each component. A textbook calculation shows that the Hubble distance at the surface of last scattering, when the CMB was formed, corresponds to approximately one degree on the sky today.

The horizon problem, then, is simple. We look at widely separated parts of the sky, and observe radiation left over from the early universe. As an empirical matter, the temperatures we observe in different directions are very nearly equal. But the physical locations from which that radiation has traveled were separated by distances larger than the Hubble distance (and therefore, the horizon size, if the universe is dominated by matter and radiation) at that time. They were never in causal contact; no prior influence could have reached both points by traveling at speeds less than or equal to that of light. Yet these independent regions seem to have conspired to display the same temperature to us. That seems like an unnatural state of affairs. We can formalize this as:

> **Horizon problem (causal version).** If different regions in the early universe had non-overlapping past light cones, no causal influence could have coordinated their conditions and evolution. There is, therefore, no reason for them to appear similar to us today.

If that's as far as it goes, the horizon problem is perfectly well-formulated, although somewhat subjective. The causal formulation merely points out that there is no reason for a certain state of affairs (equal temperatures of causally disconnected regions) to obtain, but it doesn't give any reason for expecting otherwise. Characterizing this as a "problem," rather than merely an empirical fact to be noted and filed away, requires some positive expectation for what we think conditions near the Big Bang *should* be like: some reason to think that unequal temperatures would be more likely, or at least less surprising.

b. Equilibration and Entropy

We can attempt to beef up the impact of the horizon problem by bringing in the notion of *equilibration* between different regions. Imagine we have a box of gas containing different components, or simply different densities, and for the moment we assume gravity is negligible. If the gas starts from an inhomogeneous (low-entropy) configuration, given time it will typically equilibrate and attain a uniform distribution. In that sense, an equal temperature across a hot plasma actually seems natural or likely. But if we think of the early universe as such a box of gas, it hasn't had time to equilibrate. That's a direct consequence of the fact that (in a matter/radiation-dominated universe) the horizon size is much smaller than the scales over which we are currently observing. Let's call this the "equilibration" version of the horizon problem:

> **Horizon problem (equilibration version).** If different regions in the early universe shared a causal past, they could have equilibrated and come to the same temperature. But since they did not, such as in a matter/radiation-dominated universe, observing equal temperatures today is puzzling.

The equilibration formulation of the horizon problem seems stronger than the causal version; it attempts to provide some reason equal temperatures across causally disconnected regions should be surprising, rather than merely noting their existence. But in fact, we will see that this version undermines the usual conclusions of the horizon problem and invites a critique that has been advanced by Sheldon Goldstein (2014). I will try to argue that Goldstein's critique doesn't rebut the claim that the early universe is fine-tuned, though it does highlight the misleading nature of the equilibration version of the horizon problem.

The problem arises when we try to be more specific about what might count as "natural" initial conditions in the first place. If we trace the Friedmann equation backward in time, we come to a singularity—a point where the scale factor is zero and the density and Hubble parameter are infinite. There is no reason to expect the equations of classical general relativity to apply in such circumstances; at the very least, quantum effects will become important, and we require some understanding of quantum gravity to make sensible statements. Absent such a theory, there would be some justification in giving up on the problem entirely, and simply saying that the issue of initial conditions can't be addressed without a working model of quantum gravity. (Presumably, this was the feeling among many cosmologists before inflation was proposed.)

Alternatively, we could choose to be a bit more optimistic and ask what kind of configurations would constitute natural initial conditions in the moments right after the initial singularity or whatever quantum phase replaces it. Given the phase space describing the relevant degrees of freedom in that regime, we can coarse-grain into macrostates defined by approximately equal values of macroscopically observable quantities and define the Boltzmann entropy as the logarithm of the volume of the macrostate of which the microstate is an element, as discussed in Section II. Calculating this volume clearly requires the use of a measure on the space of states; fortunately such a measure exists, given by Liouville in the classical case or by the Hilbert-space inner product in the quantum case. Given that machinery, states with high Boltzmann entropy seem natural or generic or likely, simply because there are many of them; states with low Boltzmann entropy seem unnatural because there are relatively few, suggesting the need for some sort of explanation.[5]

---

[5] Note that the expectation of high entropy for the early universe, whether convincing or not, certainly has a different character than our expectation that long-lived closed systems will be high-entropy in the real world. In the latter case, there is a dynamical mechanism at work: almost all initial conditions will evolve toward equilibrium. In the case of the early universe, by contrast, we are actually making a statement about the initial conditions themselves, saying that high-entropy ones would be less surprising than low-entropy ones because the former are more numerous.

Let's apply this to a matter/radiation-dominated universe, in which different regions we observe in the CMB are out of causal contact. Following Goldstein, we can draw an analogy with two isolated boxes of gas. The boxes could be different sizes, and have never interacted. All we know is that there is some fixed number of particles in the boxes, with some fixed energy. Goldstein's observation is that if we know nothing at all about the particles inside the two boxes, it should be *completely unsurprising* if they had the same temperature. The reasoning is simple: at fixed energy and particle number, there is more phase space corresponding to approximately equal-temperature configurations than to unequal-temperature ones. Such configurations maximize the Boltzmann entropy. Given the two boxes, some fixed number of particles, and some fixed total energy, a randomly chosen point in phase space is very likely to feature equal temperatures in each box, even if they have never interacted. Therefore, one might tentatively suggest, perhaps seeing equal temperatures in causally disconnected parts of the early universe isn't actually unlikely at all, and the horizon problem shouldn't be a big worry.

c. Gravity and Dynamics

For isolated boxes of gas, this logic is surprising, but valid. A randomly selected state of two isolated boxes of gas is likely to have equal temperatures in each box, even in the absence of interactions. For the early universe, however, the boxes turn out to not provide a very useful analogy, for a couple of (related) reasons: the importance of gravity, and the fact that we are considering time-dependent trajectories, not simply individual states.

Gravity plays the crucial role in explaining a peculiar feature of the early universe: it is purportedly a low-entropy state, but one in which the matter is radiating as a nearly-homogeneous black body, exactly as we are trained to expect from systems in thermal equilibrium. The simple resolution is that when the *self-gravity* of a collection of particles becomes important, high-entropy configurations of specified volume and density are inhomogeneous rather than homogeneous. Therefore, if the imaginary boxes that we consider are sufficiently large, we wouldn't expect the temperature to be uniform even inside a single box, much less between two boxes.

This can be seen in a couple of different ways. One is to consider the Jeans instability: the tendency of sufficiently long-wavelength perturbations in a self-gravitating fluid to grow. In a fluid with density ρ and speed of sound c_s, modes are unstable and will grow if they are larger than the Jeans length,

$$\lambda_J = \frac{c_s}{(G\rho)^{1/2}}.$$

If the size of a box of gas is greater than the Jeans length for that gas, the distribution will fragment into self-gravitating systems of size λ_J or smaller rather than remaining homogeneous; in cosmology, that's the process of structure formation. Given the fluid pressure p as a function of the energy density, the speed of sound is defined by $c_s^2 = dp/d\rho$. For a radiation bath, the speed of sound is $c_R = 1/\sqrt{3}$, while for collisionless, nonrelativistic particles, it is $c_M \approx 0$; a cosmological matter fluid is unstable and inhomogeneities will grow on all scales.[6] Classically, the only perfectly stable configuration of fixed energy in a fixed volume would be a black hole in vacuum, which is about as inhomogeneous as you can get.

Another way of reaching the same conclusion (that randomly chosen states of self-gravitating particles are likely to be inhomogeneous) is to examine phase-space volumes directly. Consider a collection of particles with a given energy, interacting only through the inverse-square law of Newtonian gravity. The volume of phase space accessible to such a system is unbounded. Keeping the energy fixed, we can send any number of particles to infinity (or arbitrarily large momentum) while compensating by moving other particles very close together, sending their mutual energy to minus infinity. This is a real effect, familiar to researchers in galactic dynamics as the "gravo-thermal catastrophe" (Lynden-Bell and Wood 1968; Nityananda 2009). A galaxy populated by stars (or dark matter particles) interacting through Newtonian gravity will tend to become centrally condensed without limit, while ejecting other stars to infinity. The entropy of such a system is unbounded, and there is no equilibrium configuration, but generic evolution is in the direction of greater inhomogeneity. And, indeed, in semiclassical general relativity, the highest-entropy configuration of fixed energy in a fixed volume is generally a black hole when the energy is sufficiently high.[7]

In the early universe, the Jeans length is generally less than the Hubble radius (although they are of the same order for purely-radiation fluids). High-entropy states will generally be very inhomogeneous, in stark contrast with the intuition

---

[6] This story is complicated by the expansion of the universe, since the Hubble parameter acts as a friction term; in a purely radiation-dominated Friedmann-Robertson-Walker universe, density perturbations actually shrink as the universe expands. But even when radiation dominates the energy density, there are still slowly moving matter particles. Given sufficient time in a hypothetical non-expanding universe, density perturbations in non-relativistic matter would grow very large. This leads us to the discussion of trajectories rather than states, which I undertake in the next section.

[7] When the energy is too low, any black hole will have a higher Hawking temperature than its surroundings, and it will lose energy and shrink. Its temperature will grow as the hole loses mass, and eventually it will evaporate completely away. This reflects the fact that black holes have negative specific heat.

we have from boxes of gas with negligible self-gravity, in accordance with Penrose's analysis mentioned in Section II. Hence, the equilibration version of the horizon problem is extremely misleading, if not outright incorrect. In an alternative world in which particles could still gravitationally attract each other but the universe was not expanding, so that past light cones stretched forever and the horizons of any two points necessarily overlapped, our expectation would *not* be for a smooth universe with nearly constant temperatures throughout space. It would be for the opposite: a highly inhomogeneous configuration with wildly non-constant temperatures. Whatever the fine-tuning problem associated with the early universe may be, it is not that "distant regions have not had time to come into thermal equilibrium." Indeed, the lack of time to equilibrate is seen to be a feature, rather than a bug: it would be even harder to understand the observed uniformity of the CMB if the plasma had had an arbitrarily long time to equilibrate.

From this perspective, the thermal nature of the CMB radiation is especially puzzling. It cannot be attributed to "thermal equilibrium," since the early plasma is not in equilibrium in any sense.[8] Sometimes an attempt is made to distinguish between "gravitational" and "nongravitational" degrees of freedom and to argue that the nongravitational degrees of freedom are in equilibrium, whereas the gravitational ones are not. This is problematic at best. The relevant gravitational effect isn't one of degrees of freedom (which would be independently propagating gravitational waves or gravitons), but the simple existence of gravity as a force. One might try to argue that the primordial plasma looks (almost) like it would look in an equilibrium configuration in a universe where there was no gravitational force due to gravity, but it is unclear what impact such an observation is supposed to have.[9]

The same conclusion can be reached in a complementary way, by recalling that we are considering a highly time-dependent situation, rather than two boxes of gas at rest. As mentioned in Section II, there is a great amount of freedom in choosing spatial hypersurfaces in a perturbed cosmological spacetime; for example, we could choose the temperature of the fluid as our time coordinate (as long as the fluid was not so inhomogeneous that constant-temperature surfaces were no longer spacelike). Then, by construction, the temperature is completely

---

[8] This is trivial, of course. Equilibrium states are time-independent, whereas the early universe is expanding and evolving. More formally, there is no timelike Killing vector for an equilibrium state.

[9] There has been at least one attempt to formalize this notion, by inventing a scenario in which the strength of gravity goes to zero in the very early universe (Greene et al. 2011). While intriguing, from the trajectory-centered point of view advocated in the next section, even this model doesn't explain why our observed universe exhibits such an unlikely cosmological history.

uniform at any moment of time. But we are not observing the CMB at the moment of recombination; the photons that we see have been redshifted by a factor of over a thousand. From that perspective, the question is not "Why was the temperature uniform on spacelike hypersurfaces?" but instead "Why is the redshift approximately equal in completely separate directions on the sky?"[10] This formulation suggests the importance of considering cosmological trajectories, which we turn to in Section IV.

d. Inflation

It is useful to see how inflation addresses the horizon problem. At the level of the causal version, all inflation needs to do is invoke an extended period of accelerated expansion. Then the past horizon size associated with a spacetime event becomes much larger than the Hubble radius at that time, and widely separated points on the surface of last scattering can easily be in causal contact. We can quantify the total amount of inflation in terms of the number of e-folds of expansion,

$$N_e = \int_{a_i}^{a_f} d\ln a = \int_{t_i}^{t_f} H dt.$$

where the integral extends from the beginning to the end of the period of accelerated expansion ($\ddot{a} > 0$). Generally, $N_e > 50$ e-folds of inflation are required to ensure that all the observed regions of the CMB share a causal patch.

Whether or not these regions have "equilibrated" depends on one's definitions. During inflation, the energy density of the universe is dominated by the approximately constant potential energy of a slowly rolling scalar field. The evolution is approximately adiabatic until the reheating period when inflation ends and the inflaton energy converts into matter and radiation. Perturbations generically shrink during the inflationary period, in accordance with intuition from the cosmic no-hair theorem (Wald 1983), which states that a universe dominated by a positive cosmological constant approaches a de Sitter geometry. But it is not a matter of degrees of freedom in different regions sharing energy, as in a conventional equilibration process; perturbations simply decrease locally and independently as the universe expands.

Once accelerated expansion occurs, it is important that inflation *end* in such a way that homogeneous and isotropic spatial slices persist after reheating (when

---

[10] Alternatively, and equivalently, we could define spacelike hypersurfaces by demanding that each hypersurface be at a constant redshift factor from the present time. On such surfaces, the temperature would generically be very nonuniform. In these coordinates, the question becomes: Why did distant regions of the universe reach the recombination temperature at the same time?

energy in the inflaton field is converted into matter and radiation). The success of this step is highly nontrivial; indeed, this "graceful-exit problem" was highlighted by Guth (1981) in his original paper, which relied on bubble nucleation to enact the transition from a metastable false vacuum state to the true vacuum. Unfortunately, there are only two regimes for such a process; if bubbles are produced rapidly, inflation quickly ends and does not last for sufficient e-folds to address the horizon problem; though if they are produced slowly, we are left with a wildly inhomogeneous universe with a few bubbles appearing and colliding while inflation continues in a false vacuum elsewhere. An attractive solution to this dilemma came in the form of slow-roll inflation (Linde 1982; Albrecht and Steinhardt 1982). Here, the field is not trapped in a false vacuum, but rolls gradually down its potential. Slow-roll inflation can very naturally produce homogeneous and isotropic spatial slices even if it proceeds for a very large number of e-folds. Central to the success of this model is the fact that the rolling field acts as a "clock," allowing regions that have been inflated to extreme distances to undergo reheating at compatible times (Anninos et al. 1991). It is, thus, crucially important that the universe during slow-roll inflation is *not* truly in equilibrium even though its evolution is approximately adiabatic; the evolving inflaton field allows for apparent coordination among widely separated regions.

Inflation, therefore, solves the puzzle raised by the horizon problem, in the following sense: given a sufficient amount of inflation and a model that gracefully exits from the inflationary phase, we can obtain a homogeneous and isotropic universe in which distant points share a common causal past. The success of this picture can obscure an important point: the conditions required to get inflation started in the first place are extremely fine-tuned. This fine-tuning is often expressed in terms of entropy; a patch of spacetime ready to begin inflating has an enormously lower entropy than the (still quite low-entropy) homogeneous plasma into which it evolves (Penrose 1989; Carroll and Chen 2004). In this sense, inflation "explains" the fine-tuned nature of the early universe by positing an initial condition that is even more fine-tuned. In the context of the horizon problem, this issue can be sharpened. One can show that, in order for inflation to begin, not only must we have very particular initial conditions in which potential energy dominates over kinetic and gradient energies, but the size of the patch over which these conditions obtain must be strictly larger than the Hubble radius (Vachaspati and Trodden 2000). In other words, even in an inflationary scenario, it is necessary to invoke smooth initial conditions over super-horizon-sized distances. As compelling as inflation may be, it still requires some understanding of pre-inflationary conditions to qualify as a successful model.

Contemporary discussion of inflation often sidesteps the problem of the required low-entropy initial condition by appealing to the phenomenon of eternal inflation: in many models, if inflation begins at all, it continues without end in some regions of the universe, creating an infinite amount of spacetime volume (Guth 2000). While plausible (although for some recent concerns, see Boddy, Carroll, and Pollack 2014), this scenario raises a new problem: rather than uniquely predicting a universe like the kind we see, inflation predicts an infinite variety of universes, making prediction itself a difficult problem. I won't discuss this issue here, but for recent commentary, see Ijjas, Steinhardt, and Loeb (2013); Guth, Kaiser, and Nomura (2014); Linde (2014); Ijjas, Steinhardt, and Loeb (2014).

IV. Trajectories

a. Initial Conditions and Histories

The horizon problem, as discussed in Section III, might not seem like an extremely pressing issue. The causal version is fairly weak, merely noting a state of affairs rather than offering any reason we should expect the contrary, while the equilibration version is misleading—the problem is not that distant regions are unable to equilibrate, it's that equilibration would have made things more inhomogeneous. Nevertheless, there is no question that the early universe *is* fine-tuned. A better statement of the fine-tuning problem comes from considering cosmological trajectories—histories of the universe over time—rather than concentrating on initial conditions.

The relationship between initial conditions and trajectories is somewhat different in classical mechanics and quantum mechanics. In classical mechanics, the space of trajectories of a system with a fixed phase space and time-independent Hamiltonian is isomorphic to the space of initial conditions, or indeed the space of possible conditions at any specified moment of time. We can think of the history of a classical system as a path through a phase space Γ with coordinates $\{q^i, p_i\}$, where the $\{q^i\}$ are canonical coordinates and the $\{p_i\}$ are their conjugate momenta, governed by a Hamiltonian, $\mathcal{H}(q^i, p_i)$ (which, for our purposes, is taken to be time-independent). Evolution is deterministic and reversible (information-conserving), implying that the state at any one time plus the Hamiltonian is sufficient to determine the state at any prior or subsequent time. Classically, however, trajectories (and time itself) can come to an end; that's what happens at spacelike singularities such as the Big Bang or inside a Schwarzschild black hole. This reflects the great amount of freedom that exists in choosing the geometry of phase space, including the possible existence of singular points on the manifold. It is therefore possible for some conditions to be truly "initial," if they occur at the past boundary of the time evolution.

In quantum mechanics, there are no singularities or special points in the state space. A general solution to the Schrödinger equation $\hat{H}|\Psi\rangle = i\partial_t|\Psi\rangle$ with a time-independent Hamiltonian can be written in terms of energy eigenstates as

$$|\Psi(t)\rangle = \sum_n r_n e^{i(\theta_i - E_n t)} |E_n\rangle,$$

where the constant real parameters $\{r_i, \theta_i\}$ define the specific state. Each eigenstate merely rotates by a phase, singularities cannot arise, and time extends infinitely toward the past and future (Carroll 2008).[11] In contrast to the arbitrariness of classical phase space, the geometry of pure quantum states is fixed to be $\mathbb{CP}^n$, and evolution is always smooth (Kibble 1979; Brody and Hughston 2001; Bengtsson et al. 2002). There is then no special "initial" condition; the state at any one time can be evolved infinitely far into the past and future. When we come to quantum gravity, a canonical quantization of general relativity (Wiltshire 1995) suggests that the wave function of the universe may be an exact energy eigenstate with an eigenvalue of zero, as reflected in the Wheeler-DeWitt equation: $\hat{H}|\Psi\rangle = 0$, where the Hamiltonian, $\hat{H}$, includes both gravitational and matter dynamics. In that case, there is no time evolution in the conventional sense, although time can conceivably be recovered as an effective concept describing correlations between a "clock" subsystem and the rest of the quantum state (Page and Wootters 1983; Banks 1985). It is, nevertheless, far from clear that the Wheeler-DeWitt equation is the proper approach to quantum gravity, or indeed that local spacetime curvature is the proper thing to quantize. Evidence from the holographic principle (Hooft 1993; Susskind 1995; Bousso 2002), black hole complementarity (Susskind et al. 1993), the gauge-gravity correspondence (Maldacena 1998; Horowitz and Polchinski 2006), the entanglement/spacetime connection (Ryu and Takayanagi 2006; Swingle 2012; Van Raamsdonk 2010; Maldacena and Susskind 2013), and thermodynamic approaches to gravity (Jacobson 1995; Verlinde 2011) suggests that

---

[11] The Borde-Guth-Vilenkin (BGV) theorem (Borde, Guth, and Vilenkin 2003) demonstrates that spacetimes with an average expansion rate greater than zero must be geodesically incomplete in the past (which is almost, but not quite, equivalent to saying there are singularities). This has been put forward as evidence that the universe must have had a beginning (Mithani and Vilenkin 2012); there are explicit counterexamples to this claim (Aguirre and Gratton 2002), but such examples are arguably unstable or at least nongeneric. However, the BGV theorem does not assume Einstein's equation or any other equations of motion; it only makes statements about classical spacetime. It is, therefore, silent on the question of what happens when gravity is quantized.

gravity can be thought of as emerging from nonlocal degrees of freedom, only indirectly related to curved spacetime. Given the current state of the art, then, it is safest to leave open the question of whether time is emergent or fundamental, and whether it is eternal or has a beginning.

Fortunately, this uncertainty over whether conditions can truly be initial does not prevent us from talking about cosmological fine-tuning. For most of the history of the universe, many important cosmological quantities are well-described by classical dynamics. This includes the expansion of the scale factor, as well as the evolution of perturbations, considered as modes in Fourier space of fixed comoving wavelength (i.e., expanding along with the universe).[12] On small scales, the dynamics are nonlinear and entropy-generating due to a variety of processes such as star formation, supernovae, magnetic fields, and violent gravitational relaxation. Consequently, the evolution at those wavelengths is not well-approximated by reversible equations on phase space. This leaves us, however, with the dynamics on large scales—in the present universe, more then ten million light-years across—that can be treated as the Hamiltonian evolution of an autonomous set of degrees of freedom. Therefore, we can circumvent conceptual problems raised by the idea of "initial conditions" by simply asking whether the trajectory of the large-scale universe since the aftermath of the Big Bang is natural, or fine-tuned, in the space of all such trajectories.

b. *The Canonical Measure*

Fortunately, it is possible to construct a preferred measure on the space of trajectories, which we can use to judge the amount of fine-tuning exhibited by our real universe. We start by considering the measure on phase space itself, and use that to find a measure on the space of paths through phase space that represent solutions to the equations of motion.[13]

In classical mechanics, there is a natural measure on phase space Γ, the Liouville measure. To construct it in terms of coordinates $\{q^i\}$ and momenta $\{p_i\}$, we first write down the symplectic 2-form on Γ,

---

[12] Inflation is an important exception. During inflation itself, the state of the universe has essentially only one branch. When inflation ends, reheating creates a large number of excited degrees of freedom, effectively "measuring" the state of the inflaton and causing the wave function to split into many branches (Boddy et al. 2014). This process explains how a perturbed post-inflationary universe can develop out of an unperturbed inflationary state.

[13] This notion of a cosmological measure is completely separate from that which arises in what is sometimes called "the measure problem in cosmology," which deals with the relative frequency of different kinds of observers in a multiverse. (See, e.g., Winitzki 2008; Aguirre et al. 2007; Linde et al. 2009; Schwartz-Perlov and Vilenkin 2010; Freivogel 2011; Salem 2012.)

$$\omega = \sum_{i=1}^{n} dp_i \wedge dq^i.$$

Note that the dimension of Γ is 2n. The Liouville measure is then a 2n-form given by

$$\Omega = \frac{(-1)^{n(n-1)/2}}{n!} \omega^n.$$

All that matters for our current purposes is that such a measure exists and is uniquely defined. What makes the Liouville measure special is that it is conserved under time evolution. That is, given states that initially cover some region $S \subset \Gamma$ and that evolve under Hamilton's equations to cover region S', we have

$$\int_S \Omega = \int_{S'} \Omega.$$

Classical statistical mechanics assumes that systems in equilibrium have a probability distribution in phase space that is uniform with respect to this measure, subject to appropriate macroscopic constraints. Meanwhile, in connecting cosmology with statistical mechanics, we assume that the microstate of the early universe is chosen randomly from a uniform distribution in the Liouville measure, subject to the (severe) constraint that the macrocondition has the low-entropy form given by the Past Hypothesis. Albert (2003) calls this the "Statistical Postulate."[14] Of course, one can question why probabilities should be uniform in *this* measure rather than some other one. Even if the Liouville measure is somehow picked out by the dynamics by virtue of being conserved under evolution, we are nevertheless free to construct any measure we like. For our present purposes, this kind of question seems misguided. As discussed in Section II, the point of fine-tuning arguments is to find clues that can guide us to inventing more comprehensive physical theories. We are not arguing for some metaphysical principle to the effect that the universe *should* be chosen uniformly in phase space according to the Liouville measure; merely that, given this measure's unique status as being picked out by the dynamics, states that look natural in this measure tell us very little, while states that look unnatural might reveal useful information. (See Schiffrin and Wald 2012 for a critique of the use of cosmological measures in the way I am advocating here.)

---

[14] Albert and Loewer have referred to the combined package of the dynamical laws, the Past Hypothesis, and the Statistical Postulate as the "Mentaculus," from the Coen brothers film, *A Serious Man* (see, e.g., Loewer 2012).

In general, having a measure on phase space Γ does not induce a natural measure on the space of trajectories T, which is one dimension lower. (There is a natural map $\Gamma \to T$, which simply sends each point to the trajectory it is on; however, while differential forms can be pulled back under such maps, they cannot, in general, be pushed forward (Carroll 2004).) In the case of general relativity, Gibbons, Hawking, and Stewart (GHS) (1987) showed that there is, nevertheless, a unique measure satisfying a small number of reasonable constraints: it is positive, independent of coordinate choices, respects the symmetries of the theory, and does not require the introduction of any additional structures. GHS relied on the fact that general relativity is a constrained Hamiltonian system: because the metric component g_{00} is not a propagating degree of freedom in the Einstein–Hilbert action, physical trajectories obey a constraint of the form $\mathcal{H} = \mathcal{H}_*$, where $\mathcal{H}$ is the Hamiltonian and $\mathcal{H}_*$ is a constant defining the constraint surface. (In cosmological spacetimes, we generally have $\mathcal{H}_* = 0$.) The space U of physical trajectories—those obeying the Hamiltonian constraint—is thus, two dimensions lower than the full phase space Γ. GHS constructed a measure on U by identifying the nth coordinate on phase space as time t, for which $\mathcal{H}$ is the conjugate variable. The symplectic form is then

$$\omega = \tilde{\omega} + d\mathcal{H} \wedge dt,$$

where

$$\tilde{\omega} \equiv \sum_{i=1}^{n-1} dp_i \wedge dq^i$$

GHS showed that the $(2n-2)$-form,

$$\Theta = \frac{(-1)^{(n-1)(n-2)/2}}{(n-1)!} \tilde{\omega}^{n-1},$$

is a unique measure satisfying their criteria.

The GHS measure has the attractive feature that it is expressed locally in phase space. Therefore, it can be evaluated on the space of trajectories simply by choosing some transverse surface through which all trajectories (or all trajectories of interest for some purpose) pass, and then calculating Θ on that surface; the results are independent of the surface chosen. In cosmology, for example, we might choose surfaces of constant Hubble parameter, or constant energy density, and evaluate the measure at that time. This feature has a crucial consequence: the total measure on some particular kind of trajectories, such as ones that are spatially flat or ones that are relatively smooth in some particular cosmological epoch, is completely independent of what the trajectories are doing at some other cosmological epoch. Therefore, changing the dynamics in the early universe (such as modifying the potential for an inflaton field) cannot possibly change the fraction of trajectories with certain specified properties in the late universe. (Adding additional degrees of freedom can, of course, alter the calculation of the measure.) New physics cannot change "unnatural" trajectories into "natural" ones.

At heart, there is not much conceptual difference between studying the purported fine-tuning of the universe in terms of the measure on trajectories and quantifying the low entropy of the early state. There are relatively few initial conditions with low entropy, and the trajectories that begin from such conditions will have a small measure. As discussed in Section II, in order to quantify fine-tuning, we generally need to specify a coarse-graining on the space of states as well as a measure. In the language of trajectories, this corresponds to specifying macroconditions the trajectories must satisfy. One benefit of the trajectory formalism is that it is relatively straightforward to ask questions about conditionalizing over macroconditions specified at a different time; for example, we can talk about the fraction of trajectories that are smooth at one time given that they are smooth at some other time. Another benefit, and a considerable one, is that we can look at features of cosmic evolution that we truly understand without claiming to have full control over the space of states (as would be necessary to completely understand the entropy). An objection to Penrose's argument is sometimes raised that we don't know enough about how to calculate the entropy in quantum gravity to make any statements at all; using the measure on classical trajectories allows us to make fine-tuning arguments while remaining wholly in a regime where classical general relativity should be completely valid.

c. Flatness

An interesting, and surprisingly, nontrivial application of the GHS measure is to the flatness problem—as we will see, it doesn't really exist. (In this section and the next, I am drawing on work from Carroll and Tam 2010.) Consider a Robertson-Walker universe with scale factor $a(t)$ and curvature parameter κ, obeying the Friedmann equation, with an energy density from components ρ_i such that each scale as a power law, $\rho_i = \rho_{i0} a^{-n_i}$, for some fixed n_i. This includes the cases of non-relativistic matter, for which we have $n_M = 3$, and radiation, for which $n_R = 4$. Then we can define the corresponding density parameters Ω_i, as well as an "effective density parameter for curvature," via

$$\Omega_i \equiv \frac{8\pi G \rho_{i0} a^{-n_i}}{3H^2}, \quad \Omega_\kappa \equiv -\frac{\kappa}{a^2 H^2}.$$

The Friedmann equation then implies that $\sum_i \Omega_i + \Omega_\kappa = 1$. The ratio of the curvature to one of the densities evolves with time as

$$\frac{\Omega_\kappa}{\Omega_i} \propto a^{n_i - 2}.$$

Whenever $n_i > 2$, as for matter or radiation, the relative importance of curvature grows with time. The conventional flatness problem is simply the observation that because the curvature is not very large today, it must have indeed been extremely small at early times. Roughly speaking (since details depend on the amounts of matter, radiation, and vacuum energy, as well as their evolutions), we must have had $\Omega_\kappa / \Omega_{matter/radiation} < 10^{-55}$ in the very early universe in order for the curvature to not dominate today.

As we have argued, however, such a statement has impact only if the set of trajectories for which $\Omega_\kappa / \Omega_{matter/radiation} < 10^{-55}$ in the very early universe is actually small. It *seems* small, since 10^{-55} is a small number. But that just means that it would be small if trajectories were chosen uniformly in the variable $\Omega_\kappa / \Omega_{matter/radiation}$, for which we have given no independent justification. Clearly, this is a job for the GHS measure.

To make things quantitative, consider a Robertson-Walker universe containing a homogeneous scalar field $\phi(t)$ with potential $V(\phi)$. (A scalar field is more directly applicable than a matter or radiation fluid, since the scalar has a Hamiltonian formulation; one can, however, consider scalar field theories that mimic the behavior of such fluids, so no real generality is lost.) The scale factor will obey the Friedmann equation, with the energy density given by

$$\rho_\phi = \frac{1}{2}\dot\phi^2 + V(\phi).$$

The dynamical coordinates for this model are $a(t)$ and $\phi(t)$, with conjugate momenta p_a and p_ϕ, as well as a Lagrange multiplier N (the lapse function) that enforces the Hamiltonian constraint. The lapse function is essentially the square root of the 00 component of the metric; it is non-dynamical in general relativity, since no time derivatives of g_{00} appear in the action. Setting $8\pi G = 1$ for convenience, the Einstein-Hilbert Lagrangian for the scale factor coupled to the scalar field is

$$\mathcal{L} = 3\left(Na\kappa - \frac{a\dot a^2}{N}\right) + a^3\left[\frac{\dot\phi^2}{2N} - NV(\phi)\right].$$

The canonical momenta, defined as $p_i = \partial\mathcal{L}/\partial\dot q^i$, are

$$p_N = 0, \quad p_a = -6N^{-1}a\dot a, \quad \text{and} \quad p_\phi = N^{-1}a^3\dot\phi.$$

Performing a Legendre transformation, the Hamiltonian is

$$\mathcal{H} = N\left[-\frac{p_a^2}{12a} + \frac{p_\phi^2}{2a^3} + a^3 V(\phi) - 3a\kappa\right].$$

The equation of motion for N sets it equal to an arbitrary constant, which we can choose to be unity. Varying the action with respect to N gives the Hamiltonian constraint, $\mathcal{H}_* = 0$, which is equivalent to the Friedmann equation. Setting N = 1, the remaining phase space Γ is four-dimensional. The Hamiltonian constraint surface is three-dimensional, and the space of physical trajectories U is two-dimensional. The GHS measure can be written as the Liouville measure subject to the Hamiltonian constraint:

$$\Theta = (dp_a \wedge da + dp_\phi \wedge d\phi)|_{\mathcal{H}=0}.$$

The measure can be evaluated in expanding Friedmann-Robertson-Walker universes by integrating over a specified transverse surface, for example by setting $H = H_*$. The answer works out to be

$$\mu \equiv \int_{H=H_*} \Theta$$

$$= \int_{H=H_*} \Theta_{a\phi}\, da\, d\phi$$

$$= -6 \int_{H=H_*} \frac{3a^3 H_*^2 - a^3 V + 2ak}{(6a^2 H_*^2 - 2a^2 V + 6k)^{1/2}}\, da\, d\phi.$$

We can make this expression look more physically transparent by introducing the variable

$$\Omega_V = \frac{V(\phi)}{3H^2},$$

as well as the curvature density parameter defined earlier. The scale factor is strictly positive, so integrating over all values of Ω_κ is equivalent to integrating over all values of a(t). The measure is then

$$\mu = 3\sqrt{\frac{3}{2}} H_*^{-2} \int_{H=H_*} \frac{1 - \Omega_V - \frac{2}{3}\Omega_\kappa}{|\Omega_\kappa|^{\frac{5}{2}} (1 - \Omega_V - \Omega_\kappa)^{1/2}}\, d\Omega_\kappa\, d\phi.$$

This integral is divergent; it blows up as $\Omega_\kappa \to 0$ since the denominator includes a factor of $|\Omega_\kappa|^{5/2}$. By itself, the divergence isn't surprising; the set of classical trajectories is non-compact. The more interesting fact is *where* it diverges—for universes that are spatially flat ($\Omega_\kappa = 0$), which is certainly a physically relevant region of parameter space.

This divergence was noted in the original GHS paper (Gibbons et al. 1987), where it was attributed to "universes with very large scale factors" due to a different choice of variables. That characterization isn't very useful because a "large scale factor" doesn't pick out any particular type of trajectory; rather, it is a feature along the trajectory of any open universe. Later works correctly described the divergence as arising from nearly flat universes (Hawking and Page 1988; Coule 1995; Gibbons and Turok 2008). Gibbons and Turok (2008) advocated dealing with the infinity by discarding all flat universes by fiat and concentrating on the non-flat universes. Tam and I (Carroll and Tam 2010) took the opposite view: what the divergence is telling us is that almost every Robertson-Walker cosmology is spatially flat. Rather than throw such trajectories away, we should throw all of the others away and deal with flat universes.

What one wants, therefore, is a measure purely on the space of flat universes. The procedure we advocated in Carroll and Tam (2010) for obtaining such a measure was faulty, as our suggested regularization gave a result that was not invariant under a choice of surface on which to evaluate the measure. This problem was later solved in Remmen and Carroll (2013), which derived a measure on the phase space for flat universes by demanding that it obey Liouville's theorem; in Remmen and Carroll (2014), we used this phase-space measure to derive a measure on the space of trajectories. Applying this result to the case of inflation, we showed that one generically expects a very large amount of inflation for unbounded potentials such as $V(\phi) \propto \phi^2$, and relatively few e-folds for "natural" inflation, in which $V(\phi) \propto \cos(\phi)$ (Freese et al. 1990).

From the point of view of fine-tuning, using the GHS measure completely alters our picture of the flatness problem. We noted that the conventional formulation of the problem implicitly assumes a measure that is uniform in Ω_κ, which seemed intuitively reasonable. But in fact, the measure in the vicinity of flat universes turns out to be proportional to $1/|\Omega_\kappa|^{5/2}$, which is a dramatic difference. Rather than sufficiently flat universes being rare, they are actually generic. We take this result to indicate that the flatness problem really isn't a problem at all; it was simply a mistake, brought about by considering an informal measure rather than one derived from the dynamics.

d. Smoothness

The surprising result that almost all universes are spatially flat might raise the hope that a careful consideration of the measure might also explain the smoothness of the universe: perhaps almost all cosmological trajectories describe universes that are extremely smooth (homogeneous and isotropic) at early times. Sadly, the opposite is true, as can be seen by extending the GHS measure to perturbed spacetimes (Carroll and Tam 2010). This might seem like a difficult task, considering how many ways the universe can be perturbed. But as long as the perturbations are small, every Fourier mode evolves independently according to a linear equation of motion. Therefore, we can consider the measure on a mode-by-mode basis.

For linear scalar perturbations, the coupled dynamics of a matter/radiation fluid and the spacetime curvature in a background spacetime can be described by a single degree of freedom, the cosmological perturbation field $u(\vec{x},t)$, as discussed by Mukhanov, Feldman, and Brandenberger (1992). Given the action for this field, we can isolate the dynamical variables and construct the symplectic 2-form on phase space, which can then be used to compute the measure on the set of solutions to Einstein's equations. Various subtleties arise along the way, but the final answer is relatively straightforward. Here I will just quote the results; calculations can be found in Carroll and Tam (2010).

It is convenient to switch to conformal time,

$$\eta = \int a^{-1} dt.$$

Derivatives with respect to η are denoted by primed (′) and double-primed (″) superscripts; $\tilde{H} \equiv a'/a$ is related to the Hubble parameter $H = \dot{a}/a$ by $\tilde{H} = aH$. In Fourier space, the cosmological perturbation field is a function $u(\vec{k},t)$, where $\vec{k}$ is the co-moving wave vector. It is essentially a scaled version of the 00 component of the metric perturbation, which is just the Newtonian gravitational potential:

$$\Phi = (\bar{\rho} + \bar{p})^{1/2} u$$

where $\bar{\rho}$ and $\bar{p}$ are the background energy density and pressure. From that, we can express the energy density perturbation in conformal Newtonian gauge,

$$\delta\rho = \frac{1}{4\pi G a^2}[\nabla^2 \Phi - 3\tilde{H}(\Phi' + \tilde{H})].$$

The cosmological perturbation field obeys an equation of motion

$$u'' - c_s^2 \nabla^2 u - \frac{\theta''}{\theta} u = 0.$$

Here, c_s is the speed of sound in the matter/radiation fluid, and $\theta(\eta)$ is a time-dependent parameter given by

$$\theta = \frac{1}{a}\left[\frac{2}{3}\left(1 - \frac{\hat{H}'}{\hat{H}^2}\right)\right]^{-1/2}.$$

This equation of motion can be derived from an action

$$S_u = \frac{1}{2}\int d^4x \left(u'^2 - c_s^2 \sum_i \partial_i u \partial_i u + \frac{\theta''}{\theta}u^2\right).$$

Defining the conjugate momentum, $p_u = \frac{\partial \mathcal{L}}{\partial u'} = u'$, we can describe the dynamics in terms of a Hamiltonian for an individual mode with wavenumber k,

$$\mathcal{H} = \frac{1}{2}p_u^2 + \frac{1}{2}\left(c_s^2 k^2 - \frac{\theta''}{\theta}\right)u^2.$$

This is simply the Hamiltonian for a single degree of freedom with a time-dependent effective mass $m^2 = c_s^2 k^2 - \theta''/\theta$.

One convenient hypersurface in which we can evaluate the flux of trajectories is $\eta = \eta_* =$ constant. A straightforward calculation shows that the measure evaluated on such a surface is simply

$$\mu = \int_{\eta=\eta_*} du dp_u.$$

In other words, the measure on a perturbation mode is completely uniform in the $\{u, p_u\}$ variables, much as we might have naïvely guessed, and in stark contrast to the flatness problem. All values for u and p_u are equally likely; there is nothing in the measure that would explain the small observed values of perturbations at early times. Hence, the observed homogeneity of our universe does imply considerable fine-tuning.

We can use this measure to roughly quantify how much fine-tuning is involved in the conventional assumption of a smooth universe near the Big Bang. For purposes of convenience, Carroll and Tam (2010) asked a simple question: Assuming that the universe had the observed amount of uniformity at the surface of last scattering ($\delta\rho/\bar{\rho} \leq 10^{-5}$), what fraction of the allowed trajectories were also smooth in the very early universe, say near the scale of grand unification when the energy density was $\rho \sim (10^{16} \text{ GeV})^4$? The answer is, unsurprisingly, quite small. For each individual mode, the chance that it was small at the GUT scale given that it was small at recombination is on the order of 10^{-66}. But there are many modes, and if any one of them is large, then the universe is not truly smooth. The total fraction of universes that were smooth at early times is just the product of the fractions corresponding to each mode. Choosing reasonable bounds for the largest and smallest modes considered, the total fraction of trajectories that are smooth at early times works out to be

$$f(\text{smooth at GUT scale}|\text{smooth at recombination}) \approx 10^{-6.6 \times 10^7}.$$

This represents a very conservative estimate for the amount of fine-tuning involved in the standard cosmological model.

It might seem strange to ask a conditional question that assumes the universe was smooth at the time of last scattering, rather than directly inquiring about the fraction of universes that were smooth at recombination. But that fraction is ill-defined, since the phase space of perturbations is unbounded. We could have calculated the fraction of universes that are relatively smooth today that were also smooth at recombination and obtained a similarly tiny number. More importantly, we are interested in the fine-tuning necessary for the universe to obey the Past Hypothesis. The reason that fraction is such a small number is that most trajectories that are smooth at last scattering contain modes that were large at earlier times but decayed. That is morally equivalent to trajectories that start with relatively high entropy, but that start with delicate correlations that cause the entropy to decrease as time passes. All of conventional cosmology assumes that the early universe was not like that; 10^{-66} quantifies the amount of fine-tuning implied by this assumption.

We can, therefore, conclude that the smoothness of the early universe does indeed represent an enormous amount of fine-tuning. In reaching this conclusion, we made no reference to the causal structure nor to any thwarted attempts at equilibration. Those considerations, which play a central role in formulating the horizon problem, are red herrings. The real sense in which the early universe was fine-tuned is extremely simple: the overwhelming majority of cosmological trajectories, as quantified by the canonical measure, represent highly nonuniform universes at early times, and we don't think the real universe was like that. Clearly, the specific numerical value we obtain is not of central importance; what is certain is that the history of our actual universe does not look anything like it was chosen randomly.

V. Discussion

I have argued that the traditional discussion of the fine-tuning of the early universe in terms of the horizon and flatness problems is misguided. The difficulty with the horizon problem is not that it is incorrect, but that it is inconclusive, since we don't have a clear picture of how the situation would change if distant regions on the surface of last scattering actually had been in causal contact. The flatness problem is based on an implicit use of an unjustified measure on the space of initial conditions; when a more natural measure is used, we see that almost all Robertson-Walker universes are spatially flat. It is much more sensible to quantify fine-tuning in terms of the measure on the space of cosmological trajectories. From that perspective, it is clear that the overwhelming majority of such trajectories, conditionalized on some reasonable requirement in the late universe, are wildly inhomogeneous at early times. The fact that the actual early universe was not like that, as specified by the Past Hypothesis, gives us a clear handle on the kind of fine-tuning we are faced with.

As with most discussions of fine-tuning, in this paper I have prejudiced the discussion somewhat by assuming throughout that the universe is an approximately Robertson-Walker cosmology with a certain (large) amount of matter and radiation. While we have seen that the universe is fine-tuned even given that framework, such an assumption is much stronger than what is required, for example, by the anthropic principle. Life can certainly exist in universes with far fewer stars and galaxies than what we observe. In the presence of a stable vacuum energy, the highest-entropy configuration for the universe to be in is empty de Sitter space (Carroll and Chen 2004). The worry there is that vacuum fluctuations give rise to an ensemble of freak "Boltzmann brain" observers (Dyson et al. 2002; Albrecht and Sorbo 2004; Bousso and Freivogel 2007).[15] As argued in Boddy et al. (2014), however, quantum fluctuations in de Sitter space don't actually bring into existence decohered branches of the wave function containing such freak observers. Nevertheless, it seems reasonable to think that the space of trajectories containing one person or one galaxy in an otherwise empty background has a much greater measure than the kind of universe in which we live, with over a hundred billion galaxies; at least, such a situation has a much higher entropy. We are,

---

[15] The problem is not that a de Sitter scenario predicts that "we should be Boltzmann brains." Rather, we should consider ourselves to be one of any of the many observers that find themselves in precisely our current cognitive situation. For most such observers, that cognitive situation—e.g., my current belief that there is a person named "David Albert" who wrote a book entitled *Time and Chance*—is completely uncorrelated with the reality of their actual environment. Such a situation is cognitively unstable, for reasons explained in Albert (2003).

therefore, still left with the fundamental cosmological question: Why don't we live in a nearly empty de Sitter space?

Formulating cosmological fine-tuning in the language of a measure on trajectories puts the inflationary-universe scenario in its proper context. A major original motivation of inflation theory was to solve the horizon and flatness problems. Reformulating the issue raised by the horizon problem as a matter of the measure on cosmological trajectories brings the problem with inflation into sharp focus: the fact that most trajectories that are homogeneous at late times are highly nonhomogeneous at early times is completely independent of physical processes in the early universe. It depends only on the measure evaluated at relatively late times. Inflation, therefore, cannot solve this problem all by itself. Indeed, the measure reinforces the argument made by Penrose, that the initial conditions necessary for getting inflation to start are extremely fine-tuned, more so than those of the conventional Big Bang model it was meant to help fix. Inflation does, however, still have very attractive features. It posits an initial condition that, while very low-entropy, is also extremely simple, not to mention physically quite small. (With inflation, our observable universe could have been one Planck length across at the Planck density; without inflation, the same patch was of order one centimeter across at that time. That is an incredibly large volume, when considered in Planck units, over which to have initial homogeneity.) Therefore, while inflation does not remove the need for a theory of initial conditions, it gives those trying to construct such a theory a relatively reasonable target to shoot for.

Of course, all of this discussion about fine-tuning and the cosmological measure would be completely pointless if we did have a well-formulated theory of initial conditions (or, better yet, of our cosmological history considered as a whole). Ultimately the goal is not to explain why our universe appears unnatural; it's to explain why we live in this specific universe. Making its apparent unnaturalness precise is hopefully a step toward achieving this lofty ambition.

Acknowledgments

It is a pleasure to thank David Albert for inspiration and conversations over the years, Barry Loewer for his patience, Shelly Goldstein for useful discussions, my collaborators Heywood Tam and Grant Remmen for their invaluable insights, and Tim Maudlin for the interactions that proximately inspired this paper. This research is funded in part by DOE grant DE-SC0011632, and by the Gordon and Betty Moore Foundation through Grant 776 to the Caltech Moore Center for Theoretical Cosmology and Physics.

References

Aaronson, Scott, Sean M. Carroll, and Lauren Ouellette. 2014. Quantifying the rise and fall of complexity in closed systems: The coffee automaton.

Ade, P. A. R., and others. 2014. Planck 2013 results. I. Overview of products and scientific results. *Astronomy & Astrophysics* 571.

Aguirre, Anthony, and Steven Gratton. 2002. Steady-state eternal inflation. *Physical Review D* 65.

Aguirre, Anthony, Steven Gratton, and Matthew C Johnson. 2007. Hurdles for recent measures in eternal inflation. *Physical Review D* 75.

Albert, D. 2003. *Time and Chance*. Cambridge, MA: Harvard University Press.

Albrecht, Andreas, and Lorenzo Sorbo. 2004. Can the universe afford inflation? *Physical Review D* 70.

Albrecht, Andreas, and Paul J. Steinhardt. 1982. Cosmology for grand unified theories with radiatively induced symmetry breaking. *Physical Review Letters* 48: 1220–1223.

Anninos, P., R. A. Matzner, T. Rothman, and M. P. Ryan. 1991. How does inflation isotropize the universe? *Physical Review D* 43: 3821–3832.

Banks, Tom. 1985. TCP, quantum gravity, the cosmological constant and all that... *Nuclear Physics B* 249: 332–360.

———. 2000. Cosmological breaking of supersymmetry or little lambda goes back to the future II. https://arxiv.org/abs/hep-th/0007146.

Banks, Tom, and W. Fischler. 2001. M-theory observables for cosmological space-times. https://arxiv.org/abs/hep-th/0102077.

Bengtsson, I., J. Braennlund, and K. Zyczkowski. 2002. CP^n, or, entanglement illustrated. *International Journal of Modern Physics A* 17: 4675–4696.

Boddy, Kimberly K., Sean M. Carroll, and Jason Pollack. 2014. De Sitter space without quantum fluctuations. *Foundations of Physics* 46: 702–735.

Borde, Arvind, Alan H. Guth, and Alexander Vilenkin. 2003. Inflationary spacetimes are incomplete in past directions. *Physical Review Letters* 90: 151301.

Bousso, Raphael. 2002. The holographic principle. *Reviews of Modern Physics* 74: 825–874.

Bousso, Raphael, and Ben Freivogel. 2007. A paradox in the global description of the multiverse. *Journal of High Energy Physics* 06: 018.

Brawer, Roberta. 1995. Inflationary cosmology and horizon and flatness problems: The mutual constitution of explanation and questions. PhD diss., Massachusetts Institute of Technology.

Brody, D. C., and L. P. Hughston. 2001. Geometric quantum mechanics. *Journal of Geometry and Physics* 38: 19–53.

Bunn, E. F., and D. W. Hogg. 2009. The kinematic origin of the cosmological redshift. *American Journal of Physics* 77: 688–694.

Carroll, Sean M. 2004. *Spacetime and Geometry: An Introduction to General Relativity*. Cambridge: Cambridge University Press.

———. 2008. What if time really exists? https://arxiv.org/abs/0811.3772.

Carroll, Sean M., and Jennifer Chen. 2004. Spontaneous inflation and the origin of the arrow of time. https://arxiv.org/abs/hep-th/0410270.

Carroll, Sean M., and Heywood Tam. 2010. Unitary evolution and cosmological fine-tuning. https://arxiv.org/abs/1007.1417.

Coule, D. H. 1995. Canonical measure and the flatness of a FRW universe. *Classical and Quantum Gravity* 12: 455–470.

Dicke, Robert H., and P. J. E. Peebles. 1979. The big bang cosmology—Enigmas and nostrums. In S. W. Hawking and W. Israel, eds., *General Relativity: An Einstein Centenary Survey*, 504–517. Cambridge: Cambridge University Press.

Dyson, Lisa, Matthew Kleban, and Leonard Susskind. 2002. Disturbing implications of a cosmological constant. *Journal of High Energy Physics* 10: 011.

Earman, John. 2006. The "past hypothesis": Not even false. *Studies in History and Philosophy of Science Part B: Studies in History and Philosophy of Modern Physics* 37: 399–430.

Egan, Chas A., and Charles H. Lineweaver. 2010. A larger estimate of the entropy of the universe. *Astrophysical Journal* 710: 1825–1834.

Freese, Katherine, Joshua A. Frieman, and Angela V. Olinto. 1990. Natural inflation with pseudo Nambu-Goldstone bosons. *Physical Review Letters* 65: 3233–3236.

Freivogel, Ben. 2011. Making predictions in the multiverse. *Classical and Quantum Gravity* 28: 204007.

Gibbons, G. W., and Neil Turok. 2008. The measure problem in cosmology. *Physical Review D* 77: 063516.

Gibbons, G. W., S. W. Hawking, and J. M. Stewart. 1987. A natural measure on the set of all universes. *Nuclear Physics B* 281: 736–751.

Goldstein, Sheldon. 2014. "Unpublished."

Greene, Brian, Kurt Hinterbichler, Simon Judes, and Maulik K. Parikh. 2011. Smooth initial conditions from weak gravity. *Physics Letters B* 697: 178–183.

Guth, Alan H. 1981. The inflationary universe: A possible solution to the horizon and flatness problems. *Physical Review D* 23: 347–356.

———. 2000. Inflation and eternal inflation. *Physics Reports* 333: 555–574.

Guth, Alan H., David I. Kaiser, and Yasunori Nomura. 2014. Inflationary paradigm after Planck 2013. *Physics Letters B* 733: 112–119.

Hawking, S. W., and Don N. Page. 1988. How probable is inflation? *Nuclear Physics B* 298: 789–809.

Hooft, Gerard 't. 1993. Dimensional reduction in quantum gravity. https://arxiv.org/abs/gr-qc/9310026.

Horowitz, Gary T., and Joseph Polchinski. 2006. Gauge/gravity duality. https://arxiv.org/abs/gr-qc/0602037.

Ijjas, Anna, Paul J. Steinhardt, and Abraham Loeb. 2013. Inflationary paradigm in trouble after Planck2013. *Physics Letters B* 723: 261–266.

———. 2014. Inflationary schism after Planck2013. *Physics Letters B* 736: 142–146.

Jacobson, Ted. 1995. Thermodynamics of spacetime: The Einstein equation of state. *Physical Review Letters* 75: 1260–1263.

Kibble, T. W. B. 1979. Geometrization of quantum mechanics. *Communications in Mathematical Physics* 65: 189–201.

Linde, Andrei D. 1982. A new inflationary universe scenario: A possible solution of the horizon, flatness, homogeneity, isotropy and primordial monopole problems. *Physics Letters B* 108: 389–393.

Linde, Andrei D. 2014. Inflationary cosmology after Planck 2013. https://arxiv.org/abs/1402.0526.

Linde, Andrei D., Vitaly Vanchurin, and Sergei Winitzki. 2009. Stationary measure in the multiverse. *Journal of Cosmology and Astroparticle Physics* 01: 031.

Loewer, Barry. 2012. The emergence of time's arrows and special science laws from physics. *Interface Focus* 2: 13–19.

Lynden-Bell, D., and R. Wood. 1968. The gravo-thermal catastrophe in isothermal spheres and the onset of red-giant structure for stellar systems. *Monthly Notices of the Royal Astronomical Society* 138: 495.

Maldacena, Juan Martin. 1998. The large N limit of superconformal field theories and supergravity." *Advances in Theoretical and Mathematical Physics* 2: 231–252.

Maldacena, Juan, and Leonard Susskind. 2013. Cool horizons for entangled black holes. *Fortschritte der Physik* 61: 781–811.

Misner, Charles W. 1969. Mixmaster universe. *Physical Review Letters* 22: 1071–1074.

Mithani, Audrey, and Alexander Vilenkin. 2012. Did the universe have a beginning? https://arxiv.org/abs/1204.4658.

Mukhanov, Viatcheslav F., H. A. Feldman, and Robert H. Brandenberger. 1992. Theory of cosmological perturbations. Part 1. Classical perturbations. Part 2. Quantum theory of perturbations. Part 3. Extensions. *Physics Reports* 215: 203–333.

Nityananda, Rajaram. 2009. The gravitational dynamics of galaxies. *Pramana* 73: 193–214.

Page, Don N., and William K. Wootters. 1983. Evolution without evolution: Dynamics described by stationary observables. *Physical Review D* 27: 2885.

Penrose, R. 1989. Difficulties with inflationary cosmology. *Annals of the New York Academy of Sciences* 571: 249–264.

Price, Huw. 1997. Cosmology, time's arrow, and that old double standard. In Steven F. Savitt, ed., *Time's Arrows Today: Recent Physical and Philosophical Work on the Direction of Time*, 66–94. Cambridge: Cambridge University Press.

Remmen, Grant N., and Sean M. Carroll. 2013. Attractor solutions in scalar-field cosmology. *Physical Review D* 88: 083518.

———. 2014. How many e-folds should we expect from high-scale inflation? *Physical Review D* 90: 063517.

Ryu, Shinsei, and Tadashi Takayanagi. 2006. Holographic derivation of entanglement entropy from AdS/CFT. *Physical Review Letters* 96: 181602.

Salem, Michael P. 2012. Bubble collisions and measures of the multiverse. *Journal of Cosmology and Astroparticle Physics*.

Schiffrin, Joshua S., and Robert M. Wald. 2012. Measure and Probability in Cosmology." *Physical Review D* 86.

Schwartz-Perlov, Delia, and Alexander Vilenkin. 2010. Measures for a transdimensional multiverse. *Journal of Cosmology and Astroparticle Physics* 1201: 021.

Susskind, Leonard. 1995. The world as a hologram. *Journal of Mathematical Physics* 36: 6377–6396.

Susskind, Leonard, Larus Thorlacius, and John Uglum. 1993. The stretched horizon and black hole complementarity. *Physical Review D* 48: 3743–3761.

Swingle, B. 2012. Entanglement renormalization and holography. *Physical Review D* 86: 065007.

Vachaspati, Tanmay, and Mark Trodden. 2000. Causality and cosmic inflation. *Physical Review D* 61: 023502.

Van Raamsdonk, Mark. 2010. Building up spacetime with quantum entanglement. *General Relativity and Gravitation* 42: 2323–2329.

Verlinde, Erik P. 2011. On the origin of gravity and the laws of Newton. *Journal of High Energy Physics* 1104: 029.

Wald, Robert M. 1983. Asymptotic behavior of homogeneous cosmological models in the presence of a positive cosmological constant. *Physical Review D* 28: 2118.

Wallace, David. 2011. The logic of the past hypothesis. http://philsci-archive.pitt.edu/8894/.

Wiltshire, David L. 1995. An introduction to quantum cosmology. In B. Robson, N. Visvanathan, and W. S. Woolcock, eds., *Cosmology: The Physics of the Universe*, 473–531. Singapore: World Scientific.

Winitzki, Sergei. 2008. "Predictions in eternal inflation." In Martin Lemoine, Jerome Martin, and Peter Patrick, eds., *Inflationary Cosmology: Lecture Notes in Physics*, 157–191. Heidelberg: Springer.

Chapter Five

The Meta-Reversibility Objection

▶ CHRISTOPHER J. G. MEACHAM

1. Introduction

Statistical mechanics is an inherently probabilistic theory. It explains thermodynamic phenomena—the diffusion of milk poured into a cup of hot cocoa, the cooling of the cup of hot cocoa when it's carried out into the snow—by assigning high probabilities to such events. And it explains the absence of antithermodynamic behavior—the separation and ejection of all of the milk from a cup of hot cocoa, the spontaneous warming of a cup of hot cocoa left outside—by assigning small probabilities to such events.

But how to understand these probabilities is a contentious issue. The recent proposals in the literature can be roughly divided into two camps.[1] One view of statistical mechanical probabilities, recently defended by Albert (2000) and Loewer (2001), understands them as *chances*—i.e., physical probabilities given by the laws

---

[1] Those who defend a "typicality" approach to statistical mechanics might not seem to fit into either of these camps (see Goldstein, 2001, and Maudlin, 2007, for a description of such views). As I understand it, however, this approach is not a competitor to the ones described here. Rather, the typicality approach crosscuts the divide between these two camps, and could be adopted (or rejected) by proponents of both nomic and indifference approaches.

of nature.[2] I'll call this the *nomic approach* to statistical mechanics (and accounts of statistical mechanics in this vein *nomic accounts*), since it maintains that statistical mechanical probabilities are lawful or nomic features of the world.

The other view of statistical mechanical probabilities understands them as measures of rational indifference. On this view, the statistical mechanical probabilities represent the credences that an ideally rational agent would adopt about a system, given only certain information about what that system is like. These probabilities don't come from the laws, they come from *a priori* constraints on rational belief (often called "Indifference Principles"). I'll call this the *indifference approach* to statistical mechanics (and accounts of statistical mechanics in this vein *indifference accounts*), since it maintains that statistical mechanical probabilities are the credences prescribed by Indifference Principles.

Initial formulations of nomic and indifference accounts face "reversibility worries"—while these accounts yield the right results regarding future events, they yield the wrong results regarding past events (see §3). But in both cases it seems one can overcome these reversibility worries by modifying these accounts to include something like the Past Hypothesis—a law requiring a particular low-entropy initial condition to obtain—as discussed by Albert (2000).

In this paper, I argue that this parity between nomic and indifference accounts is illusory. For while appealing to a Past Hypothesis suffices to free nomic accounts from reversibility worries, it doesn't suffice to free indifference accounts from these worries.

More precisely, I show that both nomic and indifference accounts can escape the standard reversibility worries—what I'll call the Reversibility Objection—by appealing to the Past Hypothesis. But by positing a Past Hypothesis, indifference accounts become susceptible to another kind of reversibility worry—what I'll call the Meta-Reversibility Objection. And I argue that there is no easy way for indifference accounts to escape the Meta-Reversibility Objection. Thus, reversibility considerations give us a strong reason to favor nomic accounts over indifference accounts.

The rest of the paper will proceed as follows. In §2 I provide some background regarding Bayesianism and statistical mechanics, present some assumptions, and describe the accounts of statistical mechanics I'll focus on in more detail.

In §3 I discuss the Reversibility Objection to the initial formulations of the nomic and indifference accounts. I begin the section with a rough description of

---

[2] A number of other people, including Winsberg (2008), Callender and Cohen (2009), and Frigg and Hoefer (2010), have followed Albert and Loewer in advocating something like the nomic approach to statistical mechanics.

the worries motivating the Reversibility Objection. In §3.1 I provide a more rigorous characterization of the Reversibility Objection. This allows us to see the possible replies to the Reversibility Objection, and provides us with some of the tools we need to set up the Meta-Reversibility Objection. In §3.2 I assess these replies, and show why there's pressure to follow Albert in adopting a lawful constraint on initial conditions, the Past Hypothesis. In §4 I turn to the Meta-Reversibility Objection. I begin with a rough description of the worries motivating the Meta-Reversibility Objection, and explain why similar worries don't arise for nomic accounts. In §4.1 I provide a more rigorous characterization of the Meta-Reversibility Objection. In §4.2 I assess the possible replies to the Meta-Reversibility Objection. I argue that none of the available options provide the proponent of the indifference approach with a satisfying reply. I conclude with some brief remarks in §5.

For those already familiar with these issues, here is a more detailed sketch of the dialectic that leads to the Meta-Reversibility Objection. The initial formulations of the nomic and indifference accounts, which impose no constraints on initial conditions, are unsatisfactory because they assign the wrong values to past events. This is the Reversibility Objection. We can escape this objection by adopting something like the Past Hypothesis (*PH*), a lawful constraint that requires a particular low-entropy initial condition to obtain. With this addition, the nomic and indifference accounts will assign the right values to past events.

But note that there are other lawful constraints on initial conditions one could adopt, such as the "Past Hypothesis*" (*PH**) which requires that some *high*-entropy initial condition obtain. Given something like *PH**, the nomic and indifference accounts will again assign the wrong values to past events. I'll argue in §4 that there is pressure on the indifference approach (though not the nomic approach) to take *PH** to be much more likely than *PH*. And if the indifference approach takes *PH** to be much more likely than *PH*, then the wrong values it assigns to past events given *PH** will swamp the right values it assigns to past events given *PH*. So, when all is said and done, it'll again assign the wrong values to past events. This, in a nutshell, is the Meta-Reversibility Objection.

2. Background

2.1. Rational Agents

For the purposes of this paper, I'll be restricting my attention to rational agents. I assume that the belief states of such agents can be represented with a *credence function, cr*, which takes propositions as arguments, and spits out real numbers between 0 and 1, which represent the agent's degree of confidence that the proposition is true. I assume that rational agents are *Bayesian*, in that their credences satisfy the following pair of constraints:

Probabilism: An agent's credences should satisfy the probability axioms.[3]

Conditionalization: If an agent with credences cr receives E as evidence, then her new credences cr^+ should be:

$$cr^+(A) = cr(A|E), \text{ if defined.}[4] \quad (1)$$

Note that Conditionalization doesn't care about how you break up evidence; conditionalizing on E and then conditionalizing on F yields the same result as conditionalizing on $E \wedge F$. Thus, we can reformulate Conditionalization in terms of an agent's initial credence function ic and her total evidence E as follows:[5]

Conditionalization: If an agent with initial credences ic has total evidence E, then her credences should be:

$$cr_E(A) = ic(A|E), \text{ if defined.} \quad (2)$$

I further assume that rational agents satisfy something like Lewis's (1986) Principal Principle, a principle that links their beliefs about propositions to their beliefs about the chances of those propositions. In particular, let T be a complete chance theory, let K be some background information that T requires in order to produce a chance distribution, and let $ch_{TK}(A)$ be the chance that T assigns to A given K. Then I assume that rational agents satisfy the following constraint:

Chance-Credence Principle: An agent's initial credences ic should be such that:

$$ic(A|T \wedge K) = ch_{TK}(A), \text{ if defined.}[6] \quad (3)$$

---

[3] That is, an agent's credence function cr, defined over an algebra $\mathcal{A}$ over Ω, should be such that: (1) $cr(\Omega) = 1$, (2) $\forall A \in \mathcal{A}$, $cr(A) \geq 0$, (3) $\forall A, B \in \mathcal{A}$, if $A \cap B = \emptyset$, then $cr(A) + cr(B) = cr(A \cap B)$.

[4] Where $cr(A|B) := \dfrac{cr(A \wedge B)}{cr(B)}$.

[5] Though, of course, this extension only works for agents who *have* initial credence functions.

[6] This principle is essentially that proposed by Lewis (1986), with one slight amendment—I've replaced Lewis's "complete history up to a time" H with a more general background argument K, in order to allow for statistical mechanical chances. (For some considerations in favor of this kind of formulation of the Chance-Credence Principle, see Arntzenius, 1995; Meacham, 2005; Nelson, 2009; and Meacham, 2010.) As Lewis and others have noted, this principle is problematic if one

Proponents of the indifference approach to statistical mechanics assume that rational agents also satisfy a further constraint, an *Indifference Principle* of some sort. There are many Indifference Principles, but all of the Indifference Principles compatible with Bayesianism take a similar form. We can formulate such principles in terms of an agent's initial credences ic and a measure of rational indifference μ, as follows (where different choices of μ yield different Indifference Principles):

Indifference Principle: An agent's initial credences ic should be such that:

$$ic(A) = \mu(A). \tag{4}[7]$$

An Indifference Principle and Bayesianism together will determine the unique rational credence function that an agent with total evidence E should have (i.e., $cr_E(A) = \mu(A|E)$).

Not all Indifference Principles are compatible with Bayesianism. A number of authors, such as Friedman and Shimony (1971) and Dias and Shimony (1981), have argued that some popular Indifference Principles conflict with Bayesianism.[8] But I take conflicting with Bayesianism to be a *reductio* of a proposed Indifference Principle. So I restrict my attention here to Indifference Principles that are compatible with Bayesianism.

2.2. Statistical Mechanics

Although the issues discussed in this paper are relevant to both classical and quantum statistical mechanics, I restrict my attention here to classical statistical mechanics.[9]

Consider a classical world with n particles and three spatial dimensions. The *phase space* of this world is a $6n$ dimensional space representing all of the possible

---

adopts a Humean account of chance (see Lewis, 1994; Hall, 1994; and Thau, 1994); for the purposes of this paper, I'll put these issues aside.

[7] This assumes we can treat μ as a probability function. Strictly speaking, this isn't correct (see Meacham, 2005), and accommodating this fact would require formulating the Indifference Principle in terms of (say) an equation relating a pair of primitive conditional probabilities. I'll ignore these complications here.

[8] Whether these arguments are correct depends on how these principles are understood (see Uffink, 1996).

[9] For more on classical statistical mechanics, see Tolman (1979); for more on the philosophical issues that arise, see Sklar (1993) and Albert (2000); for a helpful road map and source of references regarding the issues, see North (2011).

positions and velocities of these particles.[10] Each dimension of the phase space corresponds to the location or velocity of one of the particles in one of the three spatial directions. (Thus, the locations and velocities of each particle correspond to six dimensions.) And each point in the phase space corresponds to a possible arrangement of particle positions and velocities, with the point's position along each dimension encoding the value of the corresponding location or velocity.

Let a *microstate* be a complete specification of the locations and velocities of each particle—this corresponds to a particular point in phase space. Let a *macrostate* be an incomplete specification of the locations and velocities of each particle—this corresponds to the region of phase space containing all and only those microstates compatible with that specification.[11]

To fix on a particular phase space, we need to fix the values of certain parameters, such as the number of particles, the number of spatial dimensions, and the spatial extension of each dimension. But these details have little bearing on the points at issue. So, to simplify my presentation I'll assume that the phase space we're working with has been fixed. I describe in the appendix the somewhat clunkier versions of the central principles and arguments that take these parameters into consideration.

The standard statistical mechanical probabilities are given by the Liouville measure—i.e., the Lebesgue measure over the canonical representation of the phase space.[12] In particular, the statistical mechanical probability of the world being in macrostate A, given that it's in macrostate B, is equal to the proportion (according to the Liouville measure) of B's microstates that are compatible with A.

Note that we're assuming deterministic dynamics—given the microstate of the world at one time, the dynamics determine the microstate of the world at any other time.[13] And note that these dynamics preserve the Liouville measure—the Liouville measure of any macrostate A is the same as the Liouville measure of any macrostate A^* that we can get to by evolving A forward or backward in time in

---

[10] Following Albert, I assume that it's legitimate to apply statistical mechanics to the world as a whole. Some have resisted this suggestion, arguing that we should restrict the scope of statistical mechanics; for example, see Leeds (2003).

[11] For convenience, I'll use the term "macrostate" to refer to the property of a system, to the region of phase space occupied by microstates with that property, and to the proposition that the system has that property; context will make it clear how it's being used. Likewise, I'll use the name of a macrostate A to refer to the property, the corresponding region of phase space and the proposition of the system having that property.

[12] Or, if we're holding fixed features of the system (such as its total energy) that reduce the dimensionality of the macrostates we're considering, we employ the Lebesgue measure over the canonical representation of the appropriate sub-space.

[13] There are exceptions to the claim that classical mechanics is deterministic (see Earman, 1986; Xia, 1992; and Norton, 2003), but we can ignore these exceptions for the purposes at hand.

accordance with the dynamics. Thus, we can think of any questions about statistical mechanical probabilities in terms of probabilities over initial conditions: the answer to the question "What is the probability of A given B (at t)?" is always the same as the answer to the question "What is the probability of the initial conditions having been such that they would evolve into macrostate A at t, given that we know they're such that they'll evolve into macrostate B at t?"[14]

For the purposes of this paper, I take an *account of statistical mechanics* to be an account which provides at least: (a) a complete specification of the laws, and (b) a characterization of statistical mechanical probabilities. The discussion in this paper will focus on four accounts: the initial formulations of the nomic and indifference accounts (N1 and I1), and the revised formulations of the nomic and indifference accounts that posit a Past Hypothesis (N2 and I2). The discussion of the Reversibility Objection in §3 will focus on the first pair of accounts, N1 and I1. The discussion of the Meta-Reversibility Objection in §4 will focus on the second pair, N2 and I2.

My characterizations of the nomic accounts largely follow Albert (2000). I take the two nomic accounts, N1 and N2, to be:

N1:
a. The complete laws are (i) the laws of classical mechanics, and (ii) a *Statistical Postulate,* which assigns chances to the (lawfully permitted) initial conditions that are proportional to the Liouville measure.
b. The statistical mechanical probability of A given B is the chance of A given B assigned by the Statistical Postulate.

N2:
a. The complete laws are (i) the laws of classical mechanics, (ii) a *Statistical Postulate,* which assigns chances to the (lawfully permitted) initial conditions that are proportional to the Liouville measure, and (iii) a *Past Hypothesis,* which permits all and only the initial conditions belonging to some particular low-entropy macrostate.
b. The statistical mechanical probability of A given B is the chance of A given B assigned by the Statistical Postulate.

---

[14] Assuming, of course, that there *are* initial conditions. At worlds in which there are not—say, worlds that are infinitely temporally extended in both directions—one would have to replace this talk of initial conditions with talk about the conditions of some "early" time slice. (Similar remarks apply with respect to the Past Hypothesis presented below.)

I take the two indifference accounts, I1 and I2, to be:

I1:
a. The complete laws are (i) the laws of classical mechanics.
b. An Indifference Principle constrains rational belief. The statistical mechanical probability of A given B is what this Indifference Principle assigns as the rational initial credence in A conditional on the macrostate B and the laws.

I2:
a. The complete laws are (i) the laws of classical mechanics, and (ii) a *Past Hypothesis*, which permits all and only the initial conditions belonging to some particular low-entropy macrostate.
b. An Indifference Principle constrains rational belief. The statistical mechanical probability of A given B is what this Indifference Principle assigns as the rational initial credence in A conditional on the macrostate B and the laws.

These characterizations of I1 and I2 are underspecified, since I haven't identified what these Indifference Principles take μ to be. In order to keep my discussion as general as possible, I won't assume much about what μ is like. But I will assume that the Indifference Principles these accounts employ yield statistical mechanical probabilities that roughly match those assigned by the corresponding nomic account (so that I1 yields roughly the same statistical mechanical probabilities as N1, and I2 yields roughly the same statistical mechanical probabilities as N2). That is, I will assume that:

$$\mu(A|I1 \wedge E) \approx ch_{N1E}(A), \tag{5}$$

$$\mu(A|I2 \wedge E) \approx ch_{N2E}(A). \tag{6}$$

Note that these two constraints on μ are compatible. I1 and I2 are disjoint, since I2 entails that there's a lawful constraint on the initial conditions beyond those imposed by the dynamics (the Past Hypothesis), while I1 entails that there is not. Thus, equations (5) and (6) impose constraints on different parts of logical space, and so won't conflict.

I assume (5) and (6) in order to focus on the key differences between the nomic and indifference approaches. The core disagreement between nomic and indifference approaches isn't over which probabilities should be assigned in typical statistical mechanical cases, but instead over how these probabilities should be understood. For mere disagreement over which probabilities should be assigned won't pull the two camps apart, since for any probability assignment, one can find both a nomic account and an indifference account that yield that assignment (the nomic account saying that these values obtain as a matter of natural law, the indifference account saying they obtain in virtue of *a priori* constraints on rational belief).

3. The Reversibility Objection

Although initially attractive, the first formulations of the nomic and indifference accounts given above, N1 and I1, run into trouble.

Consider N1. This account yields the right predictions regarding what the future will be like. It predicts, for instance, that the partially diffused milk in a cup of hot cocoa will become completely diffused. This is because, according to the Liouville measure, the overwhelming majority of microstates compatible with the partially-diffused-milk-in-cocoa macrostate will evolve into microstates in which the milk is wholly diffused throughout the cocoa. And thus it assigns an overwhelmingly high chance to the milk becoming wholly diffused.

But N1 seems to yield the wrong predictions regarding what the past was like. For instance, it predicts that the partially diffused milk in a cup of hot cocoa was previously completely diffused. This is because, according to the Liouville measure, the overwhelming majority of microstates compatible with the partially-diffused-milk-in-cocoa macrostate evolved from microstates in which the milk is wholly diffused throughout the cocoa. And thus, again, it assigns an overwhelmingly high chance to the milk having been wholly diffused.

The problem is that N1 makes the same kinds of predictions about the future and the past. But this seems wrong—thermodynamic phenomena seem to be temporally asymmetric.

Let's take this worry one step further. We've said that N1 seems to yield the wrong predictions about the past—predictions that don't line up with our memories about the past, our written records about the past, and so on. But why think that our memories about the past, our written records about the past, and so on, are accurate? For, according to the Liouville measure, only a minuscule fraction of the microstates compatible with our beliefs and records evolved from past microstates that line up with our beliefs and records about the past. So, if we carry this worry to its logical conclusion, the concern is that if we believe the account is true, we're rationally required to believe the *Skeptical Hypothesis* (*SH*): that our beliefs about the past are wrong. I1 runs into similar problems. While I1 prescribes beliefs about the future that line up well with what actually happens, it prescribes beliefs about the past that don't line up with our memories, records, etc. So, as with N1, I1 requires us to believe that our memories, records, etc., are wrong. Thus, if we believe I1 is true, we're rationally required to believe the Skeptical Hypothesis. Call this the *Reversibility Objection* to the initial formulations of the nomic and indifference accounts.

3.1. Formalizing the Reversibility Objection

Let's formulate the Reversibility Objection more precisely. To begin, the argument needs a way to move from an account yielding skeptical consequences to the account being false. Thus, the argument must assume something like the following:

Adequacy: The correct account of classical statistical mechanics X is such that, given evidence like ours, one could rationally believe $X \wedge \neg SH$.

A couple of comments. First, the "evidence like ours" clause is intended to restrict our attention to agents who get roughly the same kinds of evidence as we do—evidence from their senses, their memories, and the like. What this is intended to exclude is the logical possibility of agents who *directly* get $X \wedge \neg SH$ as evidence. It's hard to imagine what getting such evidence would be like—perhaps a divine revelation that produces veridical certainty in $X \wedge \neg SH$ would qualify. But in any case, if we don't exclude such agents, then Adequacy becomes toothless, since such agents could rationally believe $X \wedge \neg SH$ for any X, as long as $X \wedge \neg SH$ is logically coherent. (To avoid cumbersome phrasing, I'll sometimes leave the "evidence like ours" clause implicit in what follows.)

Second, Adequacy entails that the correct account of classical statistical mechanics must be one that we could rationally believe. This is a nontrivial assumption. After all, just because we can't rationally believe something doesn't mean it's false. Still, it seems at least *prima facie* plausible that the correct account of classical statistical mechanics is one that we could rationally come to believe.[15]

---

[15] Of course, it's somewhat awkward to talk about what the "correct" account of classical statistical mechanics is, when we know that classical statistical mechanics doesn't actually obtain. One can see the discussion here as being in the same vein as discussions about whether the "correct" account of classical spacetime is Newtonian or Galilean. In all of these contexts, "correct" means something like "most attractive given the desiderata of theory choice and some subset of our actual evidence."

Third, assuming that the correct account of classical statistical mechanics is one that we could rationally believe, it's a further assumption to maintain that we could rationally believe both it and that the Skeptical Hypothesis is false. Now, once we grant that the correct account of statistical mechanics X is one we could rationally believe, one could argue that this further assumption is easy to get. For, following Albert (2000), one could maintain that the Skeptical Hypothesis undermines our evidence for believing something like classical statistical mechanics in the first place. So we'd only ever be in a position to believe X if we were also in a position to believe $X \wedge \neg SH$.[16]

Given Adequacy, one can formulate the Reversibility Objection to the initial formulations of the nomic and indifference accounts as follows. Let's understand "believe" as "having a credence of greater than 0.5 in," and let's use "E" to stand for any total body of evidence like ours. Then we can reformulate Adequacy as:

Adequacy: The correct account of classical statistical mechanics X is such that one could rationally have $cr_E(X \wedge \neg SH) > 0.5$.

Now suppose that $X = N1$. As we saw above, according to N1 the chance of the Skeptical Hypothesis being true, given that the world is in the macrostate picked out by E, is very high (i.e., $ch_{N1E}(SH) \approx 1$).[17] Given this, we can run the Reversibility Objection against N1 as follows:

The Reversibility Objection (against N1):

P1. $ch_{N1E}(SH) \approx 1$

P2. Bayesianism (Probabilism and Conditionalization)

---

[16] That is, suppose the Skeptical Hypothesis undermines our evidence for believing something like classical statistical mechanics, in the sense that for any account of classical statistical mechanics X, $cr_E(X|SH) \approx 0$. If $cr_E(X|SH) \approx 0$, it follows that $cr_E(X \wedge SH) \ll cr_E(SH)$, and thus that $cr_E(X \wedge SH) \approx 0$. And so it follows that $cr_E(X) = cr_E(X \wedge SH) + cr_E(X \wedge \neg SH) \approx cr_E(X \wedge \neg SH)$. Thus, if we're granting that the correct account of statistical mechanics X is such that we could rationally have $cr_E(X) > 0.5$, then it's arguably not much more of an assumption to grant that the correct account of statistical mechanics is such that we could rationally have $cr_E(X \wedge \neg SH) > 0.5$, since $cr_E(X) \approx cr_E(X \wedge \neg SH)$.

[17] I'm abusing notation here slightly. Although the T and K arguments that pick out a chance distribution are given by the conjunction of N1 and E, I am not identifying T with N1 (since T is supposed to just be a complete *chance* theory, whereas N1 is a complete description of all of the laws, not just the chancy ones), or K with E. Rather, decomposing N1 into a chancy component A and a non-chancy component B, the T and K arguments will be $T = A$ and $K = B \wedge E$, respectively.

THE META-REVERSIBILITY OBJECTION

P3. Chance-Credence Principle

P4. Adequacy

L5. $ic(SH|N1 \wedge E) \approx 1$ (P1 + P3)

L6. $cr_E(SH|N1) \approx 1$ (P2(cond) + L5)

L7. $cr_E(\neg SH|N1) \approx 0$ (P2(prob) + L6)

L8. $cr_E(N1 \wedge \neg SH) \ll cr_E(N1)$ (L7)

L9. $cr_E(N1 \wedge \neg SH) \approx 0$ (P2(prob) + L8)

L10. $cr_E(N1 \wedge \neg SH) \not> 0.5$ (L9)

C. $\neg N1$ (P4 + L10)

Next suppose that $X = I1$. Given (5), we can run the Reversibility Objection against I1 using an argument similar to the one we gave against N1:

The Reversibility Objection (against I1):

P1. $\mu(SH|I1 \wedge E) \approx 1$

P2. Bayesianism (Probabilism and Conditionalization)

P3. Indifference Principle

P4. Adequacy

L5. $ic(SH|I1 \wedge E) \approx 1$ (P1 + P3)

L6. $cr_E(SH|I1) \approx 1$ (P2(cond) + L5)

L7. $cr_E(\neg SH|I1) \approx 0$ (P2(prob) + L6)

L8. $cr_E(I1 \wedge \neg SH) \ll cr_E(I1)$ (L7)

L9. $cr_E(I1 \wedge \neg SH) \approx 0$ (P2(prob) + L8)

L10. $cr_E(I1 \wedge \neg SH) \not> 0.5$ (L9)

C. $\neg I1$ (P4 + L10)

3.2. Replies to the Reversibility Objection

To escape the Reversibility Objection, we need to either reject one of the premises of the argument or modify the account under attack. Thus, we have five ways to escape the argument:

  i. reject P1, and dispute that N1/I1 assigns a high probability to the Skeptical Hypothesis,
  ii. reject P2, Bayesianism,
  iii. reject P3, either the Chance-Credence Principle (given the nomic approach) or the Indifference Principle (given the indifference approach),
  iv. reject P4, Adequacy, or
  v. accept the conclusion, and modify one's account so that it no longer falls prey to the Reversibility Objection.

Option (i) is to reject P1, and dispute that $ch_{N1E}(SH) \approx 1$ or that $\mu(A|I1 \wedge E) \approx 1$. For example, one might argue that while our various memories considered individually would fail to tell against the Skeptical Hypothesis, the fact that all of them happen to agree with each other suffices to make the Skeptical Hypothesis unlikely.[18] More generally, one might argue that when "evidence like ours" is properly understood, or when the correct way to calculate the statistical mechanical probabilities in question is properly understood, we'll find that the chance/indifference measure of the Skeptical Hypothesis small.[19]

For the most part, people have found these kinds of argument to be unpersuasive.[20] And a full examination of these replies would take us too far afield. But let's consider a version of this kind of reply that challenges P1 by disputing what we can take to be evidence.

---

[18] See Schulman (1997), p. 154–155.
[19] For example, see Parker (2006), who argues that we cannot understand the content of our records in thermodynamic terms, and that considerations involving the contents of our records will allow us to escape the Skeptical Hypothesis.
[20] For example, see Albert (2000) for a reply to the kinds of considerations Schulman raises.

I've been following Albert (2000) in taking our evidence E to be something like our current memories and experiences, or the current macrostate of the world (at the appropriate level of coarse-grained description). But one might challenge this claim. For example, one might take our evidence to include more than what we currently "have access to," and so take E to consist of the sum of our memories and experiences we've had throughout our life. Alternatively, one might understand "evidence like ours" more ambitiously, so as to include things that we've taken science to have firmly established, such as "the universe began 14 billion years ago in a particular low-entropy state."[21]

One worry for these alternative conceptions of evidence, as we'll see when we discuss option (iii), is that given the skeptical worries in play regarding our memories and records, one can't take them to be evidence. But let's put this worry aside for now.

A bigger worry for the more modest extensions—such as that one's evidence is the sum of one's experiences throughout one's life—is that P1 will still be true. For when we consider events in the distant past—events before we were born—we'll again find that it's overwhelmingly more likely (according to the standard measure) that our records about the distant past coalesced from some high-entropy state than that they accurately reflect what the distant past was like. So, adopting these modest extensions won't help us to escape P1.

The more ambitious extensions—such as taking "the universe began 14 billion years ago in a particular low-entropy state" to be evidence—will allow us to escape P1. But given the Bayesian account we're assuming here, these extensions are difficult to maintain.

Few would want to maintain that a proposition like "the initial conditions of the universe 14 billion years ago were very low-entropy" is something we should remain confident about, no matter what further considerations arise. For example, we might want to lower our credence in this proposition if we found reason to think that some kind of dynamical explanation of thermodynamic asymmetries is correct (such as the proposal appealing to the GRW account of quantum mechanics discussed in Albert 2000), or if we found reason to believe that our astronomical evidence was systematically skewed in some way, and so on. And presumably some ways of changing one's credence in propositions like "the initial conditions of the universe 14 billion years ago were very low-entropy" in light of such considerations are rational, and some are irrational. And we'd like a story about which are which.

---

[21] Thanks to Barry Loewer here for pressing me to take this objection more seriously.

This story is what the Bayesian apparatus is supposed to give us. But we can use the Bayesian apparatus to help us here only if we're taking "the initial conditions of the universe 14 billion years ago were very low-entropy" to be a *hypothesis*—something that is up for grabs—not as evidence. If we take such propositions to be *evidence*—something that is settled, taken for granted, epistemological bedrock—then they're beyond rational evaluation.[22]

(Now, one might grant this and try to escape these worries by vacillating about what one takes evidence to be—perhaps treating propositions like "the initial conditions of the universe 14 billion years ago were very low-entropy" as evidence in some contexts, but not others, and so on. But such a position isn't intellectually stable. At the end of the day, we're going have to fit everything we want to say into a single comprehensive epistemology, a single story in which everything falls into place, where there's a determinate fact as to whether something is evidence or not.)

So, at the end of the day, the proponent of this more ambitious picture of evidence will face a dilemma. They can either (a) give a comprehensive story in which things like "the initial conditions of the universe 14 billion years ago were very low-entropy" do count as evidence, and give up on giving a story about what kinds of belief changes regarding such propositions are rational, or (b) give a comprehensive story in which things like "the initial conditions of the universe 14 billion years ago were very low-entropy" don't count as evidence. The first horn of the dilemma requires one to give up many of the epistemological ambitions of the philosophy of science, while the second horn requires one to accept that this way of escaping P1 is untenable.

Option (ii) is to reject Bayesianism—that is, to reject Probabilism, Conditionalization, or both. But I take rejecting Bayesianism as a way to avoid the Reversibility Objection to be a nonstarter. Although various reasons have been offered for modifying the standard Bayesian framework—in order to allow for agents with imprecise credences, for example, or to allow for different kinds of updating

---

[22] Such credal variations are straightforwardly ruled out by the Bayesian account sketched in §2, because Conditionalization doesn't allow one's credences in one's evidence to vary—one's credence in one's evidence must be (and remain) 1. But one can get around this obstacle by adopting a standard extension of the Bayesian approach, which replaces Conditionalization with Jeffrey Conditionalization, a rule that allows agents to get uncertain evidence. However, the move to Jeffrey Conditionalization does nothing to address the deeper problem with taking "the initial conditions of the universe 14 billion years ago were very low-entropy" as evidence. For if we treat our belief changes in such propositions as due to uncertain evidence, then they're beyond rational evaluation—we can't provide a story as to whether these "evidential" changes in belief are rational or not. But surely we *do* want a story here—an account of what our belief in this proposition should come to be, given certain observations and deductions about what the world is like.

with respect to certain kinds of belief—none of these modifications will help one avoid the Reversibility Objection.[23] This eliminates option (ii).

Option (iii) is to reject either the Chance-Credence Principle (given the nomic approach) or the Indifference Principle (given the indifference approach). Rejecting the Chance-Credence Principle is not a viable way to escape the Reversibility Objection. Virtually everyone holds that some kind of Chance-Credence Principle constrains rational belief. And while there's a strong case to be made in favor of the particular formulation of the Chance-Credence Principle we've employed, the Reversibility Objection would still go through if we replaced it with any of the other proposed formulations of the Chance-Credence Principle.[24]

Rejecting the Indifference Principle isn't a promising way to escape the Reversibility Objection either. Unlike Chance-Credence Principles, Indifference Principles are widely viewed with suspicion.[25] But a proponent of the indifference approach is committed to accepting some kind of Indifference Principle, so rejecting such principles entirely is not an option. And changing the Indifference Principle in order to try to escape the Reversibility Objection is problematic.

One worry is that changing the Indifference Principle for this reason seems methodologically unsound. The Indifference Principle is supposed to be justified by *a priori* considerations regarding what an ideally rational agent would believe in various evidential situations. To modify the Indifference Principle in order to escape something like the Reversibility Objection would be to modify the Indifference Principle in light of *a posteriori* considerations. And since the Indifference Principle is an *a priori* constraint on rational belief, *a posteriori* considerations shouldn't be relevant.

Here's a more severe concern about modifying the Indifference Principle in order to escape reversibility worries. Given I1 and evidence like ours, we want the Indifference Principle to yield the right results in typical cases, and yet avoid the prescriptions that lead to complaints like the Reversibility Objection. But it's not clear that any plausible *a priori* constraint on rational belief can do this.

Let's look at an example. Let *E* be the macrostate picked out by our evidence, where this evidence includes the fact that there's a half-melted ice cube in a cup

---

[23] See Levi (1985) and Joyce (2005) for discussion and references regarding imprecise credences, and Elga (2000) and Arntzenius (2003) for some discussion and references regarding self-locating beliefs.
[24] For the case in favor of this formulation, see Arntzenius (1995), Meacham (2005), Nelson (2009), and Meacham (2010); alternative formulations have been suggested by Vranas (2004), Hoefer (2007), and Ismael (2011), among others.
[25] For criticisms of such principles, see van Fraassen (1989), Howson and Urbach (2005), Weisberg (2011), and the references therein.

of warm water, let A_f be the macrostate containing the microstates in which the half-melted ice cube is completely melted in the near future, and let A_p be the macrostate containing the microstates in which the half-melted ice cube is completely melted in the recent past. We want the statistical mechanical probability of A_f given Il and E—the probability of the half-melted ice cube evolving to a completely melted state in the near future—to be high. We also want the statistical mechanical probability of A_p given Il and E—the probability of the half-melted ice cube having evolved from a completely melted state in the recent past—to be low. Thus, we want μ to be such that $\mu(A_f | \text{Il} \wedge E) \approx 1$ and $\mu(A_p | \text{Il} \wedge E) \approx 0$.

But these prescriptions are temporally asymmetric—future and past events are assigned different credences. Where does this temporal asymmetry come from?

The laws Il posits are temporally symmetric, so Il isn't the source of the asymmetry. What about our evidence, E? Well, the claim that the ice cube is in a half-melted state isn't temporally asymmetric. And if we take our evidence to be something like the current "time slice" of our experience, or the current macrostate of the world (at a certain level of coarse-grained description), then the rest of our evidence isn't temporally asymmetric either. For such evidence is synchronic, and in order to be temporally asymmetric, evidence must have temporal "breadth." For this reason, one might maintain that our evidence is diachronic rather than synchronic. For instance, perhaps our evidence is something like the sum of all of the experiences we've had throughout our lives. And this diachronic evidence is temporally asymmetric.

In some ways, this is a tricky position to maintain. Evidence, in the relevant sense—the notion of "evidence" that appears in Conditionalization—is something you ought to have a credence of 1 in, something you ought to be virtually certain of. And given the skeptical worries in play regarding our memories and our records of the past, our memories and past experiences are not something we can take to be evidence in this sense.

But more importantly, these kinds of attempts to extend the notion of evidence run into the problems we saw earlier in the discussion of option (i). The modest extensions, like the one just described above, will provide us with temporally asymmetric evidence, but this asymmetry seems unlikely to yield the dramatic asymmetry in prescriptions that we want. The ambitious extensions (which take things like "the initial conditions of the universe 14 billion years ago were very low-entropy" to be evidence), on the other hand, will give us temporally asymmetric evidence that yields the asymmetric prescriptions we desire, but as we've already seen, these accounts of evidence are problematic. So neither Il nor E can yield the temporal asymmetry we want with respect to our assignments to A_f and A_p. Thus, the temporal asymmetry must come from the indifference measure itself. But it's hard to see how any kind of *a priori* consideration could justify treating one direction in time differently from the other. Certainly, none of the standard considerations appealed to in order to justify Indifference Principles—symmetry, parsimony, information-theoretic considerations, etc.—provide grounds for such an asymmetry.

Together, these considerations tell against option (iii). What about option (iv), rejecting Adequacy? The role of Adequacy in the argument is to move us from the uncomfortable result that one can't rationally both believe the proposed account of statistical mechanics and reject the Skeptical Hypothesis, to the conclusion that the proposed account of statistical mechanics is not correct. To reject Adequacy and accept the rest of the argument is to concede the uncomfortable result but hold on to the proposed account of statistical mechanics anyway. While this technically allows one to escape the Reversibility Objection, it's not very satisfying. Certainly, this seems like a lonely position, since not many people are going to be comfortable embracing the skeptical consequences of the account and calling it a day. This leaves us with option (v): modifying the account under attack, N1/I1. Given that we're going to change N1/I1, what parts of the account should we change? There are three options: (v.a) we can change what it takes the laws to be, (v.b) we can change what it takes the statistical mechanical probabilities to be, or (v.c) we can add some further claim to the account which does neither of the above. Let's consider these options in reverse order.

The third way of modifying an account, (v.c)—adding a claim to the account that doesn't change the laws or the statistical mechanical probabilities—won't help us escape the Reversibility Objection. Let A be the claim we're adding to the account, and let our new account, the conjunction of A and N1/I1, be NA1/IA1. Since NA1/IA1 logically entails N1/I1, if follows from the probability axioms that $cr_E(\text{N1}/\text{I1} \wedge \neg SH) \geq cr_E(\text{NA1}/\text{IA1} \wedge \neg SH)$. Thus, if we're rationally required to be such that $cr_E(\text{N1}/\text{I1} \wedge \neg SH) \approx 0$, we're also rationally required to be such that $cr_E(\text{NA1}/\text{IA1} \wedge \neg SH) \approx 0$. So the Reversibility Objection will apply to NA1/IA1 as well.

The second way of modifying an account, (v.b)—changing the statistical mechanical probabilities—ends up collapsing into one of the other options. The Reversibility Objection arises due to the constraints on rational belief that the account imposes. So, to escape the Reversibility Objection, we need to change the values the account assigns that constrain rational belief, be they the chances (on the nomic approach) or the indifference measures (on the indifference approach). Changing the chances requires changing the laws that assign the chances, so modifying the nomic account in the second way, (v.b), amounts to a special case of changing the account in the first way, (v.a). And changing the indifference measure is to modify the Indifference Principle, which (on the indifference approach) is just option (iii). This leaves us with the first way of modifying an account, (v.a)—changing the laws. Unlike the other options, this option seems viable. Suppose we modify N1/I1 by adding a lawful constraint that allows all and only initial conditions that belong to some particular low-entropy macrostate. That is, suppose we modify N1/I1 by adding the *Past Hypothesis* to the laws. Then we get a new account, N2/I2. And N2/I2 is immune to the Reversibility Objection.

The Reversibility Objection doesn't apply to N2 because it isn't the case that $ch_{N2E}(SH) \approx 1$. The nomic accounts of classical statistical mechanics employ a Statistical Postulate that assigns chances to the lawfully permitted initial conditions that are proportional to the Liouville measure. And if the account includes a Past Hypothesis, as N2 does, then most of the microstates in our initial account are not lawfully permitted—the laws forbid any microstate that, when evolved backward according to the dynamics, doesn't lead to the particular low-entropy initial condition required by the Past Hypothesis. Once we rule out these possibilities and renormalize, we find that the majority of the remaining microstates compatible with our evidence *do* line up with our memories and records of what the past was like, *contra* the Skeptical Hypothesis. Thus, $ch_{N2E}(SH) \approx 0$.[26]

The Reversibility Objection doesn't apply to I2 because it isn't the case that $\mu(SH|I2 \wedge E) \approx 1$. It follows from (6) that $\mu(A|I2 \wedge E) = ch_{N2E}(A)$. And since $ch_{N2E}(SH) \approx 0$, it follows that $\mu(SH|I2 \wedge E) \approx 0$ as well.

So by adopting a Past Hypothesis, and replacing N1/I1 with N2/I2, we can escape the Reversibility Objection.

There has been some debate over whether one should understand the Past Hypothesis as a law, or as (say) a contingent generalization.[27] Now that we've given a precise characterization of the Reversibility Objection, we can see that regardless of whether we adopt the nomic approach or the indifference approach, the Past Hypothesis has to be a law in order to address the Reversibility Objection. For if the Past Hypothesis is only a contingent generalization, then adding it to one's account amounts to adopting option (v.c), not (v.a). And, as we've seen, adopting option (v.c) doesn't help us escape the Reversibility Objection.[28]

---

[26] For a more detailed description of this argument, see Albert (2000). Although the claim that the Past Hypothesis allows us to get around the Reversibility Objection is widespread, it is at least mildly contentious. For some worries regarding this claim, see Uffink (2002), Winsberg (2004a, 2004b), Parker (2005), Earman (2006), and Callender (2010).

[27] See Callender (2004) and North (2011) for discussion and references.

[28] Here's another way to see why this is so. Call the contingent version of the Past Hypothesis "PH_c," and call the account we get by adding a contingent Past Hypothesis to N1/I1, "N1$_c$/I1$_c$." A little thought shows that $cr_E(N1_c/I1_c) = cr_E(N1/I1 \wedge PH_c) \approx cr_E(N1/I1 \wedge \neg SH)$. Thus, because we're rationally required to not believe N1/I1 $\wedge \neg SH$, we're rationally required to not believe N1$_c$/I1$_c$.

4. The Meta-Reversibility Objection

Adopting the Past Hypothesis addresses the Reversibility Objection as posed. But one might wonder whether we can raise the problem in another way.

Here is one way to see why N1/I1 is susceptible to the Reversibility Objection. Conditional on a particular low-entropy initial condition obtaining, N1/I1 assigns the right values to past events. But conditional on some high-entropy initial condition obtaining, N1/I1 assigns the wrong values to past events. And since N1/I1 unconditionally assigns a much higher value to high-entropy initial conditions than to low-entropy initial conditions, the assignments given high-entropy initial conditions will swamp those given low-entropy initial conditions. Thus, when all is said and done, N1/I1 assigns the wrong values to past events.

Now, as we saw in §3.2, we can fix this by lawfully requiring the low-entropy initial condition to obtain—that is, by adding the Past Hypothesis (*PH*) to the laws. If we do that, then the resulting account, N2/I2, ends up assigning the right values to past events.

At this point, however, one could raise a different reversibility worry. To begin, note that there are other lawful constraints on initial conditions one could impose. For example, one could lawfully require that some *high*-entropy initial conditions obtain; call this the Past Hypothesis* (*PH**). If we do that, then the resulting account, call it N2*/I2*, ends up assigning the wrong values to past events. Now suppose we're rationally required to have credences over lawfully required initial conditions that mirror our credences over contingently obtaining initial conditions, so that we assign a much higher unconditional credence to lawful high-entropy initial conditions like *PH** than to lawful low-entropy initial conditions like *PH*. Then, since we'll unconditionally assign a much higher credence to the account that employs *PH** (N2*/I2*) than to the account that employs *PH* (N2/I2), the credences assigned by N2*/I2* will swamp those assigned by N2/I2. So, when all is said and done, we'll be required to adopt the wrong credences about past events. Call the worry that an account will commit one to this result the *Meta-Reversibility Objection*.

The Meta-Reversibility Objection shouldn't worry proponents of nomic accounts. The objection gets traction only if one's account requires one to assign credences over lawful constraints on initial conditions in a certain way. And nomic

---

What makes $N1_c/I1_c$ seem like it might help with the Reversibility Objection is the observation that, unlike N1/I1, believing $N1_c/I1_c$ doesn't commit you to believing the Skeptical Hypothesis. This is true. But since we're already committed to not believing $N1_c/I1_c$, this doesn't help us satisfy Adequacy.

accounts don't impose any requirements on how one assigns credences over lawful constraints on initial conditions. Nomic accounts require only that one's credences line up with the chances. And there aren't any chances over different lawful constraints on initial conditions. After all, the Statistical Postulate won't give us chances until we know what the lawful constraints on initial conditions (if any) are!

Indeed, the proponent of nomic accounts is perfectly free to assign, in good conscience, a high initial credence to N2 and a low initial credence to N2*. This may not be egalitarian, but so what? We've known that we need to do something like this in order to ground our inductive biases since Goodman's (1954) famous grue cases. If we want to end up believing the world is like we think it is, our initial credences will have to be biased toward the theories we actually believe in certain ways.

But the Meta-Reversibility Objection *is* a worry for proponents of indifference accounts. For the Indifference Principle makes prescriptions about everything. Thus, it makes prescriptions about lawful constraints on initial conditions.[29] And there is substantial pressure on the proponent of the indifference approach to maintain that I2* should be assigned a much higher credence than I2.

I'll discuss this more in §4.2, but briefly, the worry is this. In order to yield the expected prescriptions given I1, the indifference measure must assign values to initial conditions that are proportional to the Liouville measure. But if this is how we ought to be indifferent with respect to which initial conditions contingently obtain, it's hard to see why we shouldn't be indifferent in the same way with respect to which initial conditions lawfully obtain. After all, the structure of our ignorance over the lawfully required initial conditions mirrors the structure of our ignorance over the contingent initial conditions. And the standard rationales offered to justify indifference measures—symmetry considerations, information-theoretic considerations, etc.—apply equally well to both cases. If this is right, then the Indifference Principle will prescribe a much higher credence to I2* than to I2. So the Meta-Reversibility Objection poses a real threat to indifference accounts.

---

[29] One might wonder whether one could resist this worry by simply adopting a weaker Indifference Principle, which makes prescriptions only in some limited range of domains. As we'll see, this isn't a promising option. The typical rationales used to justify the Indifference Principle's prescriptions in these limited domains will apply just as well to wider domains. And, as we'll discuss in §4.2, imposing arbitrary constraints on these rationales would be fatal to the Indifference Principle project.

4.1. Formalizing the Meta-Reversibility Objection

Let's formulate the Meta-Reversibility Objection more precisely. To begin, let's spell out the motivating worry a bit more carefully.

Let I2⁻ be the proposition that the correct account of classical statistical mechanics is some modified version of I1 that adds a lawful constraint on boundary conditions. Thus, I2 is equivalent to $I2^- \wedge PH$, and I2* is equivalent to $I2^- \wedge PH^*$. Let PH_c be the contingent version of the Past Hypothesis—i.e., the claim that, as a matter of contingent fact, the initial conditions came from some particular low-entropy macrostate—and let PH_c^* be the contingent version of the Past Hypothesis*.

The worry sketched above is that the Indifference Principle will require one to assign credences to PH and PH* given I2⁻ and E that mirror those it requires one to assign to PH_c and PH_c^* given I1 and E. That is, the worry is that if the indifference measure is such that:

$$\mu(PH_c | I1 \wedge E) \ll \mu(PH_c^* | I1 \wedge E), \tag{7}$$

it will also be such that:

$$\mu(PH | I2^- \wedge E) \ll \mu(PH^* | I2^- \wedge E). \tag{8}$$

And this entails that, given evidence like ours, E, our credence in I2* should be much larger than our credence in I2.[30]

This puts us in an unhappy situation. If our credence in I2* is much larger than our credence in I2, then we can't rationally believe that I2 obtains. Thus, I2 can't satisfy Adequacy. Of course, we may be able to rationally believe something like I2*. But I2* is subject to the original Reversibility Objection—because almost all of the microstates compatible with our evidence and the Past Hypothesis* fail to line up with our memories and records of the past, our credence in $I2^* \wedge \neg SH$ will be very low. Thus, I2* won't satisfy Adequacy either. Either way, it looks like the indifference accounts of classical statistical mechanics fail to satisfy Adequacy. They are either susceptible to the Reversibility Objection described in §3, or they are susceptible to the worry just described above—the Meta-Reversibility Objection.

---

[30] We have $\mu(PH | I2^- \wedge E) \ll \mu(PH^* | I2^- \wedge E)$. Multiplying both sides by $\mu(I2^- \wedge E)$ gives us $\mu(PH \wedge I2^- \wedge E) \ll \mu(PH^* \wedge I2^- \wedge E)$ or, equivalently, $\mu(I2 \wedge E) \ll \mu(I2^* \wedge E)$. Dividing both sides by $\mu(E)$ gives us $\mu(I2 | E) \ll \mu(I2^* | E)$. Then applying the Indifference Principle gives us $ic(I2|E) \ll ic(I2^*|E)$, and applying Conditionalization gives us $cr_E(I2) \ll cr_E(I2^*)$.

Here's another way to think about these concerns. Proponents of the indifference approach face a choice between whether or not to take there to be lawful constraints on boundary conditions; i.e., a choice between accepting I1 (which entails that there are no lawful constraints on boundary conditions) or I2⁻ (which entails that there are). If they accept I1, then they're rationally required to believe the Skeptical Hypothesis—that their memories and records about the past are false. If they accept I2⁻, then they're also rationally required to believe the Skeptical Hypothesis. For although they'll believe that *if* I2 is true then it's likely that their memories and records about the past are reliable, their credence in I2 will be so low compared to its competitors that this will have little impact on their overall credence. The laws that will dominate their credence will be those of the high-entropy initial conditions, like I2*, which predict that it's much more likely that we coalesced from a high-entropy state than that we evolved from a low-entropy state. So, when all is said and done, they'll again be rationally required to believe the Skeptical Hypothesis.

Here is a formal characterization of the argument. Given (8), we can set up the Meta-Reversibility Objection against I2 as follows:

The Meta-Reversibility Objection (against I2):

P1. $\mu(PH | I2^- \wedge E) \ll \mu(PH^* | I2^- \wedge E)$

P2. Bayesianism (Probabilism and Conditionalization)

P3. Indifference Principle

P4. Adequacy

L5. $ic(PH | I2^- \wedge E) \ll ic(PH^* | I2^- \wedge E)$

L6. $cr_E(PH | I2^-) \ll cr_E(PH^* | I2^-)$ (P2(cond) + L5)

L7. $cr_E(PH \wedge I2^-) \ll cr_E(PH^* \wedge I2^-)$ (P2(prob) + L6)

L8. $cr_E(I2) \ll cr_E(I2^*)$ (L7)

L9. $cr_E(I2) \approx 0$ (P2(prob) + L8)

L10. $cr_E(I2 \wedge \neg SH) \approx 0$ (P2(prob) + L9)

L11. $cr_E(I2 \wedge \neg SH) \not> 0.5$ (L10)

C. $\neg I2$ (P4 + L11)

This gives us the Meta-Reversibility Objection against I2. What happens if we try to run the Meta-Reversibility Objection against N2?

To run the argument against N2, we would have to replace P1 with $ch_{N2^-_E}(PH) \ll ch_{N2^-_E}(PH^*)$, replace P3 with the Chance-Credence Principle, and replace "I2" and "I2⁻" with "N2" and "N2⁻" throughout (where N2⁻ is the nomic analog of I2⁻—the proposition that the correct account of classical statistical mechanics is some modified version of N1 that adds a lawful constraint on boundary conditions). But this argument won't work, because this version of P1 is false. P1 is false because N2⁻ and E don't suffice to pick out a chance distribution. It's true that N2⁻ includes the Statistical Postulate, and the Statistical Postulate posits chances. But what chances the Statistical Postulate assigns depend on what lawful constraints there are on the boundary conditions, and N2⁻ and E don't tell us what these lawful constraints are. So the ch terms on both sides of P1's inequality will be undefined, and P1 will be false.

4.2. Replies to the Meta-Reversibility Objection

As with the Reversibility Objection, we can divide the potential replies to the Meta-Reversibility Objection into five options:

i. reject P1, that $\mu(PH|I2^- \wedge E) \ll \mu(PH^*|I2^- \wedge E)$,
ii. reject P2, Bayesianism,
iii. reject P3, the Indifference Principle,
iv. reject P4, Adequacy, or
v. accept the conclusion, and modify one's account so that it no longer falls prey to the Meta-Reversibility Objection. One can try to do this by either: (v.a) changing the laws the account specifies, (v.b) changing the statistical mechanical probabilities the account assigns, or (v.c) adding some further claim to the account that does neither of the above.

The first option is to reject P1, and dispute that $\mu(PH|I2^- \wedge E) \ll \mu(PH^*|I2^- \wedge E)$. As before, one way to do this is argue that when "evidence like ours" is properly understood, or when the correct way to calculate the probabilities in question is properly understood, we'll find that P1 is false. But if this kind of reply fails in the context of the Reversibility Objection, as we're assuming, it seems it will fail here as well. So this way of rejecting P1 does not look promising.

There's another way to reject P1 here that wasn't available against the Reversibility Objection. Namely, one can reject P1 by arguing that this result comes from using the wrong Indifference Principle.[31] Given this way of pursuing option (i), options (i), (iii), and (v.b) are all versions of a single reply—they're all ways of rejecting the Indifference Principle the argument employs, and replacing it with a principle that makes different prescriptions. This is the most natural reply to the Meta-Reversibility Objection. For the argument hangs on the claim that, given I2⁻ and E, the indifference measure of PH^* is much greater than that of PH. And, one might say, it's not obvious why the proponent of the indifference approach should accept this claim.

But this claim turns out to be hard for proponents of the indifference approach to reject.[32] One reason is that they presumably want the Indifference Principle to still yield the standard probability assignments given accounts like I1 and I2. That is, they want the indifference measure to be such that:

$$\mu(PH_c|I1 \wedge E) \ll \mu(PH_c^*|I1 \wedge E). \tag{9}$$

But if the indifference measure assigns those values, it's hard to see why it won't assign similar values to the lawful versions of the Past Hypothesis and the Past Hypothesis* given I2⁻:

$$\mu(PH|I2^- \wedge E) \ll \mu(PH^*|I2^- \wedge E). \tag{10}$$

---

[31] This way of rejecting P1 was not available to opponents of the Reversibility Objection because the value I1 assigns to the relevant proposition (the value of SH given $I1 \wedge E$) is fixed. We stipulated that I1 assigns values that line up with N1's chances, so the value of SH given $I1 \wedge E$ must be equal to the chance of SH given $N1 \wedge E$. And the chance of SH given $N1 \wedge E$ is fixed by our stipulation that the chances given N1 be proportional to the Liouville measure.

By contrast, this way of rejecting P1 *is* available to opponents of the Meta-Reversibility Objection because the values that I2⁻ assigns to the relevant propositions (the values of PH and PH^* given $I2^- \wedge E$) are not fixed. It's true that the "matching values" stipulation we made in §2.2 implies that the values of PH and PH^* given $I2^- \wedge E$ should line up with the chances of PH and PH^* given $N2^- \wedge E$, if there are such chances. But, as we just saw in §4.1, there are no such chances—the chances of PH and PH^* given $N2^- \wedge E$ are not well-defined.

[32] As we saw in §3.2, there are also concerns regarding whether this way of replying to reversibility worries is methodologically sound, since reversibility worries seem *a posteriori*, while the Indifference Principle is supposed to be an *a priori* constraint. But these worries are of secondary interest relative to the main concerns I discuss in the text, so I'll put them aside.

After all, the two cases are structurally analogous. In one case we're assessing what the initial macrostate of the world is, as a matter of contingent fact, while in the other we're assessing what the initial macrostate of the world is, as a matter of lawful fact. In all other respects the two cases are the same. And all of the standard considerations that are appealed to in order to justify indifference measures— symmetry, parsimony, informational entropy, and so on—apply in the same way to both cases.

To get a feel for why this is so, let's go through an example in detail. The most popular version of the indifference approach to statistical mechanics, suggested by Jaynes (1983), maintains that the indifference measure of a macrostate should track how uninformative it is to be told that the system is in that macrostate, so that more-informative macrostates get smaller indifference measures. And Jaynes proposes that we cash out the relevant notion of informativeness here by appealing to the notion of Shannon entropy, i.e., information entropy.

Here's one way to flesh out the details of such a proposal in a Boltzmannian framework. The *Shannon Entropy S* of a random variable R is:

$$S(R) = -\sum_i p(R_i) \cdot \ln p(R_i), \tag{11}$$

where $p(R_i)$ is the probability of R yielding the ith possible outcome. Since a macrostate is not a random variable, we cannot apply S to a macrostate A directly. Instead, we need to find some way to construct a canonical random variable associated with A. Then we can take the Shannon entropy of that random variable to be the Shannon entropy associated with A. Let a $\Delta x \Delta v$-*partition*, γ, be a partition of phase space into hypercubes which uniformly have sides of dimension Δx (for the spatial dimensions) and Δv (for the velocity dimensions), in standard units. Call these hypercubes the γ-*macrostates*.

Consider some macrostate A, and some $\Delta x \Delta v$-partition, γ, that is fine-grained enough so that we can characterize A as a disjunction of γ-macrostates. Let "γA" be the random variable that has an equal probability of yielding each of the γ-macrostates that compose A.

Let the Shannon entropy associated with A (with respect to γ) be the Shannon entropy of γA. So if A can be characterized as a disjunction of n γ-macrostates, the Shannon entropy associated with A (with respect to γ) will be:

$$S(\gamma A) = -\sum_i p(\gamma A_i) \cdot \ln p(\gamma A_i) = n \cdot -\frac{1}{n} \cdot \ln \frac{1}{n} = -\ln \frac{1}{n}. \tag{12}$$

Given this, we can impose the desired constraint on μ as follows:

μ-Constraint: Let L be some (possibly incomplete) description of the laws, which includes the dynamics, and is such that the phase space associated with a system can be given a $\Delta x \Delta v$-partition. Let A and B be arbitrary macrostates in this phase space. Finally, for any macrostate X, let XL be the macrostate composed of the microstates compatible with X and the lawful constraints on boundary conditions imposed by L, if any. Then for any $\Delta x \Delta v$-partition, γ, such that AL and ABL can be characterized as disjunctions of γ-macrostates, μ must be such that:

$$\mu(A|B \wedge L) \propto \frac{e^{S(\gamma ABL)}}{e^{S(\gamma BL)}}. \tag{13}$$

A couple of comments. First, note that while the values we get for $e^{S(\gamma ABL)}$ and $e^{S(\gamma BL)}$ depend on our choice of γ, the ratio between $e^{S(\gamma ABL)}$ and $e^{S(\gamma BL)}$ does not. So this constraint does not depend on our choice of γ. Second, note that

$$e^{S(\gamma A)} = e^{-\ln\left(\frac{1}{n}\right)} = \frac{1}{e^{\ln\left(\frac{1}{n}\right)}} = \frac{1}{\frac{1}{n}} = n.$$

So for any given γ, the indifference measure of a macrostate is proportional to the number of γ-macrostates it's composed of. Third, note that for any γ, each γ-macrostate has the same Liouville measure. So for any given γ, the Liouville measure of a macrostate A is proportional to the number of γ-macrostates it's composed of. Thus, this constraint requires the indifference measure to be proportional to the Liouville measure. Fourth, this constraint is incomplete, since it does not fully specify μ. A complete version of this constraint would also tell us how to apply information-theoretic considerations to determine how to be indifferent about everything, and would yield the above constraint as a special case.

This constraint on μ yields the values we want given L=I1 and L=I2. But it also yields the values that make P1 of the Meta-Reversibility Objection true given L=I2⁻. And this is what we should expect, given a Shannon entropy approach to indifference. The Shannon entropy of a macrostate A can be seen as a measure of how much more information you need, after being told that a system's in macrostate A, to know what the system's microstate is.[33] And the only thing relevant to *that* is the *size* of the macrostate. The bigger the macrostate, the bigger the range

---

[33] Or, if we divide up A into a number of small macrostates, a measure of how much more information you need to know a system's small macrostate after being told that it's in A.

of microstates it contains, and the more you have to say to pick out a particular microstate. The smaller the macrostate, the smaller the range of microstates it contains, and the less you have to say to pick out a particular microstate. But these information considerations are the same regardless of whether we're assigning credence to the initial conditions contingently being in macrostate A, or assigning credence to the initial conditions lawfully being in macrostate A. The only thing that matters, information-wise, is the size of the macrostate A. Whether A obtains contingently or lawfully is irrelevant.

Here is a related concern for trying to escape reversibility worries by modifying the Indifference Principle. Given I2 and evidence like ours, we want the Indifference Principle to yield the desired prescriptions in typical cases, and yet avoid the prescriptions that lead to complaints like the Meta-Reversibility Objection. But it's not clear that any plausible *a priori* constraint on rational belief can do this.[34]

Consider again the example of the half-melted ice cube in warm water, from §3.2. Given that our evidence (E) includes that there is a half-melted ice cube in warm water, we want our credence that it will be fully melted in the near future (A_f), given I2 and E, to be high, and our credence that it was fully melted in the recent past (A_p), given I2 and E, to be low. Thus, we want μ to be such that $\mu(A_f|I2 \wedge E) \approx 1$ and $\mu(A_p|I2 \wedge E) \approx 0$. And if we're proponents of I2, we would like it to be permissible to both be confident in I2 and to have the appropriate credences

---

[34] I raise the temporal asymmetry worry as a general worry for attempts to avoid the Meta-Reversibility Objection by modifying the Indifference Principle. But one can also use temporal symmetry considerations to argue directly for P1. To make things manageable, let's assume that the only possible lawful constraints on boundary conditions are pairings of one of three constraints on initial conditions (the Past Hypothesis, the Past Hypothesis*, and the Past Hypothesis⁻, which imposes a vacuous constraint on initial conditions), and one of three constraints on final conditions (the Future Hypothesis (FH), which requires the world to end in a particular low-entropy state, the Future Hypothesis* (FH^*), which requires the world to end in some particular high-entropy state, and the Future Hypothesis⁻ (FH^-), which imposes a vacuous constraint on final conditions).

Now further assume that: (1) our evidence E is temporally symmetric, (2) the indifference measure is temporally symmetric, (3) given the lawful constraints on boundary conditions, the measure assigns values proportional to the Liouville measure, (4) the measure prescribes the desired values to future events given I2⁻ and E, and (5) the measure assigns a small value to PH^- given I2⁻ and E (which, given (3), is a necessary condition in order for the measure to yield the right prescriptions regarding past events, given I2⁻ and E). Then one can argue for P1 as follows:

L1. (3) and (4) entail that the measure must assign a high value to $FH^* \vee FH^-$, and a low value to FH, given E and I2⁻.

L2. Since E, I2⁻, and the measure are all temporally symmetric (from (1) and (2)), it follows from L1 that the measure must assign a high value to $PH^* \vee PH^-$, and a low value to PH, given E and I2⁻.

L3. L2 and (5) entail that the measure must assign a high value to PH^*, and a low value to $PH^- \vee PH$, given E and I2⁻.

C. Thus (from L3), the measure must assign a greater value to PH^* than to PH, given E and I2⁻.

in propositions like A_f and A_p given E. Since I2 entails I2⁻, this won't be permissible unless μ is such that $\mu(A_f | \text{I2}^- \wedge E) \approx 1$ and $\mu(A_p | \text{I2}^- \wedge E) \approx 0$.[35]

These prescriptions are temporally asymmetric. What is the source of this asymmetry?

The laws I2⁻ posits are temporally symmetric. It's true that I2⁻ also entails that there are lawful constraints on boundary conditions, and many of the possible lawful constraints are temporally asymmetric. And conditional on some such asymmetric constraint, μ will presumably make asymmetric prescriptions. But for each temporally asymmetric constraint, there is a constraint that is its temporal inverse. And without a bias in μ that favors some asymmetric constraints over their temporal inverses, the weighted average of these contributions to μ conditional on I2⁻ cancel out. So I2⁻ is not the culprit.

What about our evidence E? E cannot be the culprit either, for as we saw in §3.2, our evidence does not suffice to ground the desired asymmetry of our prescriptions.

So the temporal asymmetry must come from the indifference measure itself, either from temporally biased assignments over the different lawful constraints on boundary conditions, or temporally biased assignments given lawful constraints on boundary conditions, or both. But it's hard to believe that any kind of *a priori* constraint on rational belief could yield this kind of temporal asymmetry. Any reasons we could come to have for believing the world is temporally asymmetric would seem to be purely *a posteriori*.

It's worth taking a step back to say a bit more about the dialectical status of these problems. The characterization of the Indifference Principle we gave in §2.1 imposed few substantive constraints on what the principle was like. So it's logically possible to adopt an Indifference Principle that yields pretty much any prescriptions one wants. One could, for example, adopt an Indifference Principle that only makes prescriptions regarding how to assign values to initial macrostates that might *contingently* obtain. Or one could adopt an Indifference Principle that employs information-theoretic considerations to assign values to the initial macrostates that might contingently obtain, but employs entirely different considerations to assign values to the initial macrostates that might be lawfully required. More generally, one could simply jerry-rig an indifference measure so that it yields the desired statistical mechanical probabilities given I1 and I2, and doesn't yield the problematic assignments that lead to the Meta-Reversibility Objection.

---

[35] Since I2 entails I2⁻, $cr_E(\text{I2}^-) \geq cr_E(\text{I2})$. So if we're confident in I2 (e.g., $cr_E(\text{I2}) \approx 1$), it follows that we're confident in I2⁻ (e.g., $cr_E(\text{I2}^-) \approx 1$). Thus $1 \approx \mu(A_f | \text{I2} \wedge E) \approx \mu(A_f | E) \approx \mu(A_f | \text{I2}^- \wedge E)$; likewise, $0 \approx \mu(A_p | \text{I2} \wedge E) \approx \mu(A_p | E) \approx \mu(A_p | \text{I2}^- \wedge E)$.

But the mere logical possibility of such a response is cold comfort. The Indifference Principle is supposed to be an *a priori* constraint on rational belief. For it to be credible that such a principle is correct, one needs to make a convincing case that *a priori* plausibility considerations will pick out this particular constraint on rational belief. And it had better *not* be the case that one can provide a plausible rationale for any logically possible set of prescriptions. For if one could, then the entire project of justifying one Indifference Principle over the others would be a nonstarter.

So, insofar as one is on board with the project of finding the right Indifference Principle, one can't just choose whatever principle gets one out of trouble, and expect to be able to provide a plausible rationale for it. Rather, one has to try to determine what the plausible constraints on rational belief are, and hope that these considerations deliver an Indifference Principle which avoids the problematic results. And what the discussion above suggests is that if we employ the kinds of considerations people have so far used to justify Indifference Principles—symmetry, Shannon entropy, etc.—and we accept the constraints that they impose on rational belief—e.g., temporal symmetry—then we won't able to do this.

So proponents of I2 can, if they like, see this as a challenge: provide a plausible *a priori* rationale for the Indifference Principle they want to use to recover statistical mechanical probabilities, and demonstrate that these same considerations don't lead to the Meta-Reversibility Objection.

Thus, options (i), (iii), and (v.b)—i.e., modifying the Indifference Principle—look like unpromising ways to escape the Meta-Reversibility Objection. What about the other options?

I take option (ii), rejecting Bayesianism, to be untenable, for the same reasons as before. While there may be compelling reasons to modify the Bayesian framework, these changes to the Bayesian framework won't help the proponent of the indifference approach escape the Meta-Reversibility Objection.

Likewise, option (iv), rejecting Adequacy, is unsatisfying for the same reasons as before. To reject Adequacy and accept the rest of the argument is to just accept the uncomfortable consequences the account is alleged to have, and then bite the bullet and hold on to the account anyway.

And again, option (v.c), conjoining some further claims to the account, won't help us escape the argument. Since I2 fails to satisfy Adequacy, any account that logically entails I2—such as the conjunction of I2 and some further claim A—will fail to satisfy Adequacy as well.

This leaves us with (v.a): changing the laws I2 posits. This option provided a promising reply to the Reversibility Objection. But in this case, it's not clear how changing the laws can help. Since we're restricting our attention to accounts of *classical* statistical mechanics, we're holding the dynamics fixed. So if we want to change I2's laws, it seems we need to reject or modify the lawful constraints it imposes on boundary conditions.[36] But here the proponent of the indifference approach faces a dilemma. For accounts that impose lawful constraints on boundary conditions that get assigned a high value by the standard Indifference Principles, like I2*, assign high values to the Skeptical Hypothesis, and so don't satisfy Adequacy. And accounts that assign low values to the Skeptical Hypothesis, like I2, impose lawful constraints on boundary conditions that get assigned a low value by the standard Indifference Principles, and so don't satisfy Adequacy. So whichever way one goes, it looks like the resulting account won't satisfy Adequacy.

5. Conclusion

The initial formulations of the nomic and indifference accounts are subject to the Reversibility Objection. And proponents of both approaches can escape the Reversibility Objection by modifying their accounts to include the Past Hypothesis. But this move doesn't free the indifference approach of reversibility worries. For by adopting the Past Hypothesis to escape one reversibility worry, indifference accounts become subject to another—the Meta-Reversibility Objection. And modifying indifference accounts again in order to avoid this objection leads us back to accounts that are susceptible to the Reversibility Objection. So, taken together, the Reversibility Objection and the Meta-Reversibility Objection pose a steep challenge to the viability of the indifference approach.[37]

Appendix

In the text I make the simplifying assumption that the phase space we're working with has been fixed. Here is how the central principles and arguments go if we relax that assumption.

---

[36] One could also keep all of the laws I2 posits and add some further laws as well. (This move doesn't run into the same worry as (ii.c) because it doesn't just *conjoin* something to I2, it *contradicts* I2. I2 asserts that the complete laws are given by classical mechanics and the Past Hypothesis, whereas this alternative account would not.) But given that the dynamics are deterministic, there's only room for laws that either (i) pose lawful constraints on the boundary conditions or (ii) pose chancy constraints on the boundary conditions. The former is the case being considered in the text, and the latter moves us out of the realm of indifference approaches.

[37] I'd like to thank Phil Bricker, Maya Eddon, Nina Emery, Barry Loewer, and Travis Norsen for comments and discussion.

Let P stand for the parameters needed to pick out a particular phase space, given the laws. Once we explicitly take P into account, the chances described in the text will now all include P as a subscript; for instance, $ch_{NIE}(SH)$ will become $ch_{N1PE}(SH)$. Likewise, all the Indifference Principles discussed in the text will all need to take P into consideration. Thus, part (b) of indifference accounts I1 and I2 will read:

b. The statistical mechanical probability of A given B is what the Indifference Principle assigns as the rational initial credence in A conditional on the macrostate B, the parameters needed pick out the phase space P, and the laws.

And the desired constraint on μ will be:

μ-Constraint: Let L be some (possibly incomplete) description of the laws, which includes the dynamics, and is such that the phase space associated with a system can be given a $\Delta x \Delta v$-partition. Let P be a specification of the further parameters needed to pick out a phase space. Let A and B be arbitrary macrostates in this phase space. Let AL (ABL) be the macrostates composed of the microstates compatible with A (A and B) and the lawful constraints on boundary conditions imposed by L, if any. Then for any $\Delta x \Delta v$-partition, γ, such that AL and ABL can be characterized as disjunctions of γ-macrostates, μ must be such that:

$$\mu(A|B \wedge P \wedge L) \propto \frac{e^{S(\gamma ABL)}}{e^{S(\gamma BL)}}. \tag{14}$$

Let's turn to the Reversibility Objection and Meta-Reversibility Objection. Let $\forall_e$ be the restricted quantifier that ranges over all and only the Ps that are compatible with our evidence. To simplify things, I'll assume that these Ps are finite in number. Then we can run the Reversibility Objection against N1 as follows:

The Reversibility Objection (against N1):

P1. $\forall_e P ch_{N1PE}(SH) \approx 1$

P2. Bayesianism (Probabilism and Conditionalization)

P3. Chance-Credence Principle

P4. Adequacy

L5. $\forall_e P \, ic(SH|N1 \wedge P \wedge E) \approx 1$ (P1+P3)

L6. $\forall_e P \, cr_E(SH|N1 \wedge P) \approx 1$ (P2(cond)+L5)

L7. $\forall_e P \, cr_E(\neg SH|N1 \wedge P) \approx 0$ (P2(prob)+L6)

L8. $\forall_e P \, cr_E(N1 \wedge P \wedge \neg SH) \ll cr_E(N1 \wedge P)$ (L7)

L9. $\Sigma_{P_e} cr_E(N1 \wedge P \wedge \neg SH) \ll \Sigma_{P_e} cr_E(N1 \wedge P)$ (L8)

L10. $\Sigma_P cr_E(N1 \wedge P \wedge \neg SH) \ll \Sigma_P cr_E(N1 \wedge P)$ (P2(prob)+L9)

L11. $cr_E(N1 \wedge \neg SH) \ll cr_E(N1)$ (P2(prob)+L10)

L12. $cr_E(N1 \wedge \neg SH) \approx 0$ (P2(prob)+L11)

L13. $cr_E(N1 \wedge \neg SH) \not> 0.5$ (L12)

C. $\neg N1$ (P4+L13)

The Reversibility Objection against I1 will be similar; we just replace P1 with $\forall_e P \, \mu(SH|I1 \wedge P \wedge E) \approx 1$, replace P3 with the Indifference Principle, and substitute "I1" for "N1" throughout.

Finally, the Meta-Reversibility Objection against I2 will be:

The Meta-Reversibility Objection (against I2):

P1. $\forall_e P \, \mu(PH|I2^- \wedge P \wedge E) \ll \mu(PH^*|I2^- \wedge P \wedge E)$

P2. Bayesianism (Probabilism and Conditionalization)

P3. Indifference Principle

P4. Adequacy

L5. $\forall_e P \, ic(PH|I2^- \wedge P \wedge E) \ll ic(PH^*|I2^- \wedge P \wedge E)$ (P1+P3)

L6. $\forall_e P \, cr_E(PH|I2^- \wedge P) \ll cr_E(PH^*|I2^- \wedge P)$ (P2(cond)+L5)

L7. $\forall_e P \, cr_E \, (PH \wedge I2^- \wedge P) \ll cr_E \, (PH^* \wedge I2^- \wedge P)$ (P2(prob) + L6)

L8. $\forall_e P \, cr_E(I2 \wedge P) \ll cr_E(I2^* \wedge P)$ (L7)

L9. $\sum_{P_e} cr_E(I2 \wedge P) \ll \sum_{P_e} cr_E(I2^* \wedge P)$ (L8)

L10. $\sum_P cr_E(I2 \wedge P) \ll \sum_P cr_E(I2^* \wedge P)$ (P2(prob) + L9)

L11. $cr_E(I2) \ll cr_E(I2^*)$ (P2(prob) + L10)

L12. $cr_E(I2) \approx 0$ (P2(prob) + L11)

L13. $cr_E(I2 \wedge \neg SH) \approx 0$ (P2(prob) + L12)

L14. $cr_E(I2 \wedge \neg SH) \not> 0.5$ (L13)

C. $\neg I2$ (P4 + L14)

References

Albert, David. 2000. *Time and Chance*. Harvard University Press.
Arntzenius, Frank. 1995. "Chance and the Principal Principle: Things Ain't What They Used to Be." Unpublished manuscript.
Arntzenius, Frank. 2003. "Some Problems for Conditionalization and Reflection." *Journal of Philosophy* 100(7):356–370.
Callender, Craig. 2004. "Measures, Explanations and the Past: Should 'Special' Initial Conditions Be Explained?" *British Journal for the Philosophy of Science* 55(2):195–217.
Callender, Craig. 2010. "The Past Hypothesis Meets Gravity." In *Time, Chance and Reduction,* ed. Andreas Hutteman and Gerhard Ernst, pp. 34–58. Cambridge University Press.
Callender, Craig, and Jonathan Cohen. 2009. "A Better Best System Account of Lawhood." *Philosophical Studies* 145:1–34.
Dias, P. M., and A. Shimony. 1981. "A Critique of Jaynes' Maximum Entropy Principle." *Advances in Applied Mathematics* 2:172–211.
Earman, John. 1986. *A Primer on Determinism*. Springer.
Earman, John. 2006. "The 'Past Hypothesis': Not Even False." *Studies in History and Philosophy of Modern Physics* 37:399–430.

Elga, Adam. 2000. "Self-Locating Belief and the Sleeping Beauty Problem." *Analysis* 60:143–147.

Friedman, Kenneth, and Abner Shimony. 1971. "Jaynes' Maximum Entropy Prescription and Probability Theory." *Journal of Statistical Physics* 3:381–384.

Frigg, Roman, and Carl Hoefer. 2010. "Determinism and Chance from a Humean Perspective." In *The Present Situation in the Philosophy of Science,* ed. Dennis Dieks, Wenceslao Gonzalez, Stephen Hartmann, Marcel Weber, Friedrich Stadler, and Thomas Uebel, pp. 351–372. Springer.

Goldstein, Sheldon. 2001. "Boltzmann's Approach to Statistical Mechanics." In *Chance in Physics: Foundations and Perspectives,* ed. Jean Bricmont, pp. 39–54. Springer.

Goodman, Nelson. 1954. *Fact, Fiction and Forecast.* Harvard University Press.

Hall, Ned. 1994. "Correcting the Guide to Objective Chance." *Mind* 103:505–517.

Hoefer, Carl. 2007. "The Third Way on Objective Probability: A Skeptic's Guide to Objective Chance." *Mind* 116:549–596.

Howson, Colin, and Peter Urbach. 2005. *Scientific Reasoning: The Bayesian Approach.* 3rd ed. Open Court.

Ismael, Jenann. 2011. "A Modest Proposal about Chance." *Journal of Philosophy* 108(8):416–442.

Jaynes, Edwin. 1983. *Papers on Probability, Statistics and Statistical Physics.* Reidel.

Joyce, James M. 2005. "How Probabilities Reflect Evidence." *Philosophical Perspectives* 19(1):153–178.

Leeds, Stephen. 2003. "Foundations of Statistical Mechanics—Two Approaches." *Philosophy of Science* 70:126–144.

Levi, Isaac. 1985. "Imprecision and Indeterminacy in Probability Judgment." *Philosophy of Science* 52(3):390–409.

Lewis, David. 1986. "A Subjectivist's Guide to Objective Chance." In *Philosophical Papers,* vol. 2, pp. 83–132. Oxford University Press.

Lewis, David. 1994. "Humean Supervenience Debugged." *Mind* 103:473–490.

Loewer, Barry. 2001. "Determinism and Chance." *Studies in the History of Modern Physics* 32:609–620.

Maudlin, Tim. 2007. "What Could Be Objective About Probabilities?" *Studies in History and Philosophy of Modern Physics* 38:275–291.

Meacham, Christopher J. G. 2005. "Three Proposals regarding a Theory of Chance." *Philosophical Perspectives* 19:281–307.

Meacham, Christopher J. G. 2010. "Two Mistakes regarding the Principal Principle." *British Journal for the Philosophy of Science* 61:407–431.

Nelson, Kevin. 2009. "On Background: Using Two-Argument Chance." *Synthese* 1:165–186.

North, Jill. 2011. "Time in Thermodynamics." In *The Oxford Handbook of Philosophy of Time.* Oxford University Press.

Norton, John D. 2003. "Causation as Folk Science." *Philosophers' Imprint* 3:1–22.

Parker, Daniel. 2005. "Thermodynamic Irreversibility: Does the Big Bang Explain What It Purports to Explain?" *Philosophy of Science* 72:751–763.

Parker, Daniel. 2006. "Thermodynamics, Reversibility and Jaynes' Approach to Statistical Mechanics." PhD thesis, University of Maryland, College Park.

Schulman, Lawrence. 1997. *Time's Arrows and Quantum Measurement*. Cambridge University Press.

Sklar, Lawrence. 1993. *Physics and Chance*. Cambridge University Press.

Thau, Michael. 1994. "Undermining and Admissibility." *Mind* 103:491–503.

Tolman, R C. 1979. *The Principles of Statistical Mechanics*. Dover.

Uffink, Jos. 1996. "The Constraint Rule of the Maximum Entropy Principle." *Studies in History and Philosophy of Modern Physics* 27:47–79.

Uffink, Jos. 2002. "Time and Chance." *Studies in History and Philosophy of Modern Physics* 33:555–563.

van Fraassen, Bas. 1989. *Laws and Symmetry*. Oxford University Press.

Vranas, Peter. 2004. "Have Your Cake and Eat It Too: The Old Principal Principle Reconciled with the New." *Philosophy and Phenomenological Research* 69:368–382.

Weisberg, Jonathan. 2011. "Varieties of Bayesianism." In *Handbook of the History of Logic,* ed. Dov Gabbay, Stephen Hartmann, and John Woods, vol. 10, pp. 477–552. North Holland.

Winsberg, Eric. 2004a. "Can Conditioning on the 'Past Hypothesis' Militate Against the Reversibility Objections?" *Philosophy of Science* 71:489–504.

Winsberg, Eric. 2004b. "Laws and Statistical Mechanics." *Philosophy of Science* 71:707–718.

Winsberg, Eric. 2008. "Laws and Chances in Statistical Mechanics." *Studies in History and Philosophy of Modern Physics* 39:872–888.

Xia, Z. 1992. "The Existence of Noncollision Singularities in the N-body Problem." *Annals of Mathematics* 135:411–468.

Chapter Six

Typicality versus Humean Probabilities as the Foundation of Statistical Mechanics

▶ DUSTIN LAZAROVICI

1. From Microscopic Laws to Macroscopic Regularities

Consider the following macroscopic regularities that we observe in our universe:

(1) Apples do not spontaneously jump up from the ground onto the tree.
(2) Rocks thrown on earth fly along (roughly) parabolic trajectories.
(3) The relative frequency of *heads* in a long series of fair coin tosses comes out (approximately) 1/2.

These regularities are of a different kind. (3) is a statistical pattern. (2) is a mechanical phenomenon. (1) turns out to be an instance of the second law of thermodynamics. All three regularities strike us as lawlike; arguably, they are even among the more basic experiences founding our belief in a lawful cosmos. However, it turns out that none of them is nomologically necessary under the fundamental laws that we take to hold in our universe. In fact, given the huge number of microscopic constituents of macroscopic objects[1] and the chaotic nature of the microscopic dynamics (small variations in the initial conditions can lead to significant differences in their evolution), the fundamental laws put very few constraints on what is physically possible on macroscopic scales.

It is possible that particles in the ground move in such a coordinated way as to push an apple up in the air (we know this because the time-reversed process is common and the microscopic laws are time-reversal invariant). It is possible for a balanced coin to land on *heads* every single time it is tossed. And it is possible, as Albert (2015, p. 1) so vividly points out, that a flying rock is "suddenly ejecting one of its trillions of elementary particulate constituents at enormous speed and careening off in an altogether different direction, or (for that matter) spontaneously disassembling itself into statuettes of the British royal family, or (come to think of it) reciting the Gettysburg Address."

Assuming deterministic laws, a physical event or phenomenon is nomologically possible if and only if there exist micro-conditions of the universe that evolve under the microscopic dynamics in such a way that the event or phenomenon obtains. Given our limited epistemic access to the microstate of the universe (or any complex system, for that matter), we thus need some inferential procedure from the fundamental dynamics to salient macroscopic regularities, other than finding the exact solution trajectory that describes our universe. In fact, even if we *did* know the exact initial conditions and could predict the entire microhistory of the world deterministically, it would seem odd if lawlike regularities such as those stated above turned out to be merely accidental, contingent on the very particular microscopic configuration of our universe. In other words, even if we were Laplacian demons and could verify that dynamical laws + initial conditions make (let's say) the second law of thermodynamics true in our world, we should want some additional fact or principle that makes it counterfactually robust and gives it more nomological authority.

Some people find it preposterous to refer to initial conditions of the universe in order to account for something like the motion of a rock or the cooling of a cup of coffee. Well, in practice, we don't. In principle, however, even the best-isolated subsystem is part of a larger system with which it has, at some point, interacted. Hence, if we make postulates about initial conditions of various individual subsystems, we commit redundancy and risk inconsistency.[2] Any attempt at a conclusive and fundamental account must, therefore, talk about the universe as a whole (cf. Oldofredi et al., 2016).

---

[1] The relevant order of magnitude is given by Avogadro's constant, which is $\sim 10^{24}$ per mol.
[2] To adopt an expression from John Bell (2004, p. 166).

In the context of classical mechanics—that we shall focus on for now—there is a reasonably widespread agreement that the following holds true as a mathematical fact:

> There exists a small (low-entropy) region M_{PH} in the phase space Γ of the universe such that the uniform Lebesgue or, more precisely, Liouville measure[3] λ on M_{PH} assigns high weight to initial conditions leading to microtrajectories that instantiate the thermodynamic regularities—in particular, the *thermodynamic arrow of time*—and other salient patterns (about coin tosses, stone throws, etc.) that we observe in our universe.
>
> That is, if we denote this set of "good" initial conditions by $\tilde{M}_{PH} \subset M_{PH}$, it holds true that $\lambda(\tilde{M}_{PH})/\lambda(M_{PH}) \approx 1$.

Following recent lectures of David Albert, we shall call this the *fundamental theorem of statistical mechanics* (FTSM), although it is not literally a theorem in the sense of rigorous proof. "Statistical mechanics" here should be understood broadly as being tasked with explaining or predicting macroscopic regularities on the basis of the microscopic laws. An important question is, of course, why the above statement is so compelling, given that it is virtually impossible to prove for more than highly idealized models. This, however, is not the focus of the present paper, so suffice it to say that arguments going back to Ludwig Boltzmann strongly suggest the truth of the FTSM or a suitably close variant (see, e.g., Bricmont, 1995; Penrose, 1999; Albert, 2000; Carroll, 2010; Goldstein, 2012; and Lazarovici and Reichert, 2015, for detailed discussions). What this essay is going to focus on instead is the physical and philosophical interpretation of this "theorem," in particular the meaning and status of the measure figuring in it.

David Albert (2000, 2015) and Barry Loewer (2007, 2012b) have developed a popular and well-worked-out position in the context of the Humean Best System Account of laws (BSA), adapting David Lewis's theory of objective chance (Lewis, 1980, 1994; Loewer, 2001, 2004; Hoefer, 2019). According to their proposal, the best system of laws for our world consists of

1. The deterministic microscopic dynamics.
2. The Past Hypothesis (PH) postulating a low-entropy initial macrostate of the universe.

---

[3] The Liouville measure is the Lebesgue measure in canonical coordinates on phase space. If we can conditionalize on the constant total energy, the relevant measure is, more precisely, the induced *microcanonical* measure on the energy surface.

3. A probability measure $P = \dfrac{\lambda}{\lambda(M_{PH})}$ on the Past Hypothesis macro-region M_{PH}.

This probability measure does not refer to any intrinsic probabilities or random events in the Humean mosaic. Its inclusion into the best systematization is justified by the fact that it comes at a relatively small cost in simplicity but makes the system much more informative, precisely because it accounts—via the FTSM—for the thermodynamic laws, the entropic arrow of time, and many other macroscopic regularities.

Loewer introduced the name "Mentaculus" for this best-system candidate, a self-ironic reference to the movie *A Serious Man* in which a rather eccentric character tries to develop a "probability map of the entire universe." As a philosophical proposal, though, the Mentaculus—eccentric or not—is certainly appealing, as it attempts to provide a precise account of objective probabilities in deterministic theories. Moreover, Albert (2000, 2015) and Loewer (2007, 2012a) employ the Mentaculus in a sophisticated analysis of counterfactuals, records, and special science laws, the details of which are beyond the scope of this essay.

Another view that has been defended by some authors (e.g., Goldstein, 2012) is that the Liouville measure on the initial macro-region should be understood as a *typicality measure*. This is to say that the FTSM is interpreted not as a probabilistic statement but as the proposition that the respective macro-regularities obtain in *nearly all* or *the vast majority of* possible worlds (consistent with the dynamical laws and the PH). In this sense, macroscopic regularities—such as the second law of thermodynamics—come out as typical, while the notion of *probability* applies to typical statistical regularities but not to the fundamental measure that figures in the FTSM.

In general, a property F is *typical* among a reference class W if nearly all members of W instantiate F. In measure-theoretic terms, this can be explicated as $\mu(\{x \in W: F(x)\}) > 1 - \epsilon$ for an appropriate typicality measure μ and very small ϵ. The basic claim is, then, that when applied to a reference class of nomologically possible worlds (parametrized by initial conditions), typicality facts can serve explanatory, epistemic, and behavior-guiding functions—more convincingly so than probability statements (Goldstein, 2001; Maudlin, 2007; Volchan, 2007; Lazarovici and Reichert, 2015; Wilhelm, 2019; Hubert, 2021).

I will not provide a comprehensive exposition of typicality right away but will develop the position out of a critique of the Mentaculus. Indeed, the goal of this paper is as much to defend the typicality account as to elaborate it in greater detail. Against the Humean theory, I will then argue that it is preferable to adopt typicality even in the context of the BSA, while there are additional motivations if one holds an anti-Humean view about laws of nature.

2. Principal Principle and the Meaning of Humean Probabilities

Typicality facts are distinct from probability facts (Wilhelm, 2019). The first and most obvious difference between probability and typicality is that typicality doesn't come in numerical degrees. A feature of the world can be *typical* or *atypical* (or neither), though not "more typical" or "less typical" than another. The role of a typicality measure is only to determine "very large" and "very small" sets (of initial conditions, i.e., possible worlds)—that is, to concretize the notion of "nearly all." No physical or epistemic meaning is attached to the exact number that it assigns to a particular set.

The Humean probability measure, on the other hand, is supposed to contain much more information. In fact, it will assign a probability—or conditional probability—to any physical proposition about the world: a probability that my dog gets sick if he eats a piece of chocolate, or that your favorite football team wins the next Super Bowl, or that the United States elect a female president in 2028.

When one asks a Humean to explain the regularity theory of chance in five minutes, one will usually get an answer along the following lines: In our world, we find an irregular pattern of coin toss outcomes. Giving you a complete list of every single outcome would be very informative but not at all simple. Telling you that each coin lands either on heads or on tails would be very simple but not at all informative. Saying that the probability of *heads* and *tails* is 0.5 strikes the optimal balance between simplicity and strength. It summarizes the statistical pattern by saying that *heads* and *tails* come out in irregular order but with a relative frequency of roughly 1/2 throughout the history of the world.

So far, so good, but this is not really what probabilities mean in the Mentaculus account. First and foremost, the probability $P(A)$ of any event A is the value that the fundamental probability measure P assigns to the set A of initial microconditions in M_{PH} for which the respective event obtains (cf. Albert, 2015, p. 8). The epistemic and behavior-guiding function of these predictions is then supposed to be manifested in a normative principle, the *Principal Principle* (PP), which states that we should align our initial credences with these objective Humean chances. Formally:

$$C(A|P(A)=x)=x,$$

or, for conditional probabilities,

$$C(A|B \wedge P(A|B)=x) = x.$$

There are other variants of the PP proposed in the literature, and debates about what constitutes "admissible information" that one can conditionalize on (Hall, 1994, 2004; Lewis, 1994; Loewer, 2004), but these subtleties will not be relevant to our discussion. I am also not going to hammer the point that x is a real number, while such sharp credences—determined to infinitely many decimal places—seem like a psychological impossibility. Still, it is good to keep it in mind and note that this issue will not arise with the alternative we are proposing.

In any case, stipulating the PP does not explain why it is rational to follow it and what physical information a probabilistic prediction of the Humean best system contains. What exactly is the Mentaculus telling us by assigning, let's say, a (conditional) probability of 30 percent to the United States electing a female president in 2028? What feature of the Humean mosaic is this supposed to summarize?

It seems clear that we cannot get more out of the best system than we put in. If it is more accurate to assign a chance of 0.3 than 0.6, there must be concrete physical facts in the world that make it so. And these facts must go into evaluating the strength of the best-system candidates competing to summarize them; that is, assigning the number 0.3 rather than 0.6 must be a part of what makes the measure "best."

One may think that the relevant fact is simply whether the event in question obtains or not. Many authors have read Lewis in this vein, as suggesting that the Humean probability law tries to fit every (macro-)event individually by assigning as high a probability as possible to any event that, in fact, occurs, and as low a probability as possible to any event that doesn't. But it cannot really work that way. Being a good predictor of presidential elections does not gain you as many points for "strength" as predicting the increase of the Boltzmann entropy in our universe. At the end of the day, the best-system probability law will be one that informs us about robust regularities and global patterns in the world—"chancemaking patterns" as Lewis (1994) called them—while the fit to singular events will count little to nothing in the trade-off with simplicity. And if the number that the measure assigns to some singular event has nothing to do with that measure being part of the best systematization (or any systematization at all), it is hard to see why this number should have physical significance, let alone serve an epistemic or behavior-guiding function.

A standard Humean response takes a more holistic view and argues that single-case probabilities are nonetheless meaningful because $P(A) = x$ is true in all and only those worlds whose best system implies $P(A) = x$. Hence, while a single-case probability may not express anything about the individual event *per se*, there is something about the structure of the mosaic as a whole that makes that particular value true (Lewis, 1980).

However, as I will argue in more detail below, many different measures would predict the same regularities and statistical patterns—by assigning to them a measure close to one—yet disagree significantly on probabilities assigned to singular events. By some standard, one of these measures (viz., the Liouville measure) may fare best in terms of *simplicity*, which is why it is elevated to a Humean law. But this does not make it a more accurate predictor of the 2028 presidential election. In particular: if, given the dynamical laws and the PH, two probability measures P and $\tilde{P}$ predict the same regularities, while $\tilde{P}$ loses out in terms of simplicity, there is *no possible world* in which $\tilde{P}$ replaces P as part of the best systematization.[4] Therefore, a proposition like "according to the Mentaculus, the probability of event A is $P(A) \approx 0.3$ rather than $\tilde{P}(A) \approx 0.6$" does not restrict the set of possible worlds any further than to those which are typical models of the laws with respect to both measures.

To summarize: According to the standard Humean view, there are certain regularities (chancemaking patterns) on which a probability law supervenes, while "probabilities" for a great many other events—in fact, for all measurable subsets of phase space, most of which have no relation to relevant macro-variables and don't correspond to any particular macro-event—come out as a by-product. This by-product, to play on one of David Albert's favorite metaphors, is not a gift from God ("I give you the most efficient summary of the regularities, and you get rational credences for all possible events for free") but mostly mathematical surplus; the probabilities assigned to particular events could be very different, yet the physical content of the laws the same. Sure, we can postulate that we should live our lives according to the numbers that the Mentaculus spits out, but as so often with articles of faith, there is no rational prospect of reward in this world.

A True Regularity Theory of Probabilities

So Humean chances, in a nutshell, are supposed to be efficient summaries of statistical patterns in the world. Then they turn out to refer, first and foremost, to a measure on sets of possible microstates. What has one to do with the other? In

---

[4] That is, unless the standard for simplicity were oddly contingent in a way that depends on microscopic details or isolated macro-events.

most cases: nothing at all (is exactly my point). In some particularly nice cases, the connection between a statistical pattern instantiated by a large series of similar events A_i (e.g., a coin landing on *heads* on the *i*th trial) and the probability $P(A_i) = p$ of the individual events that make up the pattern is provided by a *law of large numbers* (LLN)—that is, a result of the form

$$P\left(x \in M_{PH} : \left|\frac{1}{N}\sum_{i=1}^{N}\chi_i(x) - p\right| \leq \epsilon\right) = 1 - \delta(\epsilon, N), \quad \text{(LLN)}$$

with $\delta(\epsilon, N) \to 0$, $N \to \infty$, e.g., $\delta(\epsilon, N) \propto \dfrac{1}{N\epsilon^2}$ from the standard weak law of large numbers. Here, χ_i is the indicator function mapping a possible initial microstate x to 1 if the event A_i occurs, and to 0 if A_i does not occur for the micro-history with initial condition x.

Hence, $m_{emp}(x) := \dfrac{1}{N}\sum_{i=1}^{N}\chi_i(x)$ is the relative frequency ("empirical mean") of the event-type A determined by the deterministic laws as a function of the initial conditions. And we can read equation (LLN) as: "For large N (many trials), the measure of the set of initial conditions for which the relative frequency of A deviates only slightly from p is very close to 1." If we use "typical"—for now—merely as a shorthand for "instantiated for initial conditions that form a set with measure ≈ 1," this is clearly a typicality result (just like the FTSM).

As a mathematical theorem stating sufficient conditions for (LLN), the law of large numbers requires that the events are in some sense independent or uncorrelated, which is often intuitively compelling but nearly impossible to verify. The standard proof would furthermore use that p comes out as the expectation value of m_{emp} with respect to the measure P. In the end, however, the role of the measure in (LLN) is merely to tell us that a particular set of initial conditions—the initial conditions for which the statistical pattern $m_{emp} \approx p$ obtains—has nearly full measure. And at this point, it doesn't matter where the number p came from, whether it corresponds to $P(A_i)$ or not, and whether we gave it any meaning as a *probability* in the first place. Its significance as a relative frequency describing a typical statistical regularity is established by, rather than assumed in, the law of large numbers result.

Philosophically, it is thus unnecessary and misleading to think of (LLN) as a *consequence* of the single-case probabilities determined by P. It is really the other way around. What a law of large numbers result does, in effect, is to reduce theoretical probabilities to *typical frequencies*. All the best system needs to do, is to predict the statistical regularity by assigning to it a measure close to one. If it predicts, for example, that *heads* and *tails* come out about equally often in a long series of coin tosses, we can take this as the *definition* of the coin-toss probabilities and begin to justify the rationality of assigning credences about individual trials accordingly. For instance by appealing to Dutch-book arguments (if I accept bets of less than 2:1 on each trial, I can be almost certain to lose money on the long run) or maybe by invoking a principle of indifference with regard to the individual event in the pattern that we are about to observe (Schwarz, 2014).

Ultimately, this view of probabilities as typical relative frequencies is in no way tied to Humean metaphysics; but when combined with the latter, it puts the regularity theory back on its feet. Probabilities are indeed summaries of statistical regularities that the best system predicts, rather than weights it assigns to sets of possible microstates.

Admittedly, this is breaking with the Mentaculus's "imperialistic" (Weslake, 2014) view of statistical mechanics—at least in the first instance. It will require more work to identify the typical regularities that ground special science laws and rational credences for everyday decision-making. I believe that this work can be done—however, not here. My point is that the idea of the Mentaculus providing a shortcut from the fundamental laws of physics to individual chance prescriptions for every conceivable event might be philosophically appealing but is both too ambitious and too simplistic to pan out.

3. Principal Principle versus Cournot's Principle

What the previous discussion has, in fact, accomplished is to reduce the Principal Principle (PP)—at least those instances of the PP that could have a basis in physics—to a version of *Cournot's principle* (CP). Cournot's principle has been somewhat forgotten in modern times (Glenn Shafer being one of the few famous probabilists to hold up its banner) but has a long tradition in the philosophy of probability, with some version being endorsed by Kolmogorov, Hadamard, Fréchet, and Borel, among others (for excellent historical discussions, see Martin, 1996; Shafer and Vovk, 2006).

One way to introduce CP is as a remedy to the following dilemma:

1. Only probabilistic facts follow from probability theory.
2. There are no genuinely probabilistic facts in the world. Any possible event either occurs or not.
3. Hence, no facts about the world follow from probability theory.

Assumption 2 could be denied by admitting something like propensities into the physical ontology, but then replace "(physical) facts" by "empirical facts," and we end up with a very similar problem: logically, no empirical facts follow from probabilistic ones. Cournot's principle can thus be understood as a "bridge principle," leading from probabilistic results to physical/empirical predictions.

An unfortunate historical fact is that the formulation provided by its namesake sounds wrong, at least to modern philosophers of science:

> A physically impossible event is one whose probability is infinitely small. This remark alone gives substance—an objective and phenomenological value—to the mathematical theory of probability. (Cournot, 1843)

It seems clear from his further writing that Cournot meant "physically impossible" in more of an FAPP ("for all practical purposes") sense, but the terminology has not helped the acceptance of his philosophy.

A more appropriate formulation was given by Kolmogorov (1933, sec. 2.1): If an event has very low probability, *"then one can be practically certain that the event will not occur."* Equivalently, we could say: If an event has a very high probability, we should expect it to occur. Other authors have cast the principle in more decision-theoretic terms, roughly: It is rational to act as if very high probability outcomes will obtain.

In the context of statistical mechanics, where we are dealing with robust phenomena and regularities, Tim Maudlin (private communication) has convinced me that the most pertinent formulation of the principle is in terms of *explanation* (see also Lazarovici and Reichert, 2015; Wilhelm, 2019). If a feature of our world turns out to be typical or very likely according to a theory, we should consider it to be conclusively explained by that theory. If we observe a phenomenon that is atypical or extremely unlikely according to our theory, we should look for further explanation—and, in the last resort, revise or reject the theory if none can be provided.

In the end, I believe, these are all just different aspects of the same rationality principle, which may be cast in terms of expectation, belief, acceptance, explanation, and so on (at least on the level of granularity on which these notions are roughly intertranslatable). Crucially, though, epistemic or doxastic notions come in only as we consider the *normative implications* of probability or typicality results, while the results themselves do not depend on anyone's knowledge or beliefs.

We should also note that while some version of CP could be regarded as a special case of PP (when very high credence is de facto certainty), it is characteristic of the view associated with Cournot to privilege statements of "very high" or "very low" probability with respect to their physical or empirical content. Thus Borel's famous (though still somewhat too strongly worded) credo that the principle that an event with very small probability will not happen is "the only law of chance" (Borel, 1948).

Although the Lewis-Loewer theory of objective chance is traditionally associated with the Principal Principle (Lewis even regarded PP as "non-negotiable"), it is very much—if not more—compatible with Cournot's principle: If the Humean probability of a phenomenon is nearly one, we can be almost certain that this phenomenon will actually obtain. Why? Because this is what the best system is trying to tell us; because the way in which the Mentaculus predicts macro-regularities in the world is to assign them a very high chance. Ironically, in some cases, a version of what Lewis (1994) considered to be the "big bad bug" of his theory of chance can serve to vindicate even the strongest form of CP. The Mentaculus will, for instance, assign a very small though positive probability to the universe evolving on an entropic-decreasing trajectory. However, if the universe did evolve on an entropy-decreasing trajectory, this Mentaculus would not be the best systematization of our world (given that so many salient features depend on its entropic history). Hence, the fact that the Mentaculus assigns a near-zero probability to anti-entropic trajectories, together with the fact that the Mentaculus is the best systematization of our world, implies that an anti-entropic evolution of the universe is, in fact, impossible. On the other hand, if we are talking about a feature of the world that the best systematization could, in principle, fail to predict, it is immaterial whether it assigns to it a probability of 10^{-100} or 10^{-1000}. Our residual uncertainty about whether that feature is instantiated after all does not come from anything the best system tells us about the world, but from the possibility that it just had to get this one wrong in the trade-off with simplicity.

In any case, the main argument so far as to why even Humeans should favor CP over PP is that all the physical information that the Mentaculus provides is to be found, first and foremost, in statements of very high or very low probability, whereas the rationality of aligning degrees of belief with any odd value of Humean chances is spurious, at best. Note that also methodologically, the only way to *test* probabilistic laws is by some form of Cournot principle—that is, by rejecting the law if we observe relevant phenomena to which it assigns a negligibly low chance (cf. Shafer and Vovk, 2006). I am not a verificationist and do not claim that single-case probabilities are meaningless just because they cannot be empirically tested. I have, however, argued that the regularity account fails to give them meaning as deterministic chances (except to the extent that they can be reduced to typical frequencies). Humeans often claim that their probability measure is "empirical," yet provides information far beyond what is empirically testable. I don't think they can have it both ways.

4. Probability Measures versus Typicality Measures

The next step from probability to typicality comes by emphasizing the following insight: If we agree that all we need, on the fundamental level, are "probabilities" close to 1 and 0, then a whole lot of different measures could do the job. If we don't like the Lebesgue measure, how about putting a (truncated) Gaussian measure on M_{PH}? In fact, we can tweak the measure in almost any way we like. Almost any absolutely continuous measure will make a statement analogous to the FTSM true, and thus imply the same thermodynamic laws and statistical regularities.[5] We cannot be too extreme, of course. A delta-measure concentrated on an anti-entropic microstate will, evidently, lead to vastly different predictions. However, as Maudlin (2007, p. 286) concludes: against this backdrop, "our concerns about how to pick the 'right' probability measure to represent the possible initial states ... or even what the 'right' measure means, very nearly evaporate."

An important observation here is that probabilities (or weights, to use a more neutral term) close to 1 or 0 are very robust against variations of the underlying measure. Think, for instance of the normalized Liouville measure as a uniform probability density over the macro-region M_{PH}. If $\lambda(A) \approx 0$ but $\mu(A) \gg 0$, then the measure μ must differ radically from λ on the small set A. (The same holds by contraposition for probabilities near one.) In contrast, for $\lambda(B) = 0.3$, while $\mu(B) = 0.4$ (let's say), μ needs to deviate only slightly from the uniform density on the larger set B (cf. Figure 6.1). "Large" and "small" are here understood with respect to the Liouville measure, but this doesn't make the argument circular. The point is that radical deviations from the uniform density would be necessary to come to different conclusions about what is typical/atypical, while relatively small variations can lead to significantly different probability assignments for other events.

At least for the sake of argument, Albert is willing to concede that we could consider best-system candidates involving an entire set or equivalence class of probability measures, with the understanding that the theory endorses all and only those probability statements on which these measures (more or less) agree (see, e.g., Albert, 2015, footnote 2). While this compromise goes in the right direction, it strikes me as a bit of a poison pill for Humeans. A whole set or equivalence

---

[5] The insight that a great many probability distributions over initial conditions lead to the same statistical predictions lies also behind the "method of arbitrary functions," see, e.g., von Plato (1994) and Myrvold (2016).

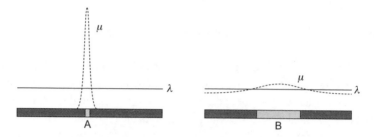

FIGURE 6.1. Schematically: λ and μ disagreeing significantly on the weight assigned to a set of small (left), respectively medium (right) λ-measure. Credit: © Dustin Lazarovici

class of measures is neither simpler nor more informative than the Liouville measure (I have only argued that it is equally informative), so these best-system candidates seem set up to fail by the standards of the BSA. Instead, we should think of it this way: We should use the Liouville measure because it is simple and natural (the natural representative of the "equivalence class," so to speak), but with the understanding that it is not a bona fide probability measure but a *typicality measure*. Its role and purpose is to designate possible properties of the world as "typical" (measure ≈ 1) or "atypical" (measure ≈ 0), or neither, while the precise numerical assignments have no physical significance.

This move, to relate CP to a concept of typicality rather than probability, is a more recent development, though there is precedent for it in the physical literature (see, e.g., Everett, 1973; and the discussion in Barrett, 2016; Dürr et al., 1992; Bell, 2004, Chapter 15, originally published in 1981; and Goldstein, 2001, on Boltzmann). There are additional motivations for this step:

(a) As argued before, Cournot's principle suggests an understanding of probabilities as typical frequencies. Evidently, this would be circular as an account of probabilities if "typical" were itself explicated in probabilistic terms.

Moreover, probabilities can mean typical frequencies when applied to events *in* the universe, but we better not refer to frequencies (not even hypothetical ones) when we speak about the universe as a whole. For this reason alone, it makes sense to distinguish two different concepts.

(b) In statistical mechanics, it is common and convenient to formalize typicality with the mathematical tool of (normalized) measures—hence the deceptive kinship to probability. There are, however, other ways to define "typical"—for instance, in terms of cardinalities, which can be relevant in other contexts and figure in the same way of reasoning. (For further differences between typicality and probability, see Wilhelm, 2019.)

(c) Some authors have argued that "typical" is a more intuitive and unambiguous notion than "probable" (see, e.g., Dürr et al., 2017). One way to spell this out is to say that the intuition associated with a typicality measure is simply one of "very large" versus "very small" sets—which does not raise further interpretive questions and makes CP immediately compelling.

If we observe a salient feature of the world and can establish that it is, in fact, a feature of *nearly all* possible worlds permitted by the laws, there is nothing left to explain. To wonder further, why our world is—in that particular respect—like nearly all possible ones, strikes me as deeply irrational.

5. The Epistemic and Metaphysical Status of Typicality

The last point leads us to the most controversial part of our discussion, or at least the one that has been most contentious between David Albert and advocates of typicality, including myself. Some Humeans, like Callender (2007), seem sympathetic to the concept of typicality as long as the relevant typicality measure supervenes, together with the dynamics and the PH, as a Humean law on the mosaic. The measure would still be an axiom of the best system, though introduced not to assign a probability to all (measurable) subsets of phase space, but to tell us which sets contain *nearly all* (or *nearly none*) of the possible initial conditions. Humeans have no reason to be offended by this proposal, even if they come to reject it.

I would, however, contend that, given the dynamical phase space, it is not a (further) empirical question—not an additional fact supervening on contingent regularities in the mosaic—if a subset exhausts "nearly all" of it. Suppose this phase space were finite. There would still be countless different ways to assign *probabilities* to its points, but that is not what we are trying to do. We just want to say that certain sets contain nearly every phase space point and don't need another physical law or postulate to tell us how to *count*. Since we are generally dealing with continuous state spaces, we use a measure to determine which sets are "very big" or "very small." Still, all reasonable measures agree on what is typical or atypical in this sense, and we are quite capable of weeding out the unreasonable ones.

David Albert (in private communications) has been particularly forceful in rejecting *this* idea, that one could appeal to some a priori notion of "nearly all," that typicality facts could come more or less for free once the rest of the theory is fixed. The deeper question here concerns the metaphysical and epistemic status of the measure (I keep using the singular, but it's important to keep in mind that many different measures—qua mathematical objects—would formalize the same notion of typicality).

Humean supervenience may seem like a ready-made answer, but it fails to capture the more subtle aspects of typicality and its use in physics. These come across if we consider the typicality measure, not as an additional law on a par with the deterministic dynamics, but as a *way of reasoning about* these laws.[6] In other words: being intimately tied to Cournot's principle—which is normative rather than descriptive—the typicality measure falls itself, at least in part, into the normative domain.

To some, this will sound both vague and pretentious. Isn't the best-system account so much more clear-cut and empirical? Alas, it is not. The Mentaculus account may suggest that the Humean mosaic is the truthmaker of a probability measure as part of the best system. However, we have seen that there are no concrete physical facts in the world that make its probability assignments *true*. The Mentaculus has de facto given up on this condition (that the axioms of the best system must be true statements about the world) and retreated to the position that there is, nonetheless, something that makes it *best*. But part of what makes a measure "best" is its relative simplicity, and there are countless ways in which one could define and compare the simplicity of measures (and theories, in general). Clearly, the standard we are meant to use could hardly supervene itself as part of the best systematization, or else it could not play the role it does in adjudicating which systematization is best, to begin with. When we appeal to simplicity, we are not making an empirical argument but also operate in the normative "space of reasons" (Sellars, 1962)—giving and taking reasons for accepting or rejecting theories. A judgment about the simplicity of probability measures is not so different, in nature or reliability, from a judgment about the appropriateness of typicality measures.

With this in mind, if someone asked about the truthmaker of the typicality measure, I would say that a choice of typicality measure can be *reasonable* or *jus-*

---

[6] Fundamentally stochastic dynamics are a somewhat different story since the dynamical laws themselves put weights on possible micro-trajectories. Nonetheless, the regularities that we should consider to be explained and predicated by the laws are those that come out as typical—i.e., that are instantiated for a set of micro-histories whose total weight is close to one. In other words, the empirical content of stochastic laws still comes from Cournot's principle.

tified, but there are no concrete physical facts in the world that make it, strictly speaking, true. (Whether there are, in addition, objective normative facts that make it true is beyond the scope of this discussion.)

Epistemically, this does not mean that the typicality measure is a priori or logically deduced from the dynamics. It is tied to and constrained by the dynamical laws—via desiderata that we will discuss in section 6—yet epistemically more robust than the latter.

While I believe that it is always the theoretical system as a whole that is challenged by empirical evidence, I cannot conceive of a situation in which it would be reasonable to revise our notion of typicality instead of adjusting the dynamical postulates (whereas the converse is common and unproblematic). As long as we can stay in the framework of Hamiltonian mechanics, we would change the Hamiltonian, not the Liouville measure. And when we are compelled to make more radical theory changes, then because the phenomena suggest new dynamical features (or a different physical ontology)—not a new concept of "typical." For instance, there are almost certainly initial conditions for a Newtonian universe which are such that particles create an interference pattern whenever they are shot through a double-slit and recorded on a screen. This and other quantum phenomena are not made *impossible* by classical mechanics; they just come out as *atypical*. However, changing the typicality measure in order to save Newtonian theory from falsification is not a serious scientific option that anyone has ever, or should ever, entertain.

Using Quine's (1951) picture of a "web of beliefs," I suggest that the dynamical laws are closer to the edges of the web than the typicality measure. Some conception of "typical"—and hence the measures suitable to express it—lies in between the dynamical laws and the logical inferences rules. While it is, in principle, possible to adjust it to new empirical evidence, the typicality measure is never the first knob to turn before making revisions in other parts of our theoretical system.

One reason is that, because the notion of typicality is so robust against variations of the measure, any revision of it would be radical—i.e., would have to correspond to extreme changes of the measure on the pertinent state space. On the flipside: Because typicality statements do not depend on the details of the measure, their explanatory work is done not by any particular choice of typicality measure but mostly by the reference class of nomic possibilities determined by the dynamical laws (and an eventual Past Hypothesis).

This leads us to a more important reason the typicality measure *should* be "less empirical" or epistemically more robust than the dynamics. We have seen that, due to the huge number of microscopic degrees of freedom, the dynamical laws put barely any constraints on what is physically possible on macroscopic scales. If the dynamics are somewhat nontrivial, the "right" initial micro-conditions could produce virtually any phenomena we like. By the same token, the "right" measure could make virtually any phenomena typical or sufficiently likely. Therefore, treating the typicality measure on the same footing as the dynamics would give us too many moving parts that can be adjusted to the data. For the Humean, this is bad because it increases the risk of a tie for the best system, at least in situations when simplicity of the dynamical and probability postulates pull in opposite directions. For the anti-Humean, it is even worse, since the more we regard the typicality measure as an independent empirical hypothesis, the less is empirical evidence able to inform us about the true laws of nature.

As an anti-Humean, I consider the laws to be essentially characterized by their modal structure. Their extensional meaning corresponds to an infinite set of nomologically possible worlds, while we have access to only one. But how then could empirical facts put any serious constraints on what the true laws might be? By the principle that the relevant phenomena of our world must come out as *typical*—not merely *possible*—with respect to the modal space carved out by the laws. This is the essence of Cournot's principle when taken to the fundamental physical level. Admittedly, we will never know with metaphysical certainty that nature complies with this principle, that our world is not an atypical model of the laws in regard to the phenomena. We can only trust that "God is subtle but not malicious," as Einstein put it.

While typicality as a *way of reasoning* plays an indispensable role in an anti-Humean epistemology, it is also very well compatible with a best-system account of laws in general. We can think of it this way: Typicality measures are not themselves law-hypotheses competing for the optimal balance between simplicity and strength, but part of the standards we use to assess the strength of law-hypotheses. For the microdynamics and the PH to be considered adequate systematizations of our world, they must determine a set of nomic possibilities that make its salient macro-regularities typical. It does not suffice that a theory reproduces them for some special and fine-tuned initial conditions. As argued before, this is not too different from the role that simplicity plays in the BSA. In particular, the "measure" of simplicity cannot be itself an axiom of the best systematization or else it could not do its job in deciding which systematization is best. Similarly, typicality could not do the job it does in scientific reasoning—in reasoning *about* theories not just *within* theories—if it were merely another empirical bookkeeping device.

Indeed, in actual scientific practice, typicality judgments do have a privileged epistemic status that the Mentaculus account seems to miss. In particular, atypicality is precisely the standard by which theories are reasonably rejected as empirically, or at least explanatorily, inadequate (think again of the double-slit experiment as a falsification of classical mechanics or the 5σ-standard commonly employed in particle physics). Interestingly, this applies in much the same way to deterministic laws as to intrinsically stochastic ones. The difference is that, in the latter case, falsifying a dynamical and a probabilistic hypothesis is one and the same, while in the deterministic case, it is primarily the dynamical laws that stand trial. And this is possible only because the typicality measure is epistemically more robust or—in a sense we shall now discuss—entailed by the dynamics. In the upshot, some concept of "typicality" and "atypicality" is part of the backdrop against which law-hypotheses are evaluated, rather than another law-hypothesis in its own right.

6. Justification of Typicality Measures

For these reasons, most advocates of typicality do not consider the typicality measure as an independent postulate of a physical theory, although it might be from a strictly logical perspective. But what then determines the "right" typicality measure and accounts for its epistemic rigidity?

One answer we have already alluded to is that the role of a typicality measure is not so much to *define* typicality but to formalize an intuitive and largely pre-theoretic notion. So what makes a set-function a *typicality* measure is not (just) that it satisfies the mathematical axioms of a normalized measure (the Kolmogorov axioms) but that it captures the semantic meaning of "typical." The notion of typicality has, in any case, a certain vagueness. Just as it is impossible, in general, to specify a *sharp* threshold value ϵ such that a set A contains "nearly all" elements if and only if $\frac{\lambda(A)}{\lambda(M_{PH})} > 1 - \epsilon$, it seems impossible to specify a set of formal criteria that make a measure convincing as part of the mathematical formalization of typicality. In the context of classical Hamiltonian mechanics, the Liouville measure is clearly a reasonable choice, while a delta-measure is clearly not, but a certain grey area in between seems unavoidable. Or consider a family of Gaussian measures with standard deviation $\sigma \to 0$, so that the distributions become more and more peaked. Again, it would be misguided to ask for a sharp threshold value of σ below which the Gaussians cease to be suitable typicality measures (or start to define a different notion of typicality). Yet this does not mean that the concept itself is ill-conceived or that the vagueness is problematic in practice. The bottom line is that a typicality statement has normative implications if and only if it is made with respect to a reasonable notion of "large" versus "small" sets. And while it is hard to state precisely what makes a measure reasonable or unreasonable, we can generally tell them apart when we see them.

Some authors put less emphasis on the intuitive content of typicality and more on the condition that the measure must be *stationary* under the dynamics. This is to say that the measure of a set of microstates at one time must correspond to the measure of the time-evolved set at any other time. In this way, the dynamical laws themselves constrain the choice of typicality measures in that they must be adapted to the dynamics. I will provide a more precise definition of stationarity and further justification for this condition below.

Fortunately or unfortunately, as long as we are dealing with classical mechanics, both approaches are perfectly consistent and complementary, since the most natural and intuitive measure—the uniform measure on phase space—is also stationary under the Hamiltonian dynamics (though not uniquely so). It could even be justified by a principle of indifference (Bricmont, 2001), although I don't find its epistemic connotations helpful.

There is, however, a notable example where stationarity and intuitiveness seem to go apart (and the principle of indifference to fail altogether). In Bohmian quantum mechanics, the natural typicality measure grounding the Born rule, and thus quantum statistics for subsystems, is given by the $|\Psi|^2$-density on configuration space, induced by the universal wave function Ψ (Dürr et al., 1992). This measure is stationary (more precisely: equivariant) under the Bohmian particle dynamics and even determined uniquely by this condition (Goldstein and Struyve, 2007). However, as Albert pointed out (private communication), we do not know what the universal wave function actually looks like; the $|\Psi|^2$-density may turn out be highly non-uniform and deviate radically from what we would have considered intuitively compelling in the classical case. In the next subsection, I will discuss how this conclusion can be avoided, and the two approaches to the typicality measure reconciled, yet again.

Stationarity, Uniformity, Symmetry

The following discussion is somewhat technical, but the basic point is simple: So far, we have talked about measures on possible initial conditions of the universe. In fact, the relevant reference class of our typicality statements—what we actually want to quantify—are not microstates but (nomologically) possible worlds. Initial conditions are just a means to parametrize the respective solution trajectories. Stationarity, in a nutshell, guarantees that a large set of solution trajectories is deemed large by the same standard if we look at the trajectories—i.e., the respective microstates $X(t)$—at any other time.

One subtlety that can lead to misunderstandings is that proponents of the Mentaculus tend to think of the fundamental probability measure on the phase space Γ as one that is uniform over the Past Hypothesis macro-region $M_{PH} \subset \Gamma$ and zero outside. This is not a stationary measure on Γ, because weight will "flow out" of the initial macro-region and disperse all over phase space. Other authors tend to think of the stationary Liouville measure on the *entire* phase space as the natural typicality measure, which is then *conditionalized* on the special initial macrostate M_{PH}. In the following considerations that focus on the solution space rather than the phase, however, this distinction will become largely immaterial.

Let S be the set of solution trajectories for the microscopic dynamics (consistent with the Past Hypothesis) in the state space $\Gamma \cong \mathbb{R}^n$. For any $t \in \mathbb{R}$, let $\epsilon_t : S \to \Gamma, X \mapsto X(t)$ be the map evaluating the trajectory X at time t. These maps can be read as charts, turning the solution set S into an n-dimensional differentiable manifold.[7] The transition maps between different charts are then $\epsilon_t \circ \epsilon_s^{-1} = \Phi_{t,s}$, where $\Phi_{t,s}$ is the flow arising as the general solution of the laws of motion (Figure 6.2).

Now, the easiest way to define a measure μ on S is in one of these charts, let's say ϵ_0. Indeed, a possible point of view is that there exists a distinguished "initial" time so that we should parametrize solutions by initial data at $t = 0$. The other point of view, emphasized by our geometric framework, is that the choice of the "time slice" is arbitrary, essentially amounting to a particular coordinatization of the solution space. Under a transition map (coordinate transformation), the measure transforms by a pullback, $\mu_t = \Phi_{t,0} \# \mu_0$, i.e., $\mu_t(A) = \mu_0(\Phi_{t,0}^{-1} A)$ for any measurable $A \subset \Gamma$, where μ_t is the measure represented in the chart ϵ_t.

A measure is *stationary* if and only if it has the same form in every time-chart, i.e.,

$$\mu_t = \Phi_{t,0} \# \mu_0 = \mu_0, \forall t \in \mathbb{R}$$

Equivariance is the next best thing if the dynamics are themselves time-dependent. Concretely, in the case of Bohmian mechanics, the particle dynamics are determined by the universal wave function Ψ_t which may itself evolve in time according to a linear Schrödinger equation. Nonetheless, we have

$$|\Psi_t|^2 d^n x = \Phi_{t,0} \# |\Psi_0|^2 d^n x, \forall t \in \mathbb{R}$$

so that the measure has the same functional form in terms of Ψ_t for all t.

---

[7] Strictly speaking, some solutions may exist only on a finite time-interval, so that the charts are only locally defined. Here, for simplicity, we assume global existence of solutions.

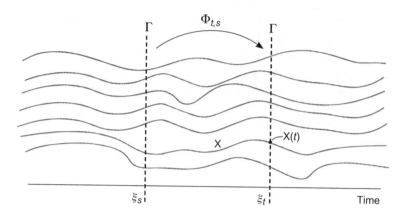

FIGURE 6.2. Sketch of the solution space and its parameterization by time slices. Credit: © Dustin Lazarovici

In conclusion, a stationary or equivariant measure on the state space Γ induces a canonical measure on the solution space S: a measure that can be defined without distinguishing a set of coordinates—i.e., a particular moment in time.

Uniformity of a measure, on the other hand, is a metric notion. It requires that

$$\mu(B(x, r)) = \mu(B(y, r)), \forall x, y \in \Gamma, r > 0,$$

where $B(x, r)$ is the ball of radius r around x. However, even if the state space Γ comes equipped with a metric, it does not, in general, induce a canonical metric on S. It is thus not clear what the metric on the solution space is supposed to be, or whether it makes sense to regard S as a metric space (Riemannian manifold) at all. But without a metric on the solution manifold, it is meaningless to ask whether a measure on it is uniform or not. From this point of view, it is indeed misleading to regard uniformity as a criterion for the typicality measure. Even the Liouville measure in classical mechanics is uniform on the "wrong space"—namely, on phase space (the space of possible initial conditions) rather than the solution space (the space of nomologically possible worlds).

The uniformity of the Liouville measure on phase space does nonetheless capture a meaningful and important feature of the typicality measure; namely, its *invariance under Galilean symmetries*. A symmetry is an isomorphism $T: \Gamma \to \Gamma$ that commutes with the flow; i.e., $\Phi_{t,s}(Tx) = T\Phi_{t,s}(x)$. This then induces a transformation $\bar{T}: S \to S$ on the solution space by $\bar{T} = \epsilon_t^{-1} \circ T \circ \epsilon_t$ which is independent of

t.[8] The most important symmetries of classical mechanics are those of Galilean spacetime, namely:

$$(q_1,\ldots,q_N; p_1,\ldots,p_N) \to (q_1+a,\ldots,q_N+a; p_1,\ldots,p_N) \quad \text{(Translation)}$$

$$(Rq_1,\ldots,Rq_N; Rp_1,\ldots,Rp_N) \quad \text{(Rotation)}$$

$$(q_1+ut,\ldots,q_N+ut; p_1+m_1u,\ldots,p_N+m_Nu) \quad \text{(Galilei boost)}$$

It is well known that the Lebesgue or Liouville measure λ is invariant under these transformations (which are just a subset of Euclidean transformations on phase space). Consequently (as is easy to check), the induced measure on S is invariant under the corresponding symmetry transformations on the solution manifold.

In Bohmian mechanics, the issue is a bit more subtle, because the wave function itself transforms nontrivially under Galilean symmetries; namely, as (see Dürr and Teufel, 2009):

$$\Psi(q_1,\ldots,q_N) \to \Psi(q_1-a,\ldots,q_N-a) \quad \text{(Translation)}$$

$$\Psi(R^{-1}q_1,\ldots,R^{-1}q_N) \quad \text{(Rotation)}$$

$$e^{\frac{i}{\hbar}\Sigma m_i\left(uq_i - \frac{1}{2}u^2\right)} \Psi(q_1+ut,\ldots,q_N+ut) \quad \text{(Galilei boost)}$$

It is, however, evident that the $|\Psi|^2$-density is invariant (more precisely, covariant) under these transformations, ensuring the invariance of the induced typicality measure under Galilean symmetries.

In the upshot, the typicality measures in both classical mechanics and Bohmian mechanics are justified and tied to the dynamics by precise mathematical features: stationarity/equivariance under the dynamics and invariance/covariance under their fundamental symmetries. However, at least in classical mechanics, these conditions are not sufficient to determine the measure uniquely or even rule out evidently inadequate choices (like a delta-measure concentrated on a stationary microstate). At the end of the day, part of what makes a measure adequate and allows it to play a normative role as it figures in typicality reasoning is that the choice doesn't seem biased or ad hoc or overly contrived. In other words, I don't think that "soft" criteria can or should be completely avoided, and they aren't, in

---

[8] Proof:. $\epsilon_t^{-1}T\epsilon_t = \epsilon_s^{-1}\Phi_{s,t}T\Phi_{t,s}\epsilon_s = \epsilon_s^{-1}\Phi_{s,t}\Phi_{t,s}T\epsilon_s = \epsilon_s^{-1}T\epsilon_s$.

fact, in scientific practice. Attempts to axiomatize typicality measures (Werndl, 2013) seem misguided, not just because the particular proposals are uncompelling, but because it is hard to see why any set of formal axioms should be more compelling than the very measures we use in physics. Come to think of it: If we couldn't rely on a form of rationality to recognize reasonable typicality measures, what should we rely upon to recognize reasonable axioms for them?

7. Conclusion

Our discussion showed that the debate between the Mentaculus and the typicality account involves at least three separate issues:

1. Is the fundamental measure of statistical mechanics a bona fide probability measure or a typicality measure?
2. What is the epistemic and metaphysical status of this measure? Is it a theoretical/empirical postulate on a par with the dynamics, or does it formalize a way of reasoning about the laws?
3. What rationality principle grounds or expresses its epistemic and behavior-guiding function, the Principal Principle or Cournot's principle?

While I have defended a view that comes down on the opposite side of the Mentaculus on every one of these points, it is important to note that the answers to the three questions are logically independent (with the exception that a typicality measure essentially collapses PP into CP). This leaves room for compromise and moderation, but also for misunderstandings since not every author defending "typicality" or "Humean chances" may have the same package deal in mind.

It is possible to maintain, for instance, that Humean laws involve a typicality measure instead of a probability measure, that a probability measure expresses a way of reasoning (e.g., a principle of indifference) rather than an empirical postulate, or that the laws include a bona fide probability measure, over and above deterministic dynamics, whose empirical and epistemic import comes from Cournot's principle.

While there is always value in compromise and moderation, the extremal positions are often the most interesting ones. Here, one of them—the Mentaculus with David Albert as one of its most prominent proponents—is attractive because of its stringency, cohesiveness, and philosophical potential, but I have argued that it is both too ambitious and simplistic in its treatment of single-case probabilities and fails to capture the different (epistemological) status that dynamical postulates and probabilistic judgments have in physics. The opposite view—which I have defended as the typicality account—fares better as an account of good scientific reasoning and of how macroscopic regularities are grounded in microscopic laws, but much of its potential remains still untapped.

References

Albert, D. Z. (2000). *Time and Chance.* Harvard University Press.
Albert, D. Z. (2015). *After Physics.* Harvard University Press.
Barrett, J. A. (2016). Typicality in Pure Wave Mechanics. *Fluctuation and Noise Letters,* 15(3):1640009.
Bell, J. S. (2004). *Speakable and Unspeakable in Quantum Mechanics.* 2nd ed. Cambridge University Press.
Borel, E. (1948). *Le hasard.* Presses universitaires de France.
Bricmont, J. (1995). Science of Chaos or Chaos in Science? *Annals of the New York Academy of Sciences,* 775(1):131–175.
Bricmont, J. (2001). Bayes, Boltzmann and Bohm: Probabilities in Physics. In J. Bricmont et al. (eds.), *Chance in Physics: Foundations and Perspectives,* pp. 3–21. Springer.
Callender, C. (2007). The Emergence and Interpretation of Probability in Bohmian Mechanics. *Studies in History and Philosophy of Science Part B: Studies in History and Philosophy of Modern Physics,* 38(2):351–370.
Carroll, S. (2010). *From Eternity to Here.* Dutton.
Cournot, A. A. (1843). *Exposition de la théorie des chances et des probabilités.* Hachette.
Dürr, D., Froemel, A., and Kolb, M. (2017). *Einführung in Die Wahrscheinlichkeitstheorie als Theorie der Typizität.* Springer.
Dürr, D., Goldstein, S., and Zanghì, N. (1992). Quantum equilibrium and the origin of absolute uncertainty. *Journal of Statistical Physics,* 67(5–6):843–907.
Dürr, D., and Teufel, S. (2009). *Bohmian Mechanics: The Physics and Mathematics of Quantum Theory.* Springer.
Everett, H. (1973). The Theory of the Universal Wave Function. In DeWitt, B. S., and Graham, N. (eds.), *The Many-Worlds Interpretation of Quantum Mechanics,* pp. 3–140. Princeton University Press
Goldstein, S. (2001). Boltzmann's Approach to Statistical Mechanics. In J. Bricmont et al. (eds.), *Chance in Physics: Foundations and Perspectives,* pp. 39–54. Springer.
Goldstein, S. (2012). Typicality and Notions of Probability in Physics. In Ben-Menahem, Y., and Hemmo, M. (eds.), *Probability in Physics,* pp. 59–71. Springer.
Goldstein, S., and Struyve, W. (2007). On the Uniqueness of Quantum Equilibrium in Bohmian Mechanics. *Journal of Statistical Physics,* 128(5):1197–1209.
Hall, N. (1994). Correcting the Guide to Objective Chance. *Mind,* 103(412):505–518.

Hall, N. (2004). Two Mistakes about Credence and Chance. *Australasian Journal of Philosophy*, 82(1):93–111.

Hoefer, C. (2019). *Chance in the World: A Humean Guide to Objective Chance*. Oxford University Press.

Hubert, M. (2021). Reviving Frequentism. *Synthese*, 199(1):5255–5284.

Kolomogorov, A. (1933). *Grundbegriffe der Wahrscheinlichkeitsrechnung: Ergebnisse der Mathematik und ihrer Grenzgebiete*. Vol. 1. Springer.

Lazarovici, D., and Reichert, P. (2015). Typicality, Irreversibility and the Status of Macroscopic Laws. *Erkenntnis*, 80(4):689–716.

Lewis, D. (1980). A Subjectivist's Guide to Objective Chance. In Jeffrey, R. C. (ed.), *Studies in Inductive Logic and Probability*, vol. 2, pp. 263–293. University of California Press.

Lewis, D. (1994). Humean Supervenience Debugged. *Mind*, 103(412):473–490.

Loewer, B. (2001). Determinism and Chance. *Studies in History and Philosophy of Science Part B: Studies in History and Philosophy of Modern Physics*, 32(4):609–620.

Loewer, B. (2004). David Lewis's Humean Theory of Objective Chance. *Philosophy of Science*, 71(5):1115–1125.

Loewer, B. (2007). Counterfactuals and the Second Law. In Price, H., and Corry, R. (eds.), *Causation, Physics, and the Constitution of Reality: Russell's Republic Revisited*, pp. 293–326. Oxford University Press.

Loewer, B. (2012a). The Emergence of Time's Arrows and Special Science Laws from Physics. *Interface Focus*, 2(1):13–19.

Loewer, B. (2012b). Two Accounts of Laws and Time. *Philosophical Studies*, 160(1):115—137.

Martin, T. (1996). *Probabilités et critique philosophique selon Cournot*. Librairie Philosophique J. VRIN.

Maudlin, T. (2007). What Could Be Objective about Probabilities? *Studies in History and Philosophy of Science Part B: Studies in History and Philosophy of Modern Physics*, 38(2):275–291.

Myrvold, W. C. (2016). Probabilities in Statistical Mechanics. In Hitchcock, C., and Hájek, A. (eds.), *The Oxford Handbook of Probability and Philosophy*, pp. 573–600. Oxford University Press.

Oldofredi, A., Lazarovici, D., Deckert, D.-A., and Esfeld, M. (2016). From the Universe to Subsystems: Why Quantum Mechanics Appears More Stochastic than Classical Mechanics. *Fluctuation and Noise Letters*, 15(03):1640002.

Penrose, R. (1999). *The Emperor's New Mind: Concerning Computers, Minds, and the Laws of Physics*, new ed. Oxford University Press.

Quine, W. V. (1951). Main Trends in Recent Philosophy: Two Dogmas of Empiricism. *Philosophical Review*, 60(1):20–43.

Schwarz, W. (2014). Proving the Principal Principle. In Wilson, A. (ed.), *Chance and Temporal Asymmetry*, pp. 81–99. Oxford University Press.

Sellars, W. (1962). Philosophy and the Scientific Image of Man. In Colodny, R. (ed.), *Frontiers of Science and Philosophy,* pp. 35–78. University of Pittsburgh Press.

Shafer, G., and Vovk, V. (2006). The Sources of Kolmogorov's Grundbegriffe. *Statistical Science,* 21(1):70–98.

Volchan, S. B. (2007). Probability as Typicality. *Studies in History and Philosophy of Science Part B: Studies in History and Philosophy of Modern Physics,* 38(4):801–814.

Von Plato, J. (1994). *Creating Modern Probability: Its Mathematics, Physics and Philosophy in Historical Perspective.* Cambridge University Press.

Werndl, C. (2013). Justifying Typicality Measures of Boltzmannian Statistical Mechanics and Dynamical Systems. *Studies in History and Philosophy of Science Part B: Studies in History and Philosophy of Modern Physics,* 44(4):470–479.

Weslake, B. (2014). Statistical Mechanical Imperialism. In Wilson, A. (ed.), *Chance and Temporal Asymmetry,* pp. 241–257. Oxford University Press.

Wilhelm, I. (2019). Typical: A Theory of Typicality and Typicality Explanation. *British Journal for the Philosophy of Science.* https://doi.org/10.1093/bjps/axz016.

Chapter Seven

The Past Hypothesis and the Nature of Physical Laws

▶ EDDY KEMING CHEN

> Therefore I think it is necessary to add to the physical laws the hypothesis that in the past the universe was more ordered, in the technical sense, than it is today—I think this is the additional statement that is needed to make sense, and to make an understanding of the irreversibility.
>
> —RICHARD FEYNMAN (1964 MESSENGER LECTURES)

1. Introduction

One of the hardest problems in the foundations of physics is the problem of the arrows of time. If the dynamical laws are (essentially) time-symmetric, what explains the irreversible phenomena in our experiences, such as the melting of ice cubes, the decaying of apples, and the mixing of cream in coffee? Macroscopic systems display an entropy gradient in their temporal evolutions: their thermodynamic entropy is lower in the past and higher in the future. But why does entropy have this temporally asymmetric tendency? Following Goldstein (2001), let us distinguish between the two parts of the problem of irreversibility:

1. The Easy Part: If a system is not at maximum entropy, why should its entropy tend to be larger at a later time?
2. The Hard Part: Why should there be an arrow of time in our universe that is governed by fundamental reversible dynamical laws?

The Easy Part was studied by Boltzmann (1964 [1896]). Crucial to Boltzmann's answer is this:

- Key to the Easy Part: States of larger entropy occupy much larger volume in the system's phase space than states of lower entropy.

So far, answering the Easy Part does not require any time-asymmetric postulates.[1] Boltzmann's program is primarily focused on closed subsystems of the universe. But its success leads us to expect that a Boltzmannian account can work at the universal level. If we model the universe as a mechanical system, we expect that, typically, the non-equilibrium state of the universe will evolve toward higher entropy at later times.

However, why is the entropy lower in the past? That is the Hard Part. A proposed answer suggests that it has to do with the initial condition of the universe:

- Key to the Hard Part: The universe had a special beginning.

We can introduce this as an explicitly time-asymmetric postulate in the theory, by using the Past Hypothesis:

Past Hypothesis (PH): At the initial time of the universe, the microstate of the universe is in a low-entropy macrostate.[2]

Given that some microstates are anti-entropic, it is standard to introduce a probability distribution over the microstates compatible with the low-entropy macrostate:

Statistical Postulate (SP): The probability distribution of the initial microstate of the universe is given by the uniform one (according to the natural measure) that is supported on the macrostate of the universe.

---

[1] Boltzmann's *Stosszahlansatz* (hypothesis of molecular chaos) is often blamed for introducing an illicit time asymmetry. But it is an innocent theoretical postulate if we understand it correctly—as a typicality or probability measure over initial conditions. See Goldstein et al. (2020), §5.5.

[2] The Past Hypothesis was originally suggested in Boltzmann (1964 [1896]) (although he seems to favor another postulate that can be called the *Fluctuation Hypothesis*) and discussed in Feynman (2017 [1965]). For recent discussions, see Albert (2000), Goldstein (2001), Callender (2004, 2011), Lebowitz (2008), North (2011), Loewer (2020), and Goldstein et al. (2020). The memorable phrase "Past Hypothesis" was coined by Albert (2000).

However, a detailed probability distribution may be unnecessary. In the typicality framework, we just need to be committed to a typicality measure:

Typicality Postulate (TP): The initial microstate of the universe is typical inside the macrostate of the universe.[3]

Unlike SP, TP is compatible with a variety of measures that agree on what is typical. PH, SP, and TP are physical postulates that have an empirical status. The answer to the Hard Part of the problem of irreversibility requires PH and either SP or TP. In fact, we also need to assume (an unconditionalized) notion of probability or typicality to answer the Easy Part. I call the answers to the Easy Part and the Hard Part the *Boltzmannian account* of the arrow of time.

How to characterize the initial macrostate of PH remains an open question. We know that the matter distribution is more or less uniform in the early universe, which is contrary to the usual conception of low entropy. However, the initial gravitational degrees of freedom are in a special state, providing a sense that the total state of the early universe has low entropy. This observation led Penrose (1979) to postulate a geometric version of PH called the *Weyl Curvature Hypothesis*: the Weyl curvature vanishes at the initial singularity. The urgent question is, of course, how to understand this in terms of quantum theory or quantum gravity. Some steps have been taken in the Loop Quantum Cosmology framework by Ashtekar and Gupt (2016). This is compatible with a Boltzmannian account, but the final details will depend on the exact theory of quantum gravity, which is currently absent.

In the philosophical literature, a number of objections have been raised against the Boltzmannian account. First, some criticize the answer to the Easy Part: the explanation is too hand-wavy and not completely rigorous (Frigg, 2007). Second, some criticize the answer to the Hard Part, claiming that PH is not even false because the entropy of the early universe is not well-defined (Earman, 2006), or that PH is not sufficient to explain the thermodynamic behaviors of subsystems (Winsberg, 2004), or that it is ad hoc, and therefore not explanatory (Price, 2004; Carroll 2010, p. 346). These are responses in the literature, and more work on these issues is certainly welcome. However, my interest here is different. I take the Boltzmannian account as a starting point. My aim is to explore the

---

[3] For more on the notion of typicality and its application in statistical mechanics, see Goldstein (2012); Lazarovici and Reichert (2015); and Wilhelm (2019).

conceptual and scientific ramifications of accepting PH and its explanation of time's arrow.

In this paper, I focus on the connection between PH and our concept of fundamental laws of nature. What is the status of PH if the Boltzmannian program turns out to be successful? Can PH be accepted as a candidate fundamental law even though it is a boundary condition of the universe? What differences does it make to our concepts of laws, chances, and possibilities? Can PH be completely expressed in mathematical language? What is the relevance of quantum theory to these issues? I argue for the following theses:

Nomic Status: The Past Hypothesis is a candidate fundamental law of nature.[4]

Axiomatic Status: The Past Hypothesis is a candidate axiom of the fundamental physical theory.[5]

Relevance: Whether the Past Hypothesis has nomic status (and / or axiomatic status) is relevant to the success of explaining time's arrows, the metaphysical account of laws, the nature of objective probability, and the mathematical expressibility of fundamental physical theory.

Some of these ideas have been defended along Humean lines, but I think we should accept them regardless of whether we think fundamental laws supervene on matter distribution or are part of the fundamental facts of the world. It turns out they are acceptable even on certain non-Humean frameworks.[6] My methodology below is naturalistic and functionalist. *Whatever plays the role of a fundamental law can be a fundamental law.* Whatever cannot be derived from more fundamental laws *and* plays the right roles in guiding our inferences about the past and the future, underlying various scientific explanations, high-level regularities, our manifest image of influence and control, and so on, is a candidate fundamental

---

[4] A *candidate* fundamental law of nature has all it takes to be a fundamental law of nature, but it may not turn out to be the true law of the actual world if it makes false predictions. For example, Newton's dynamical law $F=ma$ is a candidate fundamental law of nature, but it is not the actual fundamental law.

[5] The status of a candidate fundamental law of nature and the status of a candidate axiom of the fundamental physical theory may be equivalent. I distinguish the two theses because some people may be happy to accept one but deny the other. They may be reluctant to call PH a fundamental *law,* perhaps due to its being a boundary condition.

[6] This view is, I think, in the same spirit as the suggestion made by Demarest (2019).

law.[7] To borrow a phrase from Loewer (2020), PH "looks, walks and talks" like a fundamental law. So, we should interpret it as such. Hence, I disagree with people who think that even if PH is true and plays all the roles we suggest, it still cannot be a fundamental law—it may just be a special but *contingent* initial condition. I also disagree with people who think that whether or not PH is a fundamental law makes no substantive difference.[8]

Here is the roadmap. In §2, I provide more details of the Boltzmannian account and discuss the modifications to PH when we move from classical mechanics to quantum mechanics. The variations result in three types of physical theories: the classical Mentaculus, the quantum Mentaculus, and the Wentaculus theories. In §3, I provide positive arguments for the nomic and axiomatic status of PH. Some of these have been mentioned in the literature, but it is worth emphasizing and clarifying the exact argumentative structure. I also put forward a novel argument based on considerations about the nature of the quantum state. In §4, I discuss some apparent obstacles from recognizing PH as a fundamental law. This has to do with its nature as a boundary condition, the status of the Statistical Postulate and the Typicality Postulate, and the intrinsic vagueness in their specifications. I argue that these worries do not have much force, and they become even less worrisome in the Wentaculus theory.

2. Variations on a Theme from Boltzmann

Before we get into the philosophical and conceptual issues, let us be more explicit about what the Boltzmannian account is and how to state PH in that account. Although the Boltzmannian account is more or less the same in classical and in quantum theories, the exact form of PH is subtly different. I will exploit this difference in §4 to dissolve some of the worries about the classical version of PH. Readers familiar with the standard Boltzmannian statistical mechanics can jump to §2.3, where a new framework called the *Wentaculus* is introduced.

---

[7] I do not claim that we should reduce laws to these roles; that would be the strategy of metaphysical functionalism about laws. Rather, I am merely appealing to the methodology in naturalistic metaphysics of science that I think many people accept independently of the issue of the arrows of time.

[8] Maudlin (2007, §4) seems to regard PH as an important boundary condition but does not think of it as a fundamental law. The disagreement is based on a different view about how laws govern, which I call Dynamical Law Primitivism, which I discuss in §3.4. Carroll (2010, p. 345) suggests that there is no substantive difference between the statements "the early universe had a low entropy" and "it is a law of physics that the early universe had a low entropy." Carroll seems to be worried about the distinction between boundary conditions and laws; I discuss this in §4.1.

2.1. The Classical Mentaculus

Let us start with the Boltzmannian account in classical statistical mechanics and summarize its basic elements from the "individualistic viewpoint."[9] Let us consider a classical-mechanical system with N particles in a box of volume $\Lambda = [0, L]^3 \subset \mathbb{R}^3$ and a Hamiltonian $H = H(X) = H(\boldsymbol{q}_1, \ldots, \boldsymbol{q}_N; \boldsymbol{p}_1, \ldots, \boldsymbol{p}_N)$ that specifies the standard interactions in accord with Newtonian gravitation, Coulomb's law, and other forces obeyed by the classical system.

1. Microstate: At any time t, the microstate of the system is given by a point in a $6N$-dimensional phase space,

$$X = (\boldsymbol{q}_1, \ldots, \boldsymbol{q}_N; \boldsymbol{p}_1, \ldots, \boldsymbol{p}_N) \in \Gamma_{total} \subseteq \mathbb{R}^{6N}, \tag{1}$$

   where Γ_{total} is the total phase space of the system.

2. Dynamics: The time dependence of $X_t = (\boldsymbol{q}_1(t), \ldots, \boldsymbol{q}_N(t); \boldsymbol{p}_1(t), \ldots, \boldsymbol{p}_N(t))$ is given by the Hamiltonian equations of motion:

$$\frac{d\boldsymbol{q}_i(t)}{dt} = \frac{\partial H}{\partial \boldsymbol{p}_i}, \quad \frac{d\boldsymbol{p}_i(t)}{dt} = -\frac{\partial H}{\partial \boldsymbol{q}_i}. \tag{2}$$

3. Energy shell: The physically relevant part of the total phase space is the energy shell $\Gamma \subseteq \Gamma_{total}$ defined as:

$$\Gamma = \{X \in \Gamma_{total} : E \leq H(x) \leq E + \delta E\}. \tag{3}$$

   We only consider microstates in Γ.

4. Measure: The measure μ_V is the Lebesgue (volume) measure on $\mathbb{R}^{6N}$. The Lebesgue measure on a finite volume can be normalized to yield a probability distribution.

5. Macrostate: With a choice of macro-variables, the energy shell Γ can be partitioned into macrostates Γ_v:

$$\Gamma = \bigcup_v \Gamma_v. \tag{4}$$

   A macrostate is composed of microstates that share similar macroscopic features (i.e., similar values of the macro-variables), such as volume, density, and pressure.

---

[9] I follow the discussion in Goldstein and Tumulka (2011). These do not intend to be rigorous axiomatizations of classical statistical mechanics. What is presented below differs in emphasis from Chen (2018), as here I am explicit about the sources of vagueness, which will be discussed in §4.

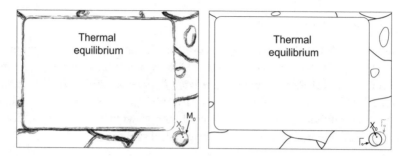

FIGURE 7.1. The partition of microstates into macrostates on phase space: (left) without exact choices of the C-parameters, (right) with exact choices of the C-parameters. X_0 represents the actual microstate of the universe at t_0. M_0 represents the vague boundaries of the PH macrostate. Γ_0 represents an admissible precisification of M_0, where Γ_0' represents another admissible precisification. The diagrams are not drawn to scale. Credit: © Eddy Keming Chen

Caveat: The partition of microstates into macrostates is exact only after we stipulate some choices of the parameters for coarse-graining (the size of the cells) and correspondence (between functions on phase space and thermodynamic quantities). We call these *C-parameters*. Without exact choices of the C-parameters, the partition is inexact and the boundaries between macrostates are vague.[10] See Figure 7.1. It is also expected that given the nature of the actual forces, some partitions will be superior to others in supporting generalizations in the special sciences.

6. Unique correspondence: Given exact choices of the C-parameters, the macrostates partition the energy shell, and as a consequence, every phase point X belongs to one and only one Γ_v. (This point is implied by #5. But I make it explicit to better contrast it with the situation in quantum statistical mechanics.)
7. Thermal equilibrium: Typically, there is a dominant macrostate Γ_{eq} that has almost the entire volume with respect to μ_V:

$$\frac{\mu_V(\Gamma_{eq})}{\mu_V(\Gamma)} \approx 1. \tag{5}$$

A system is in thermal equilibrium if its phase point $X \in \Gamma_{eq}$.

---

[10] For more discussions, see Chen (2022).

8. Boltzmann entropy: The Boltzmann entropy of a classical-mechanical system in microstate X is given by:

$$S_B(X) = k_B \log(\mu_V(\Gamma(X))), \qquad (6)$$

where $\Gamma(X)$ denotes the macrostate containing X. The thermal equilibrium state thus has maximum entropy.

Caveat: Without exact values of the C-parameters, there will be many admissible choices of the $\Gamma(X)$'s. Moreover, what is admissible is also vague. Since k_B is a scaling constant that plays no direct dynamical role, its value is also vague. Hence, the Boltzmann entropy of a microstate should be understood as a vague quantity. If we stipulate some C-parameters and the value of k_B, we can arrive at an exact boundary for the macrostate that contains X and an exact value of Boltzmann entropy for the system.

9. Low-entropy initial condition: On the assumption that we can model the universe as a classical-mechanical system of N point particles, we postulate a special low-entropy boundary condition, which Albert (2000) calls *the Past Hypothesis* (PH):

$$X_{t_0} \in \Gamma_{PH}, \mu_V(\Gamma_{PH}) \ll \mu_V(\Gamma_{eq}) \approx \mu_V(\Gamma) \qquad (7)$$

where Γ_{PH} is the PH macrostate with a volume much smaller than that of the equilibrium macrostate. Hence, $S_B(X_{t_0})$, the Boltzmann entropy of the microstate at the boundary, is very small compared to that of thermal equilibrium. Here, Γ_{PH} is underspecified; we can add further details to specify the macroscopic profile (temperature, pressure, volume, density) of Γ_{PH}.

The answer to the Easy Part of the problem of irreversibility lies in the first eight items in the numbered list, which make plausible the hypothesis of the typical tendency for a system to evolve to higher entropy toward the future. Even though microstates are "created equal," macrostates are not. Their volumes are disproportionate and uneven. Macrostates with higher entropy have much larger volume in the energy shell. Furthermore, the largest macrostate is by far that of thermal equilibrium. It is plausible that, unless the dynamics are extremely contrived, a typical microstate starting from a medium-entropy macrostate will find its way through larger and larger macrostates and eventually arrive at thermal equilibrium. That is a process in which a system's entropy gradually increases until it reaches the entropy maximum.[11] Of course, for the actual universe, there can be exceptions to the entropy increase, such as short-lived fluctuations in which entropy decreases.

However, this solves only half the problem. If typical microstates compatible with a medium-entropy macrostate will, at most times, increase in entropy toward the future, then typical microstates compatible with the same macrostate will also, at most times, increase in entropy toward the past. Hence, given the resources so far, we have shown that the medium-entropy macrostate is overwhelmingly likely at an entropy minimum produced by a thermodynamic fluctuation from equilibrium. We are led to the Hard Part of the problem: Why is the entropy so much lower in the past direction of time? Enters the Past Hypothesis. Given PH, the actual microstate starts in an atypical region of the energy shell, in a low-entropy macrostate M_0. Suppose we choose a precisification Γ_0. Given the Easy Part, typical initial microstates compatible with Γ_0 will evolve toward macrostates of higher entropy in the future direction. But there is nothing earlier than t_0, as it is stipulated to be the initial time—say, the time of the Big Bang.[12]

Therefore, if we find the universe to be in a medium-entropy macrostate Γ_t, say the state we are in right now, then the actual microstate is not like a typical microstate inside Γ_t, but a special one that is compatible with Γ_0. The reason the entropy was lower in the past is because the universe started in a special macrostate, a state of very low Boltzmann entropy. Assuming PH, it is reasonable to expect *with overwhelming probability* that entropy will be higher in the future and was lower in the past, and the sense of probability is specified by list item 4. There are two ways to understand the measure:

- A measure of probability: The natural measure picks out the correct probability measure of the initial condition. This interpretation yields the Statistical Postulate.
- A measure of typicality: The natural measure is a simple representer of a vague "collection" of measures that are equivalent as the measure

---

[11] Boltzmann's original H-theorem (Boltzmann, 1964 [1896]) is an attempt to show this. Lanford (1975) produces an exact result for a simple system of hard spheres where the Boltzmann equation is shown to be satisfied for a short duration of time, and hence, Boltzmann entropy is shown to be increasing toward the future. However, it is plausible that the equation continues to be valid and Boltzmann entropy continues to rise afterward.

[12] It can also be stipulated that t_0 is some time close to the Big Bang, in which case some antithermodynamic behavior can be displayed in the short duration before t_0.

of typicality of the initial condition. This interpretation yields the Typicality Postulate.

PH together with the Statistical Postulate supports the following classical-mechanical version of the Second Law of Thermodynamics (this is adapted from the Mathematical Second Law described in Goldstein et al., 2020, §5.2):

> **The Second Law for** X: At t_0, the actual phase point of the universe X_0 starts in a low-entropy macrostate and, with overwhelming probability, it evolves toward macrostates of increasingly higher entropy until it reaches thermal equilibrium, except possibly for entropy decreases that are infrequent, shallow, and short-lived; once X_t reaches Γ_{eq}, it stays there for an extraordinarily long time, except possibly for infrequent, shallow, and short-lived entropy decreases.

The Second Law can be stated also in the language of typicality. For simplicity, I will conduct the discussion below mostly in the language of probability.

The Second Law above is stated for the behavior of the universe, but it also makes plausible what Goldstein et al. (2020) call a "development conjecture" about isolated subsystems in the universe:

> **Development Conjecture:** Given PH, an isolated system that, at a time t before thermal equilibrium of the universe, has macrostate v appears macroscopically in the future, but not in the past, of t like a system that at time t is in a typical microstate compatible with v.

Classical mechanics with just the fundamental dynamical laws (expressed in equation (2)) is time-symmetric. Introducing the probability measure takes care of the Easy Part of the problem of irreversibility, but to solve the Hard Part of the problem—the retrodiction to the past—we need to explicitly introduce something that breaks the time symmetry. PH is a simple postulate that does the job. The number-list items about energy shell, macrostate partition, unique correspondence, and the dominance of thermal equilibrium are supposed to follow from the basic postulates about fundamental dynamics (including the structure of the Hamiltonian function) and the probability measure. Adapting the terminology of Albert (2015) and Loewer (2020), we call the collection of basic postulates the *classical Mentaculus*.

> **THE CLASSICAL MENTACULUS**
>
> 1. **Fundamental Dynamical Laws (FDL)**: The classical microstate of the universe is represented by a point in phase space that obeys the Hamiltonian equations of motion described in equations (2).
> 2. **The Past Hypothesis (PH)**: At a temporal boundary of the universe, the microstate of the universe lies inside M_0, a low-entropy macrostate that, given a choice of C-parameters, corresponds to Γ_0, a small-volume set of points on phase space that are macroscopically similar.
> 3. **The Statistical Postulate (SP)**: Given the macrostate M_0, we postulate a uniform probability distribution (with respect to the standard measure) over the microstates compatible with M_0.

This account therefore assigns probability 1 to the initial macrostate. If the probability distribution is given a status of objective probability, it delivers more than just the Second Law. It provides an exact probability for any proposition formulable in the language of phase space. This is the reason Albert and Loewer regard the Mentaculus as providing a "probability map of the world." Hence, the Mentaculus has an ambitious scope: it is possible to recover all the non-fundamental regularities, including the special science laws (such as laws of economics), and other arrows of time such as the epistemic arrow, the records arrow, the influence arrow, and the counterfactual arrow.

Whether the Albert-Loewer project can succeed in their ambitious goal of recovering all the non-fundamental regularities and arrows of time is an interesting question. Nonetheless, the classical Mentaculus as formulated provides an underpinning for the thermodynamic arrow of time. Given the universality and importance of the Second Law, the Mentaculus should be taken as a serious contender for a promising framework of the structure of a fundamental physical theory. In §2.2, I examine how to adapt the classical Mentaculus to the quantum domain. In §3, I discuss the suggestion that PH should be taken as a candidate fundamental law and some ramifications of the more ambitious project.

2.2. The Quantum Mentaculus

Let us turn to the Boltzmannian account of quantum statistical mechanics from the "individualist viewpoint." Consider a quantum-mechanical system with N fermions (with $N > 10^{20}$) in a box $\Lambda = [0, L]^3 \subset \mathbb{R}^3$ and a Hamiltonian $\hat{H}$. (Here, I follow the discussions in Goldstein et al., 2010a, and Goldstein & Tumulka, 2011.)

1. Microstate: At any time t, the microstate of the system is given by a normalized (and anti-symmetrized) wave function:

$$\psi(q_1, \ldots, q_N) \in \mathcal{H}_{total} = L^2(\Lambda^N, \mathbb{C}^k), |\psi|_{L^2} = 1 \tag{8}$$

where $\mathcal{H}_{total} = L^2(\Lambda^N, \mathbb{C}^k)$ is the total Hilbert space of the system.

2. Dynamics: The time dependence of $\psi(q_1, \ldots, q_N; t)$ is given by the Schrödinger equation:

$$i\hbar \frac{\partial \psi}{\partial t} = \hat{H}\psi. \tag{9}$$

3. Energy shell: The physically relevant part of the total Hilbert space is the subspace ("the energy shell"):

$$\mathcal{H} \subseteq \mathcal{H}_{total}, \mathcal{H} = span\{\phi_\alpha : E_\alpha \in [E, E + \delta E]\}. \tag{10}$$

This is the subspace (of the total Hilbert space) spanned by energy eigenstates ϕ_α whose eigenvalues E_α belong to the $[E, E + \delta E]$ range. Let $D = dim\mathcal{H}$, the number of energy levels between E and $E + \delta E$. We only consider wave functions ψ in $\mathcal{H}$.

4. Measure: Given a subspace $\mathcal{H}$, the measure μ_S is the surface area measure on the unit sphere in that subspace $S(\mathcal{H})$.[13]

5. Macrostate: With a choice of macro-variables,[14] the energy shell $\mathcal{H}$ can be orthogonally decomposed into macro-spaces (subspaces):

$$\mathcal{H} = \oplus_v \mathcal{H}_v, \sum_v dim\mathcal{H}_v = D \tag{11}$$

Each $\mathcal{H}_v$ corresponds to small ranges of values of macro-variables that are chosen in advance.

---

[13] For simplicity, let us assume that the subspaces we deal with are finite-dimensional. In cases where the Hilbert space is infinite-dimensional, it is an open and challenging technical question. For example, we could use Gaussian measures in infinite-dimensional spaces, but we no longer have uniform probability distributions.

[14] For technical reasons, von Neumann (1955) suggests that we round up these macro-variables (represented by quantum observables) so as to make the observables commute. See Goldstein et al. (2010b, §2.2) for a discussion of von Neumann's ideas.

Caveat: similarly to the classical case, the decomposition of Hilbert space into macrostates requires some stipulation of the exact values of the C-parameters. But in the quantum case, these parameters includes coarse-graining sizes, correspondences of functions, and also the cutoff values of how much support a quantum state needs to be inside a subspace to be counted toward belonging to the macrostate (see the next bullet point). Without the exact choices of the C-parameters, the decomposition is inexact and it is vague which microstate belongs to which macrostate. Again, it is also expected that given the nature of the actual forces, some decompositions will be superior to others for supporting generalizations in the special sciences.

6. Non-unique correspondence: Typically, a wave function is in a superposition of macrostates and is not entirely in any one of the macrostates (even if we represent macrostates with exact subspaces). However, we can make sense of situations where ψ is (in the Hilbert space norm) very close to a macrostate $\mathcal{H}_v$:

$$\langle \psi | P_v | \psi \rangle \approx 1, \tag{12}$$

where P_v is the projection operator onto $\mathcal{H}_v$. This means that $|\psi\rangle$ lies almost entirely in $\mathcal{H}_v$. In this case, we say that $|\psi\rangle$ is in macrostate v.

7. Thermal equilibrium: Typically, there is a dominant macrostate $\mathcal{H}_{eq}$ that has a dimension that is almost equal to D:

$$\frac{dim\mathcal{H}_{eq}}{dim\mathcal{H}} \approx 1. \tag{13}$$

A system with wave function ψ is in equilibrium if the wave function ψ is very close to $\mathcal{H}_{eq}$ in the sense of (12): $\langle \psi | P_{eq} | \psi \rangle \approx 1$.

8. Boltzmann entropy: The Boltzmann entropy of a quantum-mechanical system with wave function ψ that is in macrostate v is given by:

$$S_B(\psi) = k_B \log(dim\mathcal{H}_v) \tag{14}$$

where $\mathcal{H}_v$ denotes the subspace containing almost all of ψ in the sense of (12). The thermal equilibrium state thus has the maximum entropy:

$$S_B(eq) = k_B \log(dim\mathcal{H}_{eq}) \approx k_B \log(D) \tag{15}$$

where eq denotes the equilibrium macrostate.

9. Low-entropy initial condition: On the assumption that we can model the universe as a quantum-mechanical system, let us postulate a special low-entropy boundary condition on the universal wave function—the quantum-mechanical version of PH:

$$\Psi(t_0) \in \mathcal{H}_{PH}, \, dim\mathcal{H}_{PH} \ll dim\mathcal{H}_{eq} \approx dim\mathcal{H} \qquad (16)$$

where $\mathcal{H}_{PH}$ is the PH macro-space with dimension much smaller than that of the equilibrium macro-space.[15] Hence, the initial state has very low entropy in the sense of (25). More details can be added to narrow down the range of choices of $\mathcal{H}_{PH}$.

The quantum Boltzmannian account is similar to the classical one. The higher-entropy macrostates have much higher dimensions than lower-entropy ones, and the equilibrium macrostate has by far the largest dimension. It is plausible that, unless the dynamics is very contrived, a medium-entropy wave function will find its way through larger and larger subspaces and eventually arrive at the equilibrium subspace. Again, if typical wave functions in non-equilibrium macrostates evolve toward higher entropy in the future, then typical ones also come from higher entropy states in the past. The quantum mechanical Past Hypothesis blocks that inference. The reason there is a thermodynamic arrow in a quantum universe is because the universal wave function started in a special state, a subspace with very low entropy (the one described by the quantum PH). Assuming the quantum PH, it is reasonable to expect that *with overwhelming probability* the entropy is higher in the future and lower in the past, with the probability measure specified in list item 4. Again, we can understand it as a measure of probability or a measure of typicality.

Together, the probability distribution and PH support the quantum-mechanical version of the Second Law:

> **The Second Law for Ψ:** At t_0, the actual wave function of the universe Ψ_0 starts in a low-entropy macrostate, and with overwhelming probability, it evolves toward macrostates of increasingly higher entropy until it reaches thermal equilibrium, except possibly for entropy decreases that are infrequent, shallow, and short-lived; once Ψ_t reaches $\mathcal{H}_{eq}$, it stays there for an extraordinarily long time, except possibly for infrequent, shallow, and short-lived entropy decreases.

---

[15] Again, we assume that $\mathcal{H}_{PH}$ is finite-dimensional, in which case we can use the surface area measure on the unit sphere as the typicality measure for #10. In QSM it remains an open question how to formulate the low-entropy initial condition when the initial macro-space is infinite-dimensional.

(This makes plausible a similar Development Conjecture for typical isolated subsystems.)

Note again we stipulate three basic postulates in the quantum version of the Boltzmannian account: the fundamental dynamical laws, PH, and SP. Let us call this the *quantum Mentaculus*.

THE QUANTUM MENTACULUS

1. **Fundamental Dynamical Laws (FDL):** The quantum microstate of the universe is represented by a wave function Ψ that obeys the Schrödinger equation (9).
2. **The Past Hypothesis (PH):** At a temporal boundary of the universe, the wave function Ψ_0 of the universe lies inside a low-entropy macrostate that, given a choice of C-parameters, corresponds to $\mathcal{H}_{PH}$, a low-dimensional subspace of the total Hilbert space.
3. **The Statistical Postulate (SP):** Given the subspace $\mathcal{H}_{PH}$, we postulate a uniform probability distribution (with respect to the surface area measure on the unit sphere of $\mathcal{H}_{PH}$) over the wave functions compatible with $\mathcal{H}_{PH}$.

The quantum Mentaculus, as a candidate fundamental theory of physics, faces the quantum measurement problem. To solve the measurement problem, there are three promising options: Everettian quantum mechanics, Bohmian mechanics, and GRW spontaneous collapse theories. We have three distinct kinds of the quantum Mentaculus.

First, the Everettian version is completely the same as the original quantum Mentaculus in terms of the basic postulates. However, it diverges greatly from common sense: we have to give up the expectation that experimental outcomes are unique and determinate. Instead, our experiences are to be understood as experiences of agents in an emergent multiverse (Wallace, 2012).

Second, the Bohmian version posits that in addition to the wave function, which evolves unitarily according to the Schrödinger equation, particles have precise locations, and their configuration $Q = (Q_1, Q_2, \ldots, Q_N)$ follows the guidance equation, which is an additional law in the theory:

$$\frac{dQ_i}{dt} = \frac{\hbar}{m_i} \operatorname{Im} \frac{\nabla_i \Psi(q)}{\Psi(q)} (q = Q) \qquad (17)$$

Moreover, the initial particle distribution is given by the quantum equilibrium distribution:

$$\rho_{t_0}(q) = |\Psi(q,t_0)|^2 \tag{18}$$

Adding the above two postulates to the quantum Mentaculus completes the Bohmian Mentaculus.

Third, the GRW version requires revisions to the linear evolution represented by the Schrödinger equation. The wave function typically obeys the Schrödinger equation, but the linear evolution is interrupted randomly (with rate $N\lambda$, where N is the number of particles and λ is a new constant of nature of order 10^{-15} s^{-1}) by collapses:

$$\Psi_{T+} = \frac{\Lambda_k(X)^{1/2}\Psi_{T-}}{\|\Lambda_k(X)^{1/2}\Psi_{T-}\|}, \tag{19}$$

where Ψ_{T-} is the pre-collapse wave function, Ψ_{T+} is the post-collapse wave function, the collapse center X is chosen randomly with probability distribution $\rho(x) = \|\Lambda_k(x)^{1/2}\Psi_{T-}\|^2 dx$, $k \in \{1, 2, \ldots N\}$ is chosen randomly with uniform distribution on that set of particle labels, and the collapse rate operator is defined as:

$$\Lambda_k(x) = \frac{1}{(2\pi\sigma^2)^{3/2}} e^{-\frac{(Q_k - x)^2}{2\sigma^2}}, \tag{20}$$

where Q_k is the position operator of "particle" k, and σ is another new constant of nature of order 10^{-7} m postulated in current GRW theories. The GRW Mentaculus replaces the deterministic Schrödinger evolution of the wave function by this stochastic process. It still requires PH. However, as Albert (2000, §7) points out, it is plausible (though not proven) that SP is no longer needed, and the GRW collapses suffice to make anti-entropic trajectories unlikely (through the quantum probabilities stipulated by the GRW stochastic process).

2.3. The Wentaculus

In this subsection, let us consider the Boltzmannian account of quantum statistical mechanics with a very special "fundamental density matrix." This account is inspired by Dürr et al. (2005), proposed in Chen (2018), and discussed at length in Chen (2019a, 2020b). It is another variation on the same theme from Boltzmann, but it suggests some astonishing possibilities, one of which is the Initial Projection Hypothesis, which will be introduced shortly.

The density matrix can play the same *dynamical role* as the wave function does in the previous theories. In a quantum system represented by a density matrix W, W is the complete characterization of the quantum state; it does not necessarily refer to a statistical state representing our ignorance of the underlying wave function. In general, W can be a pure state or a mixed state. A density matrix $\hat{W}$ is pure if $\hat{W} = |\psi\rangle\langle\psi|$ for some $|\psi\rangle$. Otherwise it is mixed. For a spinless N-particle quantum system, a density matrix of the system is a positive, bounded, self-adjoint operator $\hat{W}: \mathcal{H} \to \mathcal{H}$ with $\text{tr}\hat{W} = 1$, where $\mathcal{H}$ is the Hilbert space of the system. In terms of the configuration space $\mathbb{R}^{3N}$, the density matrix can be viewed as a function $W: \mathbb{R}^{3N} \times \mathbb{R}^{3N} \to \mathbb{C}$. In the unitary case, $\hat{W}$ always evolves deterministically according to the von Neumann equation:

$$i\hbar \frac{d\hat{W}(t)}{dt} = [\hat{H}, \hat{W}]. \tag{21}$$

Equivalently:

$$i\hbar \frac{\partial W(q, q', t)}{\partial t} = \hat{H}_q W(q, q', t) - \hat{H}_{q'} W(q, q', t), \tag{22}$$

where $\hat{H}_q$ means that the Hamiltonian $\hat{H}$ acts on the variable q. The von Neumann equation generalizes the Schrödinger equation (9).

Importantly, now the "fundamental" quantum state can be either pure or mixed. Even when it is mixed, there is no underlying pure state that is more basic. The mixed state is completely objective. This perspective, called *Density Matrix Realism*, is in sharp contrast to the prevalent view called *Wave Function Realism*. In the density-matrix realist framework, we need to modify the Boltzmannian quantum statistical mechanics described in the earlier section. Here are the key changes:

- Microstate: At any time t, the microstate of the system is given by a density matrix $\hat{W}(t)$ that can be pure or mixed. (Macrostates are still represented by orthogonal subspaces of the energy shell.)
- Dynamics: In the unitary case, the density matrix $\hat{W}(t)$ evolves according to the von Neumann equation (21).
- Being in a macrostate: Typically, a density matrix is a superposition of macrostates and is not entirely in any one of the macro-spaces. However, we can make sense of situations where $\hat{W}(t)$ is very close to a macrostate $\mathcal{H}_\nu$:

$$\text{tr}(\hat{W}(t) I_\nu) \approx 1, \tag{23}$$

where I_ν is the projection operator onto $\mathcal{H}_\nu$. This means that almost all of $\hat{W}(t)$ is in $\mathcal{H}_\nu$. In this situation, we say that $\hat{W}(t)$ is in macrostate $\mathcal{H}_\nu$.

- Thermal equilibrium: Typically, there is a dominant macro-space $\mathcal{H}_{eq}$ that has a dimension that is almost equal to D:

$$\frac{dim\mathcal{H}_{eq}}{dim\mathcal{H}} \approx 1 \quad (24)$$

A system with density matrix $\hat{W}(t)$ is in equilibrium if $\hat{W}(t)$ is very close to $\mathcal{H}_{eq}$ in the sense of (23): $(tr\hat{W}(t)I_{eq}) \approx 1$.

- Boltzmann entropy: The Boltzmann entropy of a quantum-mechanical system with density matrix $\hat{W}(t)$ that is very close to a macrostate ν is given by:

$$S_B(\hat{W}(t)) = k_B \log(dim\mathcal{H}_\nu) \quad (25)$$

for which W is in macrostate $\mathcal{H}_\nu$ in the sense of (23).

Next, let us consider how to adapt PH in a density-matrix realist framework. The wave-function version of PH says that every initial wave function is entirely contained in the PH subspace $\mathcal{H}_{PH}$. Similarly, for density-matrix theories, we can propose that every initial density matrix is entirely contained in the PH subspace:

$$tr(\hat{W}(t_0)I_{PH}) = 1, \, dim\mathcal{H}_{PH} \ll dim\mathcal{H}_{eq} \approx dim\mathcal{H} \quad (26)$$

where I_{PH} is the projection operator onto the PH subspace. Assuming $\mathcal{H}_{PH}$ is finite-dimensional, there is also a natural probability distribution over all density matrices inside this subspace. See Chen & Tumulka (2022) for a mathematical characterization. The probability distribution and the density-matrix Past Hypothesis support a Second Law for W, which is similar to the Second Law for Ψ.

However, there is an even more natural way to implement the idea of PH in the density-matrix framework, which I favor. PH picks out a particular subspace $\mathcal{H}_{PH}$. It is canonically associated with its projection I_{PH}. In matrix form, it can be represented as a block-diagonal matrix that has a $k \times k$ identity block, with $k = dim\mathcal{H}_{PH}$, and zero everywhere else. There is a natural density matrix associated with I_{PH}, namely the normalized projection $\frac{I_{PH}}{dim\mathcal{H}_{PH}}$. Hence, we have picked out the natural density matrix associated with the PH subspace. I propose that the initial density matrix is the normalized projection onto $\mathcal{H}_{PH}$:

$$\hat{W}_{IPH}(t_0) = \frac{I_{PH}}{\dim \mathcal{H}_{PH}} \qquad (27)$$

I call this postulate the *Initial Projection Hypothesis* (IPH) in Chen (2018). Crucially, it is different from (16) and (26); while IPH picks out a unique quantum state given PH, the other two permit infinitely many possible quantum states inside the PH subspace. Remarkably, we no longer need a fundamental postulate about probability or typicality for the quantum state. We know that we can decompose a density matrix *non-uniquely* into a probability-weighted average of pure states, and in the canonical way we can decompose $\hat{W}_{IPH}(t_0)$ as an integral of pure states on the unit sphere of $\mathcal{H}_{PH}$ with respect to the uniform probability distribution:

$$\hat{W}_{IPH}(t_0) = \int_{S(\mathcal{H}_{PH})} \mu(d\psi) |\psi\rangle\langle\psi| \qquad (28)$$

The decomposition here is not an intrinsic expression of what $\hat{W}_{IPH}(t_0)$ is, as witnessed by the non-uniqueness. But the expression is something that can nonetheless be used fruitfully in statistical analysis (Chen, 2020b, §3.2.3).

By doing away with the need for an extra postulate about initial quantum states, we only need two basic postulates. I call the theory *the Wentaculus*.

THE WENTACULUS

1. **Fundamental Dynamical Laws (FDL):** The quantum state of the universe is represented by a density matrix $\hat{W}(t)$ that obeys the von Neumann equation (21).
2. **The Initial Projection Hypothesis (IPH):** At a temporal boundary of the universe, the density matrix is the normalized projection onto $\mathcal{H}_{PH}$, a low-dimensional subspace of the total Hilbert space. (That is, the initial quantum state of the universe is $\hat{W}_{IPH}(t_0)$ as described in equation (27).)

Similar to the quantum Mentaculus, the Wentaculus also suffers from the quantum measurement problem. There are three promising solutions, each of which gives rise to a distinct version of the Wentaculus.

First, there is the Everettian Wentaculus that looks exactly like the basic Wentaculus. For this theory, we need to embrace the idea that there is a (vague) multiplicity of emergent worlds that is similar to the original Everettian theory. What is interesting about the Everettian Wentaculus is that it suggests an astonishing possibility. If IPH is interpreted as a fundamental law, then the theory is *strongly deterministic,* in the sense of Penrose (1989) that the laws pick out a unique micro-history of the fundamental ontology (represented by W(t)). This theory does not postulate any objective probability. This is another step toward the Everettian aspiration of constructing a theory without any fundamental contingency.

Second, the Bohmian Wentaculus postulates that, in addition to the universal density matrix W that evolves unitarily according to the von Neumann equation, there are actual particles that have precise locations in physical space, represented by $\mathbb{R}^3$. The particle configuration $Q = (Q_1, Q_2, \ldots, Q_N) \in \mathbb{R}^{3N}$ follows the guidance equation (written for the ith particle):[16]

$$\frac{dQ_i}{dt} = \frac{\hbar}{m_i} \mathrm{Im} \frac{\nabla_{q_i} W(q, q', t)}{W(q, q', t)} (q = q' = Q), \qquad (29)$$

Moreover, the initial particle distribution is given by the density-matrix version of the quantum equilibrium distribution:

$$P(Q(t_0) \in dq) = W(q, q, t_0) dq. \qquad (30)$$

Third, the GRW Wentaculus postulates that the universal density matrix typically obeys the von Neumann equation, but the linear evolution is interrupted randomly by collapses (with rate $N\lambda$, where N is the number of particles and λ is a new constant of nature of order 10^{-15} s^{-1}):[17]

$$W_{T+} = \frac{\Lambda_k(X)^{1/2} W_{T-} \Lambda_k(X)^{1/2}}{tr(W_{T-} \Lambda_k(X))}, \qquad (31)$$

where W_{T-} is the pre-collapse density matrix, W_{T+} is the post-collapse density matrix, with k uniformly distributed in the N-element set of particle labels and X distributed by $\rho(x) = tr(W_{T-} \Lambda_k(x))$, with the collapse rate operator defined as before in (20).

In this section, I presented several versions of PH. They can all be traced back to the original Boltzmannian idea that the initial state of the universe is special and has low entropy. Differences arise when we move to the Wentaculus framework where IPH selects a unique and simple initial *microstate* of the universe. The

---

[16] This version of the guidance equation is first proposed by Bell (1980), then recast as the dynamical equation for the fundamental density matrix in Dürr et al. (2005).

[17] To my knowledge, the W-GRW equations first appear in Allori et al. (2013).

microstate is given by a mixed-state density matrix. The reason a mixed state can play the role of a microstate is because it enters directly into the fundamental dynamical equations, such as (21), (29), and (31). The different initial-condition postulates—(7), (16), (26), and (27)—form a family, and I shall continue using the generic label, the "Past Hypothesis," to refer to them and will only use specific labels when their differences are relevant.

3. Why the Past Hypothesis Is Lawlike

It is clear that PH has a special status in the Boltzmannian account. It has been suggested that PH is like a law of nature. For example, this is emphasized by Feynman (2017 [1965]) as quoted in the epigraph. Making a similar point about classical statistical mechanics, Goldstein et al. (2020) suggest that PH is an interesting kind of law:

> The past hypothesis is the one crucial assumption we make in addition to the dynamical laws of classical mechanics. The past hypothesis may well have the status of a law of physics—not a dynamical law but a law selecting a set of admissible histories among the solutions of the dynamical laws. (p. 553)

In this section I offer four types of positive arguments to support the view that PH is a candidate fundamental law of nature. These arguments also support the weaker thesis that PH is a candidate axiom in the fundamental theory. My methodology is naturalistic and functionalist. I argue for the nomic status of PH by locating the roles of the fundamental laws in our physical theories and by showing that PH plays such roles. These roles include backing scientific explanations, constraining nomological possibilities, and supporting objective probabilities.

Some of these arguments (§§3.1–3.3) are related to ideas that have appeared in the literature. I try to make the premises explicit, with the hope that the arguments are clear enough for others who disagree to examine and criticize. Usually, the arguments are made in the Humean framework, but as I argue, they can also be made on behalf of non-Humeans who have a minimalist conception of what it is for laws to really *govern*. That is the account I favor. Of course, the minimalist account is at odds with the idea about "dynamical governing":

Dynamical Governing: Only dynamical laws can be fundamental laws of nature.

In §3.4, I offer a new argument for the nomic status of PH based on considerations of the nature of quantum entanglement.

3.1. Arguments from the Second Law

The (fundamental) nomic status of PH is supported by the nomic status of the Second Law of Thermodynamics. The Second Law is a law of nature; whatever underlies a law is a law. A law that cannot be derived from other laws is a fundamental law; therefore, PH is a fundamental law. Let us spell out the argument in more detail:

- **P1** The Second Law of Thermodynamics is a law of nature.
- **P2** A law of nature can be scientifically explained only by appealing to more fundamental laws of nature and laws of mathematics.
- **P3** The Second Law of Thermodynamics is scientifically explained (in part) by PH, and PH is not a law of mathematics.
- **C1** So, PH is a law of nature and is more fundamental than the Second Law.
- **P4** PH is not scientifically explained by fundamental laws.
- **P5** A law of nature that is not scientifically explained by fundamental laws is a fundamental law.
- **C2** So, PH is a fundamental law of nature.

Comments on P1. First, the Second Law of Thermodynamics summarizes an important regularity: the tendency for things to become more chaotic and more decayed as time passes. It is part of our concept of lawhood that this irreversible tendency is lawlike. We learn about this law much more directly in our experiences than the microscopic equations of motion. Second, nature's irreversible tendency is encoded in our concept of physical necessity. For example, we learn that it is physically impossible for a metal rod to spontaneously heat up on one side and then cool down on the other; it is physically impossible to create a perpetual motion machine of the second kind (a machine that violates the Second Law), and this is impossible no matter who tries to do it—whenever and wherever.[18] Hence, the Second Law is not an accidental feature of the world. Of course, the usual formulation of the Second Law in terms of the absolute monotonic increase of entropy is too strong. It should be modified in two ways: it holds for the overwhelming majority of nomologically possible initial conditions, and for each entropic trajectory, there can be short-lived, shallow, and infrequent decreases of entropy (see Second Laws for X and for Ψ).

---

[18] This becomes more complicated if an "Albertian demon" turns out to be physically possible. See Albert (2000, §5); and Maudlin, this volume, Chapter 8.

Recognizing the importance of the Second Law, Eddington (1928) suggests:

> If someone points out to you that your pet theory of the universe is in disagreement with Maxwell's equations—then so much the worse for Maxwell's equations. If it is found to be contradicted by observation—well, these experimentalists do bungle things sometimes. But if your theory is found to be against the second law of thermodynamics I can give you no hope; there is nothing for it but to collapse in deepest humiliation. (p. 74)

That may be too strong. Nevertheless, we should accept that the Second Law is nomologically necessary. It is not merely an accidental feature of the world, such as the contingent fact that all gold spheres are less than one mile in diameter. Third, counterfactuals are backed by laws of nature; laws are what we hold fixed when evaluating counterfactuals. The Second Law backs counterfactuals about macroscopic processes that display a temporal asymmetry: if there were a half-mixed ink drop in my water cup right now, it would have been more separated in the past and more evenly mixed in the future. (See §3.2 for more on the counterfactual arrow.)

Comments on P2. The notion of scientific explanation here is not a fully analyzed notion. What is relevant to this argument is that the Second Law is supposed to be *derived as a theorem* from the basic postulates of the Mentaculus or the Wentaculus. It is expected that, assuming the laws of mathematics, the fundamental dynamical laws, PH, and SP, an initial microstate starting from the initial macrostate will, with overwhelming probability, travel to macrostates of increasingly higher entropy until it reaches thermal equilibrium (except possibly for short-lived and infrequent decreases of entropy). So, it is the Second Law's mathematical derivation from the basic postulates of the physical theory that is the relevant notion here.[19]

A standard response to P2 is that a nonfundamental law (such as those in the special sciences) can be explained (in part) by some contingent boundary conditions. Examples may include the laws of genetics and laws of economics. However, it is not clear what does the explanatory work in those cases. Let us suppose that some special science law S arise from boundary conditions B. Suppose B obtains. Then there is a high objective probability that S obtains. What is the origin of these objective probabilities? What is the physical explanation? If everything

---

[19] This is different from the notion of metaphysical explanation. See Loewer (2012), and Hicks and van Elswyk (2015). On their views, a fundamental law of nature is metaphysically (but not scientifically) explained by the matter distribution.

is ultimately physical, and the physical theory is informationally complete in so far as the motion of objects go, then it seems that the objective probabilities are ultimately backed by some postulates in physics. The probabilities in physics may supply conditional probabilities on which $Pr(S|B)$ is high. What does the explaining, then, is the probability supplied by physics, and the real law should be the high probability of S obtaining given B, which is consistent with the physicalist picture we started with.

Winsberg (this volume, Chapter 2) offers another potential counterexample to P2. He suggests that due to the near certainty of the existence of Boltzmann brains and large fluctuations in future epochs of the universe, it is important to postulate also a Near Past Hypothesis (see also Chen, 2021):

Near Past Hypothesis (NPH) We are inside the first epoch of the universe between the initial time and the first relaxation to thermal equilibrium.

Winsberg argues that NPH should have the same status as PH and SP because it is also a necessary postulate to derive the Second Law (among other special science laws). However, NPH is an indexical statement about our location in time; as such, it cannot be a candidate fundamental law of nature. So if a non-law can be part of the explanation for the Second Law, then there is no reason to think that only laws can back laws. This is an interesting insight, but I do not think Winsberg's argument undermines P2. If indexical statements cannot be laws, then the Second Law should not be stated in an indexical way. We should use the non-indexical version of the Second Law (and special science laws), such as the Second Law for X, the Second Law for Ψ, and the Second Law for W. In those versions, fluctuations are already taken into account (in a non-indexical way). Hence, we do not need to invoke NPH to derive those versions of the Second Law from PH, SP, and the dynamical laws.

Comments on P3. This premise is true if we grant the explanatory success of the Boltzmannian account, which is assumed in this paper. It is clear that PH is not a law of mathematics.

Comments on P4. P4 is an open scientific question. Perhaps some future theory (e.g., along the lines of Carroll and Chen, 2004) can explain PH using some simple and satisfactory dynamical laws. Still, it is also a scientific possibility that PH remains a fundamental law in the final theory and is not explained further. Given the openness of P4, we should accept C2 only to the degree of acknowledging PH as a *candidate* fundamental law.

Comments on P5. This follows from our concept of a fundamental law of nature.

The argument above supports the (fundamental) nomic status of PH. If nomic status implies axiomatic status, then the argument also supports the idea that PH is an axiom in the fundamental physical theory. But there is another, more straightforward argument for the axiomatic status. The predictive consequences of a physical theory should come entirely from its axioms and their deductive consequences. A good physical theory aims at capturing as many regularities as possible using simple axioms. The Second Law describes an important regularity. Therefore, we postulate PH and SP in addition to the fundamental dynamical laws. These postulates have an axiomatic status in the Mentaculus.

The Mentaculus is a good theory; it is better than "Mentaculus−," the Mentaculus minus PH and SP. The Mentaculus predicts not only the motion of planets, but also the overwhelming probability that my table will not spontaneously rearrange itself into the shape of a statue. The Mentaculus- can tell us everything about the motion of planets but is silent about many macroscopic regularities we see around us. Even so, the Mentaculus is a pretty simple theory. Someone might suggest that to achieve maximal predictive power, we can add a statement about the exact microstate of the universe at t_0 as an additional axiom to Mentaculus−. But the exact microstate is a detailed fact that complicates the Mentaculus− such that its axioms will no longer be simple enough. (In the Wentaculus, IPH pins down a quantum microstate, but its informational content and simplicity level are the same as those of PH in the Mentaculus.)

3.2. Arguments from Other Asymmetries

The thermodynamic arrow of time described by the Second Law is best explained by PH. That provides strong support for the nomic status and the axiomatic status of PH. What about other arrows of time? In this subsection, I present arguments based on the counterfactual arrow, the records arrow, the epistemic arrow, and the intervention arrow. The upshot is that they can also be traced back to the nomic status of PH, without which they would be left completely mysterious. Many of these ideas can be found in Albert (2000, 2015) and Loewer (2007), and they are also discussed in Frisch (2005, 2007), Demarest (2019), Fernandes (this volume, Chapter 12), Callender (2004), Horwich (1987), and Reichenbach (1956).

The records arrow. We have photographs and videos of World War II but no photographs of the next major world war. We have detailed accounts of the life of President Washington but no detailed accounts of the life of the sixty-fifth president of the United States. There are craters on the moon indicating past meteorite impacts but no craters indicating future meteorite impacts. Similarly, there are fossils, rocks, ice sheets, all of which tell us the state of our planet in the past, but we do not have similarly abundant records that tell us the state of our planet in the future.

What is it about our world such that there are abundant records of the past but few, if any, records about the future? One could appeal to some A-theory of time, according to which the future does not yet exist and the past has already happened. So there cannot be records about the future because there are no facts about the future. But this does not seem to provide a satisfactory scientific explanation, as the temporally asymmetric probabilistic correlations have to be accepted as brute facts. In any case, in a block-universe picture compatible with a B-theory of time, the past, present, and future are all equally real; all events exist tenselessly. There are strong probabilistic correlations between physical records (e.g., fossils) that exist at a particular time and physical systems (e.g., dinosaurs) that exist at an earlier time, but no strong correlations between physical records and events at a later time.[20]

PH offers an explanation. Albert (2000) suggests that a record is a relation between two temporal ends of a physical process. A record enables us to infer what happens inside the temporal interval. For example, in a lab, the record of an electron passing through a small slit is the relation between the "ready state" of the measuring instrument at t_1 and the "click" state of the measuring instrument at t_2. If the instrument moves from "ready" to "click," then we can infer that an electron has passed through the slit between t_1 and t_2. But if the instrument was not at "ready" at t_1, we cannot infer that. However, to know that the instrument was indeed "ready," we also need to rely on an earlier record. This seems to go back in time *ad infinitum,* to records about the lab, and to records about the larger environment, and eventually to records about earlier states of the universe. To know that the cosmic microwave background (CMB) data is reliable, we also need to postulate that there is some "ready" state at the beginning of the universe. PH, stipulated at (or around) t_0, is the "mother of all ready states." It provides the underpinning necessary for our inferences based on records to be carried out.[21]

However, it is not enough that PH be true. We also need to justify the important fact that physical records are reliable. For this, we need PH to have the status of a law. (If PH is not derived from other laws, it will have the status of a fundamental law.) If a theory predicts that it is unlikely for physical systems that look

---

[20] The notions of "earlier" and "later" here can be fleshed out in a way that does not refer to an intrinsic arrow of time. What matters here is not that there are facts about earlier times or later times, but simply that the probabilistic correlations are temporally asymmetric.

[21] More details are needed to fully explain the records asymmetry; Rovelli (2020) provides an interesting analysis that adds additional constraints on the initial condition.

like records to reliably indicate past events, then the theory would undermine the rational justification for believing in it. Such a theory would be *epistemically self-undermining* because we believe in physical theories based on records about past experiments and observations.[22] The Mentaculus without PH is such a theory. It would predict that most "records" come about from random fluctuations. If we dig out a shoe of Napoleon, most likely it came about from random fluctuations and not from a low-entropy, past state. Postulating PH as a law avoids that. The uniform probability distribution conditionalized on PH will predict that most physical systems that look like "records" will be reliable records about the past (here we set aside the problems of large future fluctuations).

The epistemic arrow. Given some information about the present, there is some sense in which our knowledge about the past is more vast, detailed, and easily gained than our knowledge about the future. We know that the sun will (likely) rise tomorrow, but we do not know who will win the US presidential election of 2028 and when exactly the stock market will crash over the next twenty years. But we know exactly who won the election of 1860, when exactly the stock market crashed in the last twenty years, and so on. Similar to the records arrow, the epistemic arrow is especially puzzling in a block-universe picture compatible with a B-theory of time. All facts about the past, present, and future are out there. Why is our knowledge so skewed toward one temporal direction?

Albert (2000) explains the epistemic arrow in terms of the records arrow, which in turn is explained by the nomic status of PH. The basic idea is this. We distinguish between inferences based on records from inferences based on predictions or retrodictions. The latter uses only the current macrostate together with the dynamical laws plus SP (construed as an unconditionalized uniform probability distribution on the energy shell of phase space) to the past (retrodictions) and to the future (predictions). Inferences based on predictions will tell us that with overwhelming probability, the sun will rise tomorrow and the ice cubes in my coffee will melt in the next hour, but inferences based on retrodictions will get most things wrong about the past. For example, retrodictions will tell us that the ice cubes in my coffee were actually smaller in the past (they spontaneously got larger in my coffee), and all the books about someone named Lincoln winning the 1860 election came about from random collisions of particles. However, inferences based on records are much more powerful and demand much less detailed information about the current macrostate of the world. We can infer to past states reliably by assuming that records are reliable. Such an inference is backed

---

[22] This is somewhat parallel to the situation of empirical adequacy and records in Everettian quantum mechanics. See Barrett (1996) on the latter.

by the assumption that the recording device was "ready" at a time before the event, and there was another recording device measuring the "ready" state of that one, and so on. As discussed before, the records arrow is explained by PH. Moreover, if the epistemic arrow has physical necessity (or high objective probability), as is required to avoid epistemic self-undermining, whatever underlies it also has physical necessity. In a way similar to the records arrow, the epistemic arrow has to be non-accidental, otherwise our beliefs about past experiments would have a low probability of being accurate.

The counterfactual arrow. I am at home right now. If I had been in my office, the future would be somewhat different but the past would have been pretty much the same. Trump did not press the nuclear button on Independence Day this year. If he had, events on Labor Day would be dramatically different but events on Memorial Day would have been pretty much the same. Why is there a temporal asymmetry of counterfactual dependence? The semantics for counterfactuals is a controversial issue. It is not clear if there is a unified theory that explains every instance of counterfactuals in ordinary language. But if we focus on the counterfactuals that are important for control, decision, and action, it is often accepted that such counterfactuals are backed by laws. This is made explicit in Lewis's (1979) metric for comparing similarity relations among worlds, but it should also be compatible with a strict-conditional approach. If the laws are temporally asymmetric, and if laws entail that changes in the present macrostate would lead to vast differences in the future but not much in the past, then the counterfactual arrow has an explanation.

However, given equation (2) in classical mechanics or equations (9) and (17) in unitary quantum mechanics, changes to the current state (such as the location of Trump's index finger and the location of the nuclear button) will lead to macroscopic differences in both directions of time. It is only by assuming PH as a fundamental law can we explain the following: most physically possible trajectories compatible with the current macrostate will be such that if Trump had pressed the button, most future trajectories would be macroscopically different from the actual ones but the past trajectories would be pretty much the same. This is also due to the records arrow. Assuming PH, there will be an abundance of records about the past. In so far as PH makes it very likely that those records are reliable, PH constrains the past histories of the trajectories even if certain macroscopic features get changed in the present state. However, few, if any, records exist about the future, so there is no such constraint on future macrostates.

The correct counterfactual semantics will no doubt involve context sensitivity and other parameters. Still, it is hard to deny that laws of nature play an important role in determining the truth values of counterfactuals.

The intervention arrow. We can exert influence toward future events but we can no longer act to bring about changes to the past. The intervention arrow is intimately connected to the counterfactual arrow, and it is not clear which is conceptually prior. Some contemporary analyses of influence and intervention are couched in the causal modeling framework. In that framework, we have directed acyclic graphs with variables representing events and arrows representing the direction of effect. But if the fundamental dynamical laws are time-symmetric, what is the scientific explanation for these arrows? Often, the arrows are taken to be primitives in the causal model, left unexplained. However, if PH explains the counterfactual arrow, then it can also explain the arrow of intervention. We can flesh out the language of intervention in terms of intervention counterfactuals, and the arrow of intervention counterfactuals can be explained in a similar way by the records arrow and PH. Loewer (2007) provides such an account.

The arguments from the entropic arrow (the Second Law) and the other arrows can be taken together as an inference to the best explanation. PH (and SP) ground these asymmetries of time. Moreover, to explain them satisfactorily, we need to postulate PH as a fundamental law and we need SP to provide objective probabilities. An opponent may take all of these arrows to be fundamental features of the world, and they can postulate them as primitives in the theory. But that would strike many as a fragmented and unsatisfactory view. Postulating PH and SP in addition to the dynamical laws is a much simpler and more unified way to think about the various arrows of time: from a set of simple axioms, we can derive all of the temporally asymmetric regularities—we get a big bang for the buck.

3.3. Arguments from Metaphysical Accounts of Laws

Humeanism provides a natural home for PH to be a fundamental law and for SP to specify objective probabilities. According to Lewis (1983), the fundamental laws and postulates of objective probabilities are the *axioms* of the best system that are true about the mosaic and optimally balances various theoretical virtues such as simplicity, informativeness, and fit. On this account, the dynamical equations such as equations (2), (9), and (17) could be axioms of their respective best systems and the GRW chances could be the objective probabilities in a GRW world. Can the classical Mentaculus, the quantum Mentaculus, and the Wentaculus count as axiomatizations of the best system? This depends on whether PH can count as a fundamental Lewisian law and whether SP can count as objective probabilities on the best-system account. Anticipating the need to add a boundary condition into the best system, Lewis (1983) writes,

A law is any regularity that earns inclusion in the ideal system. (Or, in case of ties, in every ideal system.) The ideal system need not consist entirely of regularities; particular facts may gain entry if they contribute enough to collective simplicity and strength. (For instance, certain particular facts about the Big Bang might be strong candidates.) But only the regularities of the system are to count as laws. (p. 367)

But if a statement such as PH is axiomatic in the best system, why not count it toward laws? In the same paper (p. 368), Lewis distinguishes between fundamental laws and derived laws. He suggests that fundamental laws are those statements that the ideal system takes as *axiomatic* and invokes only perfectly natural properties. But PH certainly is axiomatic in the Mentaculus and the Wentaculus. Moreover, PH can be stated in the fundamental language of the respective theory. (There will be some residual vagueness, which I discuss in §4.3.) So it seems that Lewis should be open to the idea that PH is a fundamental law according to the best-system account.[23]

Hence, if we are committed to the Humean conception that laws supervene on the mosaic in the way specified by the best-system account, we are led to accept the fundamental nomic status of PH. (Similarly, as Loewer, 2001, argues, Lewis's version of Humeanism also has room to recognize the probabilities specified by SP as objective.)

On the Humean theory, given facts about the mosaic, the theoretical virtues *metaphysically determine* what the laws are. They are constitutive of laws. Laws are just certain ideal summaries of facts in the world. Laws are nothing over and above the mosaic.

On non-Humean theories, however, laws do not supervene on the mosaic. Laws may be as fundamental as the mosaic itself. Following Hildebrand (2013), we can distinguish between two types of non-Humean theories:

1. Primitivism: Fundamental laws are primitive facts in the world.
2. Reductionism: Fundamental laws are analyzed in terms of something outside the mosaic.

Maudlin (2007) advocates a primitivist version of non-Humeanism. Hildebrand (2013) provides a survey of the reductionist versions according to which laws are further explained by relations among universals, dispositions, essences, or some other more fundamental entities. It is not clear what the further analysis buys

---

[23] For a related point, see Callender (2004).

us. It is not clear to me how to reformulate various modern physical laws and objective probabilities in terms of those entities, and it is less clear to me what advantages there are to reduce laws to something further.[24] Here I agree with Maudlin that the concept of laws seems more familiar to us than the concepts employed in the further analysis (such as in terms of dispositions, universals, and the like). Maudlin's version of primitivism is influential in contemporary discussions of the metaphysics of laws in philosophy of physics. However, Maudlin (2007) favors a more restrictive version of primitivism that I call *Dynamical Law Primitivism*:

> **Dynamical Law Primitivism:** Fundamental laws are primitive facts in the world, and only dynamical laws can be fundamental laws.

This view is connected to Maudlin's view about the intrinsic and primitive arrow of time. In contrast, the spirit of the present project is to analyze time's arrow in terms of something else. In fact, the extra commitment about the primitive arrow of time can be disentangled from the basic idea about how laws govern.

The basic non-Humean idea is simply that laws really *govern*. They metaphysically explain why the nomological possibilities are the way they are and why things are as constrained as they are. The metaphysical explanation can take the form of constraints: given $S(t_0)$, some complete specification of the state of the universe at some time, there is a constraint on what the history of the world is like. If the theory is deterministic, then there is only one microscopic history compatible with the $S(t_0)$. A fundamental dynamical law is a kind of conditional constraint. Constraints can take other forms, such as by selecting a space of possible histories. This is the form of certain equations in general relativity and Maxwellian electrodynamics. It is also true of PH, as it selects "a set of admissible histories among the solutions of the dynamical laws." There is conceptual space for a minimalist conception of primitivism that places no restriction on the form of fundamental laws and in particular, not all of them have to be dynamical laws. The basic view is this:

> **Minimal Primitivism:** Fundamental laws are primitive facts in the world; there is no restriction on the form of fundamental laws. In particular, boundary conditions can be fundamental laws.

---

[24] Hildebrand (2013) suggests that reductionist theories have an advantage over primitivist theories for answering the problem of induction. I disagree, but I set it aside for future work.

Even though fundamental laws can take on any form, we expect them to be relatively simple and informative. These theoretical virtues are no longer *constitutive* of what laws are, but they can serve as our best guides to find the primitive laws[25]:

Epistemic Guides: Even though theoretical virtues such as simplicity and informativeness are not constitutive of fundamental laws, they are good epistemic guides for discovering the fundamental laws.

Chen and Goldstein (2022) develop this idea in more detail. It seems to me that Minimal Primitivism is a good version of non-Humeanism, and it may well be one that best fits our scientific practice and the actual conception of laws. The minimal primitivist view does not commit to an intrinsic and irreducible arrow of time, making it compatible with the Boltzmannian project of analyzing time's arrow in terms of the entropy gradient. PH, as we have discussed already, is virtuous in the right ways. It provides a simple explanation for the restrictions of physical possibilities and the overwhelming probability of irreversibility. According to the Epistemic Guides on the Minimal Primitivist conception, we have found strong evidence that PH is a candidate fundamental law.

Hence, both Humeanism and non-Humeanism (in the minimalist form) support the idea that PH is a candidate fundamental law.

3.4. New Light on Quantum Entanglement

I suggest that there is a new reason to take PH as a fundamental law: it can help us solve a long-standing puzzle in the foundations of quantum mechanics. One of the chief innovations of quantum theory that has no classical analog is quantum entanglement. It is also the origin of the quantum measurement problem. If we solve the measurement problem using one of the three strategies discussed in §2— along the lines of Everett, Bohm, and GRW—we are still left with the quantum state that plays an important dynamical role in the respective theories. Hence, the puzzle of quantum entanglement can be traced to the nature of the quantum state. What does the quantum state represent physically? What is it in the world? Given the role of Ψ in formulating well-posed initial-value problems in Everettian quantum mechanics (EQM) and Bohmian mechanics (BM), and its role in dynamical collapses in GRW, it is reasonable to think that Ψ represents something

---

[25] Why is this expectation rational? I do not know of any noncircular justification. We can appeal to the success of physics and the discovery of simple and informative laws in the past. We may also appeal to a deeper "meta-law" that metaphysically constrains the physical laws.

objective. Here are some options for a realist interpretation (see Chen, 2019b, for more detail):

1. High-dimensional field: On this view, the fundamental space is isomorphic to the 3N-dimensional configuration space; the wave function represents a physical field on that space (Albert, 1996). Even the defenders acknowledge that this view has highly counter-intuitive consequences. It is also an open question whether it really succeeds in recovering the manifest image of lower-dimensional objects.
2. Low-dimensional multi-field: On this view, the fundamental space is the physical space(time); the wave function represents a "multi-field" that assigns physical quantities to every spatial region composed of N points (Forrest, 1988; Belot, 2012; Chen, 2017; Hubert & Romano, 2018). However, this view also appears to have undesirable consequences. In the multi-field interpretation of Everettianism, since entanglement relations are still in the 4-dimensional mosaic, its Lorentz-invariance comes at a surprising cost—the failure of what Albert (2015) calls *narratability*. This also arises in Wallace and Timpson's (2010) spacetime state realist interpretation of Everett. For the Bohmian framework, the multi-field guides particles, but there is no influence (back-reaction) of the particles on the multi-field, even though both the particles and the multi-field are fundamental material entities.
3. Nomological interpretation: On this view, the fundamental space is the physical space(time) and the fundamental ontology consists in particles, fields, or flashes on that space; the wave function represents a physical law (Dürr et al., 1996; Goldstein & Teufel, 2001; Goldstein & Zanghì, 2013. Given the problems of the other two approaches, the nomological interpretation remains a promising solution. However, it faces problems of a different kind. First, Ψ_t is time-dependent. Can nomological entities change in time? I do not see why not. Moreover, as defenders of this view have long recognized, if the universal quantum state obeys the Wheeler-DeWitt equation $\hat{H}\Psi = 0$ then Ψ will be time-independent and not changing. Second, the universal wave function is a very detailed function and may be too complicated to be a law. This problem seems much more serious. Call this the problem of complexity.

Taking PH to be a fundamental law provides a solution to the problem of complexity in the nomological interpretation of the quantum state. This solution works in the Wentaculus framework, where IPH is given a fundamental nomic status. For IPH, the normalized projection onto $\mathcal{H}_{PH}$ contains no more and no less information than $\mathcal{H}_{PH}$, specified by PH in the Mentaculus. If $\mathcal{H}_{PH}$ is simple and informative enough to be nomological, then so is its normalized projection, which is $W_{IPH}(q, q', t_0)$. That is, we can afford the same status of a fundamental law to $W_{IPH}(q, q', t_0)$. In the Everettian Wentaculus, we can interpret $W_{IPH}(q, q', t_0)$ as a law that determines the "local beables," such as a matter-density field in physical space. In the Bohmian Wentaculus, we can interpret $W_{IPH}(q, q', t_0)$ as similar to the Hamiltonian function $\hat{H}(p, q)$: providing a velocity field of particle trajectories. In the GRW Wentaculus, we can interpret $W_{IPH}(q, q', t_0)$ as providing conditional probabilities for the configurations of "local beables," such as a matter-density field or flashes in spacetime.

However, to solve the complexity problem, it is not sufficient for IPH to be a contingent initial condition; it is crucial that IPH has the status of a fundamental law. If it is nomologically possible for W_0 to differ from the state specified by IPH, then the initial quantum state could well be (i.e., physically possible) too complicated to be regarded as nomological. Hence, only by assuming the nomic status of IPH do we obtain a solution to the problem of complexity, thereby arriving at a satisfactory way to understand the nature of the quantum state. In contrast to the first two interpretations according to which quantum entanglement relations are in the material ontology, the nomological interpretation of the quantum state locates the origin of entanglement in the laws. And by taking a nomological interpretation of the quantum state in the Wentaculus framework, we see a unified solution to two problems in the foundations of physics: the problem of irreversibility and the nature of the quantum state. Again, this is compatible with both Humeanism and (the minimal form of) non-Humeanism about laws. (For more detail, see Chen 2018, 2020a.)

4. Apparent Conflicts with Our Concept of Laws of Nature

In §3, I have provided positive arguments for the fundamental, nomic status of PH. In this section, I discuss three apparent conflicts between these arguments and the ordinary conceptions of laws of nature. These apparent conflicts may explain some people's hesitation in accepting the fundamental, nomic status of PH. However, as I argue, these conflicts are merely apparent if we adopt the Mentaculus framework, and at any rate, they become even less worrisome in the Wentaculus framework.

4.1. Boundary Condition Laws?

It is often said that PH is merely a boundary condition. The contrast between boundary conditions and fundamental laws can be seen in the differences between the paradigm cases of dynamical laws—such as equations (2), (9), (17), (21), and (29)—and boundary conditions that select subclasses of the dynamically possible trajectories. Boundary conditions do not *directly* play a dynamical role.

It is not clear how to make this worry precise. First, it is unclear why every law has to be dynamical. Second, by restricting the possible initial conditions, a boundary-condition law can be an important ingredient in the theory, as in the case of the Mentaculus. In fact, in the Wentaculus framework, the boundary condition IPH plays a *direct* dynamical role, akin to the dynamical role of the Hamiltonian function in classical mechanics. For example, in the Bohmian version, IPH (27), the von Neumann equation (21), and the guidance equation (29) can be combined into one equation:

$$\frac{dQ_i}{dt} = \frac{\hbar}{m_i} \text{Im} \frac{\nabla_{q_i} W_{IPH}(q,q',t)}{W_{IPH}(q,q',t)}(Q)$$
$$= \frac{\hbar}{m_i} \text{Im} \frac{\nabla_{q_i} \langle q|e^{-i\hat{H}t/\hbar}\hat{W}_{IPH}(t_0)e^{i\hat{H}t/\hbar}|q'\rangle}{\langle q|e^{-i\hat{H}t/\hbar}\hat{W}_{IPH}(t_0)e^{i\hat{H}t/\hbar}|q'\rangle}(q=q'=Q) \quad (32)$$

Hence, in the Bohmian version, IPH does not just select a subclass of velocity fields; it pins down a unique velocity field. In the Everettian version, IPH does not just select a subclass of possible multiverses; it pins down a unique one. In the GRW version, IPH is directly involved in setting the chances of collapses.

There is a related worry about admitting boundary condition laws. Typically we distinguish between dynamical laws and initial conditions. If initial conditions can be laws, then how can we distinguish between laws and contingent data? It seems that the distinction would collapse. However, that is not the case. Some boundary conditions such as PH have the elite status as fundamental laws, but it does not follow that all boundary conditions are similarly elite. This is because not all boundary conditions exemplify the optimal balance of required theoretical virtues (either as constitutive of what laws are or as epistemic guides to primitive laws), including simplicity, informativeness, and fit.

One may worry that it is odd to have a fundamental law that refers to a particular time (t_0). Our most familiar fundamental laws are general statements that do not refer to a particular time or place. But why is it a requirement that laws cannot refer to particular events? Suppose some places or times are in fact physically distinguished; then it seems appropriate for laws to refer to them. Think about the Aristotelian tendency for things to move toward the center of the (geocentric) universe. If that is indeed the case, then we should have a fundamental law describing that motion, and a simple candidate would just be to state that the center of the world, C_0, is the place toward which things evolve. Similarly, if the initial time, t_0, is indeed special (as the initial state accounts for a great many regularities), then it is appropriate to postulate a law that refers to t_0.

4.2. Non-Dynamical Chances?

Another worry concerns the nature of objective probabilities. The Mentaculus postulates both PH and SP. PH and SP share the explanatory burden, so they should have the same status. It is important that the probabilities of SP be objective. However, if the dynamics are deterministic (as in the Bohmian and the Everettian versions), objective probabilities seem to be either 0 or 1. How can non-trivial probabilities be objective and represent something beyond subjective credences? (In so far as typicality plays a similar explanatory role as probability, TP may face the same *prima facie* problem as SP.)

Humeanism has the resources to solve this problem. Loewer (2001) suggests that deterministic "chances" can gain entry in the Lewisian best system by the informativeness they bring and the simplicity of the postulate, such as the uniform measure specified by SP. On non-Humeanism, how to understand deterministic "chances" is an open problem. On Minimal Primitivism, perhaps the notion of primitive constraining (of the initial state) can come in degrees that can be represented by probabilities. Another way to allow deterministic "chances" is to use the notion of typicality specified by TP, which can be interpreted as only allowing the initial conditions that are *typical*. On this understanding, abnormal, anti-entropic initial conditions are not nomologically possible. (A similar problem arises in Bohmian mechanics, which requires a quantum equilibrium distribution in addition to the deterministic dynamical laws, to deduce the usual Born rule for subsystems.)

In the Wentaculus framework, it is clear that there is an objective anchor for SP (and TP). Since IPH selects a unique initial quantum state of the universe, we no longer need a probability distribution over initial quantum states. However, as a purely mathematical fact, the $W_{IPH}(t_0)$ induces a "uniform" probability distribution over pure states. This is not a fundamental postulate of probability in the theory, unlike SP or TP in the Mentaculus. It arises as a mathematical consequence of the objective quantum microstate of the universe. This emergent probability distribution, though not fundamental, can play the same role in statistical analysis about typical behaviors. For example, the usual conjecture about typical pure states approaching equilibrium can be translated into the following: most parts of the density matrix will be very close to the equilibrium subspace, which is equivalent to the claim that the density matrix will approach equilibrium. Hence, the Wentaculus framework provides an objective anchor for SP and TP.[26]

4.3. Nomic Vagueness?

The final worry about the fundamental nomic status of PH has to do with the fact that PH is vague, and an exact version of PH would be arbitrary in an unprecedented way. In Figure 7.1 of §2.1, I made clear that the boundaries of macrostates are fuzzy, and the macrostates form an exact partition only if we stipulate some arbitrary choices of the parameters for coarse-graining. These are the C-parameters: the size of coarse-graining cells, the exact correspondence between macroscopic quantities and functions on phase space, and (in the quantum case) the cutoff value for macrostate inclusion. There are better or worse ways to choose the C-parameters. But it is implausible that there are some exact values of C-parameters as known to Nature. The vagueness comes up in the classical Mentaculus and the quantum Mentaculus in how PH selects an initial macrostate that constrains the initial microstate. In the quantum case, PH only selects an exact subspace in Hilbert space when we choose some arbitrary C-parameters.

In the quantum Mentaculus, can we stipulate that there is an exact subspace $\mathcal{H}_{PH}$ as known to Nature? This leads to what I call *untraceable arbitrariness*. There is an infinity of *admissible* changes[27] to the boundary of $\mathcal{H}_{PH}$ that do not change the nomological status of most microstates compatible with $\mathcal{H}_{PH}$. This is unlike the kind of arbitrariness of natural constants or other fundamental laws. For example, any change to the value of the gravitational constant in Newtonian theory will make most worlds (compatible with the original Newtonian theory) impossible. What about the case of stochastic theories? Are the dynamical chances traceable? Yes they are, but not in the way of changing status from physically possible to physically impossible. The traceable changes are reflected in the probabilistic likelihood of most worlds.

I discuss this in more detail in Chen (2022). Here I provide a simple illustration by considering mechanisms for flipping a coin (see Figure 7.2a). Suppose we have a stochastic coin and it is flipped three times. It landed heads, tails, and heads.

---

[26] However, it does not answer the related question about the nature of the quantum equilibrium distribution in Bohmian theories. But that is different from the issue of the status of SP and TP.

[27] Admissibility here is vague, and rightly so. It can be interpreted as some measure of simplicity of theories. We want PH to be simple enough, and different ways of carving out the boundary will lead to different exact versions of PH. But we want to consider only those versions that are sufficiently simple (and not too gerrymandered).

THE PAST HYPOTHESIS AND PHYSICAL LAWS 241

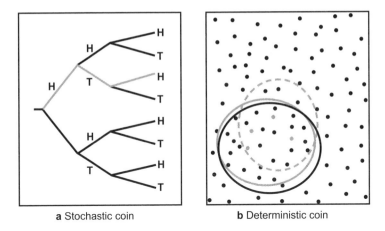

a Stochastic coin   **b** Deterministic coin

FIGURE 7.2. Illustration of nomic vagueness using the contrast between a stochastic coin and a deterministic coin. What is actually observed is this sequence: Heads, Tails, Heads. Credit: © Eddy Keming Chen

In Figure 7.2a, the possible sequences are marked in black and the actual sequence is marked in grey. The simplest chance hypotheses are going to be:

- H_α: The chance of landing heads at each flip is the same, and it is α.

So, this is a one-parameter family of chance hypotheses. $H_{0.5}$ is the hypothesis that the coin is fair, H_1 is the hypothesis that the coin always lands Heads, and so on. Among these hypotheses, a sequence of coin flips will select exactly one of the chance hypotheses from this class based on which one receives the highest likelihood value. In this case, it is $H_{2/3}$ with the highest likelihood value being $4/27$. Conversely, $H_{2/3}$ will assign a determinate chance to every sequence of coin flips. If Nature stochastically acts according to $H_{2/3}$, then changing the chance of Heads even slightly, say to 0.65, will change the chance of every sequence. So, it is in this sense that stochastic theories are traceable. Of course, we can imagine a more gerrymandered chance hypothesis according to which half of the time, the coin operates with a $5/6$ chance of landing Heads, and the other half of the time, the coin operates with a $3/6$ chance of landing Heads. This will predict the same chance as $H_{2/3}$ to every sequence, but it is far less simple. The gerrymandered chance hypothesis is not a serious competitor to $H_{2/3}$. The actual hypothesis is *by far* simpler and more fit than any competitor.

Traceability is lost in the case of the deterministic coin. In this case, there are no dynamical transition chances. The objective probabilities come from probabilities over initial conditions. Suppose a deterministic coin is flipped and it lands heads, tails, and heads. In Figure 7.2b, the grey dots represent the initial conditions of the coin (which also include details about the flipping mechanism) that deterministically lead to the sequence HTH, and the black dots represent initial conditions that lead to other outcomes (such as HHH, TTT, and so on). To simplify things, suppose the sample space is finite, so there are only finitely many initial conditions to consider. Then we can draw different probabilistic hypotheses as different "circles" over initial conditions. Suppose the black circle encloses 4 grey dots and 23 black dots. Then it represents a probabilistic hypothesis that all and only the dots within the black circle are possible initial conditions and each dot has equal probability. The black circle has the highest likelihood given the data of HTH. If the dashed circle encloses 4 grey dots and 25 black dots, then the dashed circle is less likely than the black circle given HTH. However, it is easy to have a nearby circle, say the grey circle, that (like the black circle) has 4 grey dots and 23 black dots. This is possible if the grey dots are sufficiently localized in state space such that it is easy to preserve their proportion to the black dots while changing the boundary of the circle.

Moreover, specifying the grey circle need not be more complicated than specifying the black circle. They are the same kind of probabilistic hypotheses, and there is no reason to think that one is more gerrymandered than the other, unlike the situation with the stochastic coin where to recover the same likelihood, one has to resort to time-dependent chances. This reasoning can be generalized if the state space gets richer and the sequence of coin flips gets longer. There may be infinitely many ways to slightly change the boundary of the black circle and keep the relative proportions *constant*. This means that there will be a large class of probabilistic hypotheses that have the same likelihood given a particular history of coin flips. No particular hypothesis is far simpler and more fit than all competitors. Hence, super-empirical virtues such as simplicity will become more relevant. Furthermore, since the variation of the boundary is incremental, comparing simplicity can generate a sorites series: is there a determinate class of hypotheses that are simple enough? It is implausible for there to be a sharp line. Hence, we have a vague "collection" of hypotheses that pass the simplicity bar.

The contrast between the stochastic coin and the deterministic coin is analogous to the comparison between GRW theories and a vague law such as PH. GRW chances are traceable, but the boundary of the PH macrostate and the exact probability distribution are untraceable. Hence, there are reasons to think that if PH is a fundamental law in the Mentaculus, then it is a vague law. I call this phenomenon *nomic vagueness*.

However, it is not clear why vagueness disqualifies a statement from being a fundamental law. After all, we should be led by empirical evidence and scientific practice to consider what the laws are, and our metaphysical commitments to precision and exactness should not be given absolute priority. It is a surprising consequence that a fundamental law can be vague. This gives us reason to think that perhaps the final theory of the world will not be completely mathematical expressible, in so far as vagueness and higher order vagueness defy classical logic and set-theoretic mathematics. This is a radical consequence about nomic vagueness that deserves more attention (Chen, 2020c).

Nevertheless, the situation is different in the Wentaculus. It gets rid of nomic vagueness without introducing untraceable arbitrariness. According to IPH, the initial macrostate and the initial microstate is represented by the quantum state of the universe—$W_{IPH}(t_0)$. It enters directly into the fundamental microdynamics. Hence, $W_{IPH}(t_0)$ will be traceable from the perspective of two realist interpretations of the quantum state (Chen, 2019b):

1. $W_{IPH}(t_0)$ is ontological: If the initial density matrix represents something in the fundamental material ontology, IPH is obviously traceable. Any changes to the physical values $W_{IPH}(t_0)$ will leave a trace in every world compatible with IPH.
2. $W_{IPH}(t_0)$ is nomological: If the initial density matrix is on a par with the fundamental laws, then $W_{IPH}(t_0)$ plays the same role as the classical Hamiltonian function or fundamental dynamical constant of nature. It is traceable in the Everettian version with a matter-density ontology as the initial matter-density is obtained from $W_{IPH}(t_0)$. It is similarly traceable in the GRW version with a matter-density ontology. For the GRW version with a flash ontology, different choices of $W_{IPH}(t_0)$ will, in general, lead to different probabilities of possible macro-histories. In the Bohmian version, different choices of $W_{IPH}(t_0)$ will lead to different velocity fields such that for typical initial particle configurations (and hence typical worlds compatible with the theory), they will take on different trajectories.

The traceability of $W_{IPH}(t_0)$ is due to the fact that we have connected the low-entropy macrostate (now represented by $W_{IPH}(t_0)$) to the microdynamics (where $W_{IPH}(t_0)$ appears). Hence, $W_{IPH}(t_0)$ is playing a dual role at t_0 (and only at that time): it is both the microstate and the macrostate. In contrast, the untraceability of Γ_0 in classical mechanics is due to the fact that classical equations of motion directly involve only the microstate X_0, not Γ_0. Similarly, $\mathcal{H}_{PH}$ in the standard wave-function formulation is untraceable because the Schrödinger equation directly involves only the wave function, not $\mathcal{H}_{PH}$. Many changes could be made to Γ_0 and $\mathcal{H}_{PH}$ that would not trickle down whatsoever in typical worlds compatible with these postulates. The Mentaculus, but not the Wentaculus, faces a dilemma between nomic vagueness and untraceable arbitrariness.

5. Conclusion

I have argued that, in the Boltzmannian framework, PH is a candidate fundamental law of nature. Such a view is supported by the theoretical roles PH plays in the theory. In arguing for the nomic status and the axiomatic status of PH, we see that whether it is a law makes a difference to many other issues in the foundations of physics. Moreover, its nomic status calls for some rethinking about the nature of physical laws. I suggest that, according to Humeanism and a minimal version of non-Humeanism, boundary conditions can be fundamental laws, SP and TP can be objective, and fundamental laws and chances can be vague. The conflicts with our concept of laws of nature are merely apparent, and in any case, they become much less worrisome if we adopt the Wentaculus framework. Hence, the view that PH is a candidate fundamental law should be more widely accepted than it is now.

Acknowledgments

My ideas in this paper have been influenced by discussions with many people over the years. I am especially grateful to David Albert, Craig Callender, Sheldon Goldstein, Barry Loewer, and Roderich Tumulka. I would also like to thank Eugene Chua, Saakshi Dulani, Veronica Gomez, Ned Hall, Tim Maudlin, Kerry McKenzie, Elizabeth Miller, Jill North, Charles Sebens, Ted Sider, Cristi Stoica, Anncy Thresher, David Wallace, Brad Weslake, Isaac Wilhelm, Eric Winsberg, and the participants in my graduate seminar on the arrows of time at UCSD in spring 2020.

References

Albert, David Z. 1996. Elementary quantum metaphysics. In Cushing, J. T., Fine, A., & Goldstein, S. (eds.), *Bohmian Mechanics and Quantum Theory: An Appraisal*, pp. 277–284. Kluwer Academic.

Albert, David Z. 2000. *Time and Chance*. Harvard University Press.
Albert, David Z. 2015. *After Physics*. Harvard University Press.
Allori, Valia, Goldstein, Sheldon, Tumulka, Roderich, & Zanghì, Nino. 2013. Predictions and primitive ontology in quantum foundations: A study of examples. *British Journal for the Philosophy of Science*, 65(2), 323–352.
Ashtekar, Abhay, & Gupt, Brajesh. 2016. Initial conditions for cosmological perturbations. *Classical and Quantum Gravity*, 34(3), 035004.
Barrett, Jeffrey A. 1996. Empirical adequacy and the availability of reliable records in quantum mechanics. *Philosophy of Science*, 63(1), 49–64.
Bell, John S. 1980. De Broglie-Bohm, delayed-choice, double-slit experiment, and density matrix. *International Journal of Quantum Chemistry*, 18(S14), 155–159.
Belot, Gordon. 2012. Quantum states for primitive ontologists. *European Journal for Philosophy of Science*, 2(1), 67–83.
Boltzmann, Ludwig. 1964 [1896]. *Lectures on Gas Theory*. University of California Press.
Callender, Craig. 2004. Measures, explanations and the past: Should "special" initial conditions be explained? *British Journal for the Philosophy of Science*, 55(2), 195–217.
Callender, Craig. 2011. Thermodynamic asymmetry in time. In Zalta, Edward N. (ed.), *The Stanford Encyclopedia of Philosophy*, Fall 2011 ed. Stanford University. https://plato.stanford.edu/.
Carroll, Sean. 2010. *From Eternity to Here: The Quest for the Ultimate Theory of Time*. Penguin.
Carroll, Sean M., & Chen, Jennifer. 2004. Spontaneous inflation and the origin of the arrow of time. arXiv preprint hep-th/0410270.
Chen, Eddy Keming. 2017. Our fundamental physical space: An essay on the metaphysics of the wave function. *Journal of Philosophy*, 114:7.
Chen, Eddy Keming. 2018. Quantum mechanics in a time-asymmetric universe: On the nature of the initial quantum state. *British Journal for the Philosophy of Science*, doi.org/10.1093/bjps/axy068.
Chen, Eddy Keming. 2019a. Quantum states of a time-asymmetric universe: Wave function, density matrix, and empirical equivalence. MA thesis, Department of Mathematics, Rutgers University, New Brunswick. arXiv:1901.08053.
Chen, Eddy Keming. 2019b. Realism about the wave function. *Philosophy Compass*, 14(7).
Chen, Eddy Keming. 2020a. From time asymmetry to quantum entanglement: The Humean unification. *Noûs*, 1–29.
Chen, Eddy Keming. 2020b. Time's arrow in a quantum universe: On the status of statistical mechanical probabilities. In Allori, Valid (ed.), *Statistical Mechanics and Scientific Explanation: Determinism, Indeterminism and Laws of Nature*. World Scientific.
Chen, Eddy Keming. 2020c. Welcome to the fuzzy-verse. *New Scientist*, 247(3298), 36–40.
Chen, Eddy Keming. 2021. Time's arrow and self-locating probability. *Philosophy and Phenomenological Research*, forthcoming.
Chen, Eddy Keming. 2022. Fundamental nomic vagueness. *Philosophical Review*, 131(1).

Chen, Eddy Keming, & Goldstein, Sheldon. 2022. Governing without a fundamental direction of time: Minimal primitivism about laws of nature. In Ben-Menahem, Yemima (ed.), *Rethinking the Concept of Law of Nature*. Springer.

Chen, Eddy Keming, & Tumulka, Roderich. 2022. Uniform probability distribution over all density matrices. *Quantum Studies: Mathematics and Foundations*. https://doi.org/10.1007/s40509-021-00267-5.

Demarest, Heather. 2019. Mentaculus laws and metaphysics. *Principia: An International Journal of Epistemology*, 23(3), 387–399.

Dürr, Detlef, Goldstein, S., & Zanghì, N. 1996. Bohmian mechanics and the meaning of the wave function. In *Experimental Metaphysics: Quantum Mechanical Studies in Honor of Abner Shimony*. Kluwer Academic Publishers.

Dürr, Detlef, Goldstein, Sheldon, Tumulka, Roderich, & Zanghì, Nino. 2005. On the role of density matrices in Bohmian mechanics. *Foundations of Physics*, 35(3), 449–467.

Earman, John. 2006. The "past hypothesis": Not even false. *Studies in History and Philosophy of Science Part B: Studies in History and Philosophy of Modern Physics*, 37(3), 399–430.

Eddington, Arthur Stanley. 1928. *The Nature of the Physical World*. Macmillan.

Feynman, Richard. 2017 [1965]. *The Character of Physical Law*. MIT press.

Forrest, Peter. 1988. *Quantum metaphysics*. Blackwell.

Frigg, Roman. 2007. A field guide to recent work on the foundations of thermodynamics and statistical mechanics. In *The Ashgate Companion to the New Philosophy of Physics*, pp. 99–196. Taylor & Francis.

Frisch, Mathias. 2005. Counterfactuals and the past hypothesis. *Philosophy of Science*, 72(5), 739–750.

Frisch, Mathias. 2007. Causation, counterfactuals, and entropy. In Price, Huw, & Corry, Richard (eds.), *Causation, Physics, and the Constitution of Reality: Russell's Republic Revisited*. Oxford University Press.

Goldstein, Sheldon. 2001. Boltzmann's approach to statistical mechanics. In Bricmont, J., et al. (eds.), *Chance in Physics*, pp. 39–54. Springer.

Goldstein, Sheldon. 2012. Typicality and notions of probability in physics. In Ben-Menahem, Yemima, and Hemmo, Meir (eds.), *Probability in Physics*, pp. 59–71. Springer.

Goldstein, Sheldon, & Teufel, Stefan. 2001. Quantum spacetime without observers: Ontological clarity and the conceptual foundations of quantum gravity. In Callendar, Craig, and Huggett, Nick (eds.), *Physics Meets Philosophy at the Planck Scale*, pp. 275–289. Cambridge University Press.

Goldstein, Sheldon, & Tumulka, Roderich. 2011. Approach to thermal equilibrium of macroscopic quantum systems. In *Non-Equilibrium Statistical Physics Today: Proceedings of the 11th Granada Seminar on Computational and Statistical Physics, AIP Conference Proceedings*, vol. 1332, pp. 155–163. American Institute of Physics.

Goldstein, Sheldon, & Zanghì, Nino. 2013. Reality and the role of the wave function in quantum theory. In Ney, Alyssa, and Albert, David Z. (eds.), *The Wave Function: Essays on the Metaphysics of Quantum Mechanics*, pp. 91–109. Oxford University Press.

Goldstein, Sheldon, Lebowitz, Joel L., Mastrodonato, Christian, Tumulka, Roderich, & Zanghì, Nino. 2010a. Approach to thermal equilibrium of macroscopic quantum systems. *Physical Review E*, 81(1), 011109.

Goldstein, Sheldon, Lebowitz, Joel L., Tumulka, Roderich, & Zanghì, Nino. 2010b. Long-time behavior of macroscopic quantum systems: Commentary accompanying the English translation of John von Neumann's 1929 article on the quantum ergodic theorem. *European Physical Journal H*, 35(2), 173–200.

Goldstein, Sheldon, Lebowitz, Joel L., Tumulka, Roderich, & Zanghì, Nino. 2020. Gibbs and Boltzmann entropy in classical and quantum mechanics. In Allori, Valid (ed.), *Statistical Mechanics and Scientific Explanation: Determinism, Indeterminism and Laws of Nature*. World Scientific.

Hicks, Michael Townsen, & van Elswyk, Peter. 2015. Humean laws and circular explanation. *Philosophical Studies*, 172(2), 433–443.

Hildebrand, Tyler. 2013. Can primitive laws explain? *Philosophers' Imprint*, 13(15), 1–15.

Horwich, Paul. 1987. *Asymmetries in Time: Problems in the Philosophy of Sciences*. MIT Press.

Hubert, Mario, & Romano, Davide. 2018. The wave-function as a multi-field. *European Journal for Philosophy of Science*, 8(3), 521–537.

Lanford, Oscar E. 1975. Time evolution of large classical systems. In Moser, J. (ed.), *Dynamical Systems, Theory and Applications*, pp. 1–111. Springer.

Lazarovici, Dustin, & Reichert, Paula. 2015. Typicality, irreversibility and the status of macroscopic laws. *Erkenntnis*, 80(4), 689–716.

Lebowitz, Joel L. 2008. Time's arrow and Boltzmann's entropy. *Scholarpedia*, 3(4), 3448.

Lewis, David. 1979. Counterfactual dependence and time's arrow. *Noûs*, 13, 455–476.

Lewis, David. 1983. New work for a theory of universals. *Australasian Journal of Philosophy*, 61, 343–77.

Loewer, Barry. 2001. Determinism and chance. *Studies in History and Philosophy of Science Part B: Studies in History and Philosophy of Modern Physics*, 32(4), 609–620.

Loewer, Barry. 2007. Counterfactuals and the second law. In Price, Huw, & Corry, Richard (eds.), *Causation, Physics, and the Constitution of Reality: Russell's Republic Revisited*. Oxford University Press.

Loewer, Barry. 2012. Two accounts of laws and time. *Philosophical Studies*, 160(1), 115–137.

Loewer, Barry. 2020. The Mentaculus. Manuscript.

Maudlin, Tim. 2007. *The Metaphysics within Physics*. Oxford University Press.

North, Jill. 2011. Time in thermodynamics. In *The Oxford Handbook of Philosophy of Time*, 312–350. Oxford University Press.

Penrose, Roger. 1979. Singularities and time-asymmetry. In Hawking, S. W., & Israel, W. (eds.), *General Relativity*, pp. 581–638. Cambridge University Press.

Penrose, Roger. 1989. *The Emperor's New Mind: Concerning Computers, Minds, and the Laws of Physics.* Oxford University Press.

Price, Huw. 2004. On the origins of the arrow of time: Why there is still a puzzle about the low-entropy past. In Ayala, F. J., and Arp, R. (eds.), *Contemporary Debates in Philosophy of Science,* 219–239. Wiley-Blackwell.

Reichenbach, Hans. 1956. *The Direction of Time.* University of California Press.

Rovelli, Carlo. 2020. Memory and entropy. arXiv preprint arXiv:2003.06687.

Von Neumann, John. 1955. *Mathematical Foundations of Quantum Mechanics.* Princeton University Press.

Wallace, David. 2012. *The Emergent Multiverse: Quantum Theory according to the Everett Interpretation.* Oxford University Press.

Wallace, David, & Timpson, Christopher G. 2010. Quantum mechanics on spacetime I: Spacetime state realism. *British Journal for the Philosophy of Science,* 61(4), 697–727.

Wilhelm, Isaac. 2019. Typical: A theory of typicality and typicality explanations. *British Journal for the Philosophy of Science,* forthcoming.

Winsberg, Eric. 2004. Can conditioning on the "past hypothesis" militate against the reversibility objections? *Philosophy of Science,* 71(4), 489–504.

Chapter Eight

On the Albertian Demon

▸ TIM MAUDLIN

Historical Prelude

It is fitting for any discussion of Maxwell's demon to recall the exact point that Maxwell was making with his example. His concern was the means by which the heat energy in a system can be converted into work. It would evidently be very convenient if heat energy could be extracted from a system in the form of work without any other attendant thermodynamic changes: we could, for example, run machinery by cooling the ocean, and not thereby heat anything else up. We can, of course, extract heat energy from the ocean as work by running a heat engine between the ocean and the arctic ice floes. But what the second law of thermodynamics, as Maxwell understood it, forbids is the conversion of the heat in a body at equilibrium into work without either doing work on the body or transferring heat to a cooler body.

It would be pleasing to have a crisp, exact formulation of the second law from which this result could be derived. Maxwell himself did not provide such a formulation. Indeed, his presentation of the law suggests that a rigorous verbal formulation is not the proper way to grasp the content of the law:

Admitting heat to be a form of energy, the second law asserts that it is impossible, by the unaided action of natural processes, to transform any part of the heat of a body into mechanical work, except by allowing heat to pass from that body into another at lower temperature. Clausius, who first stated the principle of Carnot in a manner consistent with the true theory of heat, expresses this law as follows:—

It is impossible for a self-acting machine, unaided by any external agency, to convey heat from one body to another at higher temperature.

Thomson gives it a slightly different form:—

It is impossible, by mean of inanimate material agency, to derive mechanical effect from any portion of matter by cooling it below the temperature of the coldest of the surrounding objects.

By comparing together these statements, the student will be able to make himself master of the fact which they embody, an acquisition which will be of much greater importance to him than any form of words on which a demonstration may be more or less compactly constructed. (Maxwell 1902, p. 153)

Maxwell's insouciance about a precise verbal formulation, which could serve as an axiom for demonstrations, may annoy modern formalistic sensibilities. But perhaps it is the height of wisdom. After all, the very distinction between "mechanical energy" and "heat energy" that underlies the engineering of heat engines is not itself precise. Some contemporary discussions of Maxwell's demon focus on "gases" supposedly consisting of a *single atom* confined to a box or a piston. But what useful distinction could there possibly be between the "heat energy" and the "mechanical energy" of a single atom, bouncing around in a container? What could it mean to say that such a system is, or is not, in "thermal equilibrium"? The basic notions that concern Maxwell are inherently vague. The distinction between heat energy and mechanical energy in a steam engine is clear enough for practical purposes, but there is no exact line that separates this case from the single atom. And if these concepts are inherently vague, it's a fool's errand to seek a perfectly precise statement of the second law.

Not only did Maxwell not seek exactitude in the statement of the law, he immediately recognized that that whatever validity it has does not follow from fundamental physical principles. The passage cited above continues:

Suppose that a body contains energy in the form of heat, what are the conditions under which this energy or any part of it may be removed from the body? If heat in a body consists in a motion of its parts, and if we were able to distinguish these parts, and to guide and control their motions by any kind of mechanism, then by arranging our apparatus so as to lay hold of every moving part of the body, we could, by a suitable train of mechanism, transfer the energy of the moving parts of the heated body to any other body in the form of ordinary motion. The heated body would thus be rendered perfectly cold, and all its thermal energy would be converted into the visible motion of some other body.

Now this supposition involves a direct contradiction to the second law of thermodynamics, but is consistent with the first law. The second law is therefore equivalent to a denial of our power to perform the operation just described, either by a train of mechanism, or by any other method yet discovered. Hence, if the heat of a body consists in the motion of its parts, the separate parts which move must be so small or so impalpable that we cannot in any way lay hold of them to stop them. (Maxwell 1902, pp. 152–154)

In sum, according to the kinetic theory of heat, heat just *is* the kinetic energy of the small parts of a body, and kinetic energy is a paradigm form of mechanical energy. So there cannot possibly be any fundamental physical principle limiting the conversion of "heat energy" into "mechanical energy." The kinetic theory of heat itself is incompatible with ascribing some inviolable status to the second law, as Maxwell understands it.

Maxwell's demon is introduced much later in the book, in a rather offhand fashion. This is appropriate, since the point being made there is essentially the same point as is made in the passage above. The additional bit of information Maxwell now has to hand is the velocity distribution formula for a gas at equilibrium. The spread of the distribution permits an especially simple way to imagine a violation of the second law:

LIMITATION OF THE SECOND LAW OF THERMODYNAMICS

Before I conclude, I wish to direct attention to an aspect of the molecular theory which deserves consideration.

One of the best established facts in thermodynamics is that it is impossible in a system enclosed in an envelope which permits neither change of volume nor passage of heat, and in which both the temperature and the pressure are everywhere the same, to produce any inequality of temperature or of pressure without the expenditure of work. This is the second law of thermodynamics, and it is undoubtedly true as long as we can deal with bodies only in mass, and have no power of perceiving or handling the separate molecules of which they are made up. But if we conceive a being whose faculties are so sharpened that he can follow every molecule in its course, such a being, whose attributes are still as essentially finite as our own, would be able to do what is at present impossible to us. For we have seen that the molecules in a vessel full of air at uniform temperature are moving with velocities by no means uniform, though the mean velocity of any great number of them, arbitrarily selected, is almost exactly uniform. Now let us suppose that such a vessel is divided into two portions, A and B, by a division in which there is a small hole, and that a being, who can see the individual molecules, opens and closes this hole, so as to allow only the swifter molecules to pass from A to B, and only the slower ones to pass from B to A. He will thus, without expenditure of work, raise the temperature of B and lower that of A, in contradiction to the second law of thermodynamics.

This is only one of the instances in which conclusions which we have drawn from our experience of bodies consisting of an immense number of molecules may be found not to be applicable to the more delicate observations and experiments which we may suppose made by one who can perceive and handle the individual molecules which we deal with only in large masses. (Maxwell 1902, pp. 338–339)

It seems worthwhile, and only fair, to point out that the claim that Maxwell makes in this passage, and the use that the imaginary demon is put to in the argument, *is perfectly correct and accurate.* In particular, modern discussions about whether the entropy of the demon must inevitably rise if he is to do what is described *are irrelevant to the point Maxwell is making.* Let the entropy of the demon rise as much as you like: still, he accomplishes the task of changing the thermodynamic state of the gas from equilibrium to disequilibrium without doing any work on the gas, changing the volume of the container, or allowing heat to flow in or out of the container. Thermal energy in the gas would become available, after the demon is done, to be converted into mechanical energy by a normal heat engine, and in the process no heat would have flowed from the gas to a surrounding body at lower temperature. After the action of such an engine, the new equilibrium state of the gas would have a lower temperature, and work would have been extracted from the system without heating a cooler external reservoir. *This is a violation of the second law as Maxwell has stated it,* and nothing in the laws of physics prevents the existence of such a demon. So the second law, as Maxwell understands it, cannot be a logical consequence of the basic dynamical laws. Period. End of story.

All of the tortured modern discussion of Maxwell's demon has exactly zero relevance to Maxwell's point. These discussions arise only when one decides to mean something else by "the second law of thermodynamics" than Maxwell meant, and hence when one decides that what counts as a *violation* of the second law is something different from what Maxwell had in mind. For example, this new second law might be stated as follows:

The Entropy of a Closed System Never Decreases

Let's call this the *Modern Entropic Version* (MEV) of the second law. There are nearby cousins of the MEV one might also consider, such as the claim that the entropy of a closed system not at equilibrium (i.e., not in its maximum entropy state) always increases. It is not worthwhile to quibble over these variations. One thing to notice is that the modern version is stated in terms of *entropy*, rather than in terms of heat and work and mechanical energy. So in order to even understand the MEV, one needs a disquisition on entropy. Maxwell's version never so much as mentions entropy. And one needs to discuss "closed system," although the application of that term seems clear enough.

For example, Maxwell's box of gas, with the demon sealed inside, is a closed system. And the total entropy of the system includes the entropy of the demon. So the question of the entropy increase of the demon is relevant to the MEV even though it is irrelevant to Maxwell's understanding of the second law. And Maxwell's example is not *obviously* a counterexample to the MEV, even though it is obviously a counterexample to his second law. So one wonders if the MEV is on firmer foundations, with respect to the basic dynamical laws, than Maxwell's version.

David Albert points out in *Time and Chance* that the MEV, taken literally at its word, is no better off than Maxwell's version, and is equally subject to counterexample-by-demon.[1] The trick is to take the demon out of the box. In fact, the trick is to employ the services of *Laplace's* demon (who has much greater abilities than Maxwell's ever needed). The Laplacian demon can arrange the exact phase point (or near enough) of a box of gas in such a way that, *after the box is sealed up again and isolated,* it will evolve, in a predictably short period, from equilibrium to disequilibrium. We know that such phase points exist, and nothing in

---

[1] Albert 2000, pp. 103–104.

the laws of nature prevents such an external demon. So the MEV, taken at face value, cannot be even a statistical consequence of the fundamental laws.

Albert comments:

> These (however) are plainly not quite the sorts of demons that either Maxwell or his critics have in mind. The demons we've just been talking about (after all)—notwithstanding that they are statistically reliable producers of violations of the literal prohibitions of the second law of thermodynamics—will not necessarily leave the entropy of the world any lower (once the whole business is over) than it was when we decided to put them to work. And so they aren't the sorts of demons than can make you any *money*—they aren't the sorts of demons (that is) with whose help one can reliably increase that portion of the total energy of the world which is available for our exploitation by gross mechanical procedures. Let's call them *pseudo*-Maxwellian demons, then. (Albert 2000, p. 105)

I would like to register a historical objection here. Whatever his critics have in mind, this is *exactly* the sort of demon that Maxwell was concerned with. The perceptive and calculational and manipulative powers of this demon vastly outstrip Maxwell's: that is the price of taking him out of the box. But this demon makes *Maxwell's* point just the same. Recall: Maxwell would not care if the entropy of the demon goes up even if he is *inside* the box.

Prospects for the MEV

In order to discuss the MEV, one needs an account of entropy. Albert adopts what is now called the Boltzmann entropy: the entropy of a system is proportional to the logarithm of the volume of phase space occupied by the macrostate of the system. This definition inherits any vagueness inherent in "macrostate," so one should not be under the illusion that the MEV is somehow perfectly precise. Albert tries to finesse this point by boundless generosity. Let the division of the phase space into macrostates be *anything you like:* still he will prove the possibility of a demon that, even when included as part of the system, systematically and predictably reduces the entropy of the system (see Albert 2000, p. 110). Hence, the MEV cannot follow from the microdynamics.

Now if one becomes *too* generous here, the existence of such spontaneously entropy-decreasing systems can be rendered uninteresting. For example: Let the system be a box of gas with a single visible dust mote in it. And let the volume of the box be divided, by a set of thin wires, into a collection of regions. There is one central region containing one tenth of the volume of the whole box, and 9,000 smaller regions arrayed around the large one, each containing one ten-thousandth of the volume of the whole box. And let the macrostate of this system, by stipulation, be the usual thermodynamic parameters together with a specification of *which region the dust mote is in*. Clearly, the maximal entropy state of such a system occurs when the gas is in thermal equilibrium and the dust mote is in the large central region. If the dust mote wanders off into one of the smaller regions, the entropy, as we have defined it, will go down by $k_B \ln(1000)$.

Suppose we start this system off in its maximal-entropy state, i.e., in thermal equilibrium with the dust mote in the center. By random Brownian motion, the mote will almost certainly soon wander off into some small region or other, and the entropy (as we have defined it) will go down. So, by definition, we have created a system that violates the MEV.

Now this really is a demon that does not illustrate what Maxwell had in mind. The decrease in entropy, as we have defined it, has nothing at all to do with any physical change that makes heat energy available for conversion into mechanical energy. This demon absolutely does not allow you to "make money" by converting heat energy into work, and not because the entropy decrease in one system is overbalanced by an entropy increase elsewhere. The point of the example is to show that not just any old definition of "macrostate," together with the corresponding definition of "entropy," renders an MEV that makes contact with Maxwell's concerns.

The dust-mote-in-a-box example may seem perverse: no one would spontaneously suggest such a division into macrostates. I have invented the example for pedagogical purposes soon to be revealed. But exactly the same point can be made about absolutely standard examples of "macrostates." Suppose our gas is composed of two species of particle, the same in all respects save color. One form of the particle is yellow, the other blue, and there are equal numbers of each. Then the macroscopic color distribution of the gas would standardly be accepted as part of its macrostate. The maximum entropy state would be one in which the gas appears to be uniformly green. If all the blue particles were on one side of the box and all the yellow on the other, or if the blues were so distributed in the yellows to spell out *mene, mene, tekel, upharsin*, the macrostate would be of astonishingly low-entropy. One would expect either of these "low-entropy" states to spontaneously relax to equilibrium, and never to spontaneously emerge from equilibrium.

Still, suppose—*mirabile dictu*—such a thing did spontaneously happen. Or suppose that Albert's Laplacian demon arranged for it, so you could expect it to happen. Still, you couldn't *make any money* from this spectacular reduction in "entropy": not an erg of heat energy would become available for mechanical purposes. The irreversible mixing of the segregated colors into a uniform green is a prototypical example of the MEV, but it makes no contact at all with Maxwell's concerns.

Maxwell's own account of the second law, of course, has no problems in this regard, because his versions are all *stated* in terms of "making money"—that is, in terms of converting heat energy into mechanical energy. So there is no question whether a violation of Maxwell's version of the second law would be of a certain practical use: it would by definition.

It might seem that Albert's perfect generosity in how to define macrostates circumvents this problem. The generosity does not *prevent* you from making a stupid choice of macrostates, but is certainly *permits* you to make an intelligent choice. But matters are not quite so simple. For this generosity does not mean that we don't have to figure out what "stupid" and "intelligent" mean here. Which choices of macrostate make the appropriate contact with "making money" (i.e., converting heat to mechanical energy) to render the resulting MEV interesting enough to even bother with? And what, exactly, determines the distinction between "intelligent" and "stupid"? Albert never addresses these questions. I think they lay at the heart of a certain puzzle, or lacuna, in his argument.

The Albertian Demon

One advantage of the MEV, if one defines entropy in terms of volumes of phase space, is that one can immediately identify a feature of the fundamental dynamics that has fairly direct consequences for the MEV—namely, Liouville's theorem. Since Liouville's theorem states that the microdynamics preserves volumes in phase space, the definition of entropy in terms of those volumes means that the theorem has some consequences for the dynamics of entropy. Let's derive one such consequence.

Once we have divided the phase space up by a macrotaxonomy, every microtrajectory—i.e., every trajectory of a phase point—determines a unique macrotrajectory. This raises the question of the possibility for a useful macrodynamics: Given only the macrostate of a system, can there be reliable generalizations about how that macrostate is likely to evolve? We can say that a system is *macropredictable* if such a macrodynamics exists: the macrostate reliably evolves in accordance with a dynamics formulated in terms of the macrostates. To take the hackneyed example, an ice cube in a glass of hot water, isolated from the rest of the universe, is macropredictable: in such cases the ice cube reliably melts. Given details about the size and shape of the ice cube, the temperature of the water, and so on, this can be refined. The melting will take place at a predictable rate, etc. Note that to *prepare* such a macropredictable system in its initial state, we only need to be able to control its macrocondition. As Maxwell points out, we don't need access to the fine microscopic details.

The explanation is simple. Such a system is macropredictable if and only if the vast majority of the microstates compatible with the initial macrostate evolve, via the microdynamics, in such a way as to produce the same macrotrajectory. But that means that the overwhelming volume of phase points consistent with the initial macrostate must generate that macrotrajectory. So if the macrotrajectory implies that the system will be in some given macrostate at a given time, that macrostate must occupy at least as much volume of phase space as the initial macrostate did. If it didn't, Liouville's theorem would be violated.

In short, Liouville's theorem implies that no macropredictable system can macropredictably evolve so that its entropy goes down.

The dust-mote-in-a-box system does not violate this theorem. Although it is macropredictable *that* the entropy of the system will go down, the system is not, by our definition, macropredictable. The dust mote undergoes Brownian motion, and we cannot macropredict exactly which macrostates it will go through, in what order, as time goes on. This inability to predict the macroevolution from the initial macrostate is *essential* to the dust-mote-in-a-box's ability to reliably reduce its "entropy."

The main point of Albert's discussion of Maxwell's demon is exactly to emphasize the gap between

> (P) No closed macropredictable system can exist whose entropy macropredictably goes down.

and

> (Q) No closed system can exist whose entropy macropredictably goes down.[2]

P follows from Liouville's theorem plus definitions, and Q does not. But one could, in principle, make money with a system that violates Q and satisfies P.

---

[2] By "macropredictably" here, I mean the same thing as Albert does when he uses "robustly"—that the behavior can be predicted with very high probability from the initial macrostate in the usual way. That is, such behavior is typical.

The critical additional qualifier "macropredictable" is entailed if one requires that the demon act in a cycle and also that the precise final state of the target system be macropredictable. The essential issue for the demon is not the cyclical property *per se:* *any* sort of macropredictabilty of the demon would work just as well, so long as the entropy of its macropredictable final macrostate D_f is no larger than the entropy of its initial macrostate D_i. For suppose it is macropredictable at the initial time that (1) that the demon will end up, after a given period of time, in the particular macrostate D_f, and (2) the target system will end up at that same time in a certain final macrostate T_f. Then the whole system will macropredictably evolve from $T_i + D_i$ to $T_f + D_f$. So the entropy of $T_f + D_f$ cannot be less than the entropy of $T_i + D_i$. Hence if the entropy of D_f equals that of D_i, then the entropy of T_f cannot be less than that of T_i. And if the demon works in a (macroscopic) cycle, $D_i = D_f$, and *a fortiori* the entropy of D_f equals that of D_i. So the entropy of the target system cannot have gone down.

Albert blocks this inference by eliminating the requirement that the demon be macropredictable. One could achieve the same end by eliminating the requirement that the state of the target system be macropredictable: one might be able to predict that the target system will end up in *one or another* of many distinct low-entropy macrostates, but not be able to predict which. In either case, the link to Liouville's theorem is broken, and so nothing evident stands in the way of constructing a demon. Albert concludes that nothing *does* stand in the way, or at least that we have, at present, no reason to think anything fundamental stands in the way. But the situation is not so simple.

How to Make a Demon

Time and Chance does not contain working blueprints for an Albertian demon. The closest one comes is the abstract diagram on page 107, which illustrates a particular flow on phase space (Figure 8.1). The idea is that the large phase volume of the initial state of the demon + target system flows into a collection of disjoint, macroscopically distinct regions of phase space. Liouville's theorem is respected because the total volume of all the final regions taken together is the same as the volume of the initial state, but each of the N possible final states has a volume which is 1/N of the original. So the final entropy will be $k_B \ln(N)$ less than the original. This is obviously the same idea as the dust-mote-in-a-box.

The problem with the dust-mote-in-a-box was that the macrotaxonomy is stupid: reduction of entropy, defined in this way, cannot make energetic money.

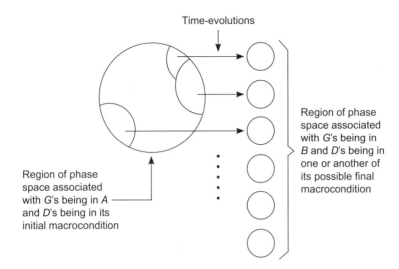

FIGURE 8.1.

Now one may harbor the same suspicion about Albert's abstract demon. No doubt, *given a Hamiltonian*—i.e., *given a flow on phase space*—one can *invent* a macrotaxonomy that has the features illustrated in Albert's diagram. Just take the initial macrostate, let it evolve for some period, then partition the resulting region of phase space into N equal-volume pieces. *Given this macrotaxonomy,* one would have created an Albertian demon. But, of course, there is no reason at all to think that a reduction of entropy, so defined, connects up to issues of heat energy or mechanical energy at all.

Albert's generosity with the macrotaxonomy, though, seems to answer this. His pitch appears irresistible:

> Here's the deal: you tell me how you want to carve the phase space up into macroconditions, and I will be able to come back to with a design for a Maxwellian demon which will be robustly capable of lowering the entropy of (as *you* calculate it) of a larger isolated system of which it forms a part. (Albert 2000, p. 110)

Who could refuse such an offer? The issue of stupid macrotaxonomies seems to be moot: one doesn't invent the macrotaxonomy to suit the flow on phase space, as suggested above, but rather designs the demon to suit the macrotaxonomy.

But like most offers that seem too good to be true, Albert's contains a hitch. What, exactly, does Albert have in mind by "I will be able to come back with a design"? As I have discovered by asking, he has in mind this: for any fixed macrotaxonomy *there exists a Hamiltonian that produces the right kind of flow.* That is, instead to seeing where the dynamics takes the initial macrostate and dividing up that patch, Albert proposes to take the target final macrostates as given and then specifies a dynamics that splits the initial macrostate among them. The one constraint is Liouville's: the sum of the volumes of the possible final macrostates must be at least as great as the volume of the initial macrostate.

But if there one thing in the universe that is not amenable to our intervention and design, it the fundamental Hamiltonian! *That* is handed to us by nature, with no avenue of appeal. So the observation that, as an abstract matter, a convenient Hamiltonian exists that splits up the initial volume in the right way does us no good at all. That is not a "design," it is an idle wish.

And that's not the worst of it. Suppose I grant Albert the fantastic ability to specify the fundamental Hamiltonian of the universe as he wishes. Even that handsome gift does not solve our problems. For Albert's seeming generosity—you give me whatever macrotaxonomy you like and I'll design a Hamiltonian to suit your needs—is smoke and mirrors. It is like offering your opponent in rock-scissors-paper the free choice of any move at all, so long as he provides it to you first and you can premise your own move on his. Recall, one thing we want to avoid are stupid macrotaxonomies—macrotaxonomies that do not connect in the right way with concepts like *heat energy* and *mechanical energy.* Reducing the entropy relative to a stupid macrotaxonomy will not allow you to make money. *But the properties of energy in a system are determined by its Hamiltonian.* Indeed, the Hamiltonian typically provides a measure of the total energy of a system. So I *can't* settle on a smart macrotaxonomy—a macrotaxonomy in terms of which the entropy measures usefully available energy—unless the Hamiltonian is *fixed.* No matter how hard I try, I may not end up with what I want. If Albert's "design" changes the Hamiltonian of the system, then it can change the relevance of the macrotaxonomy I settled on. The seemingly generous offer is empty.

It's not just that Albert's offer seems too good to be true: once the Hamiltonian is fixed, it is provably is too good to be true. That is, in the sort of case we are concerned with, it is provably *impossible* to design the sort of demon that is suggested by Figure 8.1. This is not to say that some less efficient, and hence less desirable, Albertian demon cannot be made. But it does show that there are more considerations at play than Albert's discussion suggests.

The Perfectly Efficient Albertian Demon

The official definition of an Albertian demon is given in the passage above: the Albertian demon is "robustly capable of lowering the entropy (as *you* calculate it) of a larger isolated system of which it forms a part." We have already seen how the idea of such a demon avoids Liouville's theorem by a divide-and-conquer strategy: almost all of the total volume of the initial macrostate flows into N disjoint possible final macrostates, each of which has a lower volume and hence a lower entropy than the initial one. Now the most efficient use of these N possible final macrostates occurs when each state gets *filled up* under the Hamiltonian flow. That is, the *smallest volume* (and hence the *lowest entropy*) possible for these final states (assuming them all to be of equal size) obtains when each is exactly 1/N of the volume of the initial state, and hence when (almost all) of the volume of each final state is occupied by a phase point that originates in the initial macrostate. If the final macrostates are not so efficiently filled under the flow, then the overall ability of the demon to reduce entropy will suffer. As an obvious example, if only half of the available volume of each final macrostate comes to be occupied under the flow from the original macrostate, then the volumes of the final states must be 2/N of the initial state, and the entropy is only reduced by $k_B \ln(N/2)$. The perfectly efficient Albertian demon gets the most bang for the division into N possible macroscopically different end states by using all their phase volume. If we were able to freely specify the Hamiltonian, and hence the flow on phase space, without undercutting the significance of the macrotaxonomy, then such an Albertian demon would be possible.

There is a second sort of perfect efficiency we can ask of our demon. Consider Maxwell's own description above: the demon checks every particle that approaches his door and determines whether its velocity is greater or less than the initial mean velocity of the gas. If it is greater, he allows the particle to pass from side A to side B, and if it is less he allows the particle to pass from side B to side A. Slow particles on side A and fast particles on side B are not allowed to pass. This is clearly the most efficient way to produce a temperature difference between the sides: allowing slow particles to go from A to B or fast particles from B to A would undo the effect he is trying to achieve. So let's build this sort of efficiency into the definition of the perfectly efficient Albertian demon as well. If there is nothing to prevent the construction of an Albertian demon, it is hard to see what could prevent constructing one that is perfectly efficient in this sense.

Here's the rub: a perfectly efficient Albertian demon, as defined, is provably impossible. Let's take Maxwell's own design, but make its program discrete in the following way. Divide time into equal, finite, discrete segments Δt_i. For each Δt_i, the demon must decide whether to have the door open or closed for that interval. He only opens the door during a particular interval if that will increase the temperature difference in the right way. He continues to operate until some target temperature difference has been achieved, and then he stops. At the end of every possible run of the demon, the door has been open or closed during each Δt_i, so the net effect of the demon on the gas in each run is determined by a particular sequence of openings and closings.

Clearly, the demon must operate differently for each distinct sequence of openings and closings that he can produce. Albert's trick for keeping the entropy of the demon from rising as it operates is to associate each possible different sequence of actions of the demon with a different possible *final macrostate* of the demon. In this way, even though there may be a huge collection of possible final macrostates for the demon, *each one* can have an entropy no larger than the demon's initial entropy. There is no reason, from purely entropic considerations, that the physics of the demon cannot be thermodynamically reversible, and hence not increase its entropy.

Recall: we want a perfectly efficient demon to *completely fill* the phase space of the possible final macrostates of the demon + target system. That is, if we track how the whole phase space of the initial macrostate evolves under the Hamiltonian flow, it divides up and completely fills the phase space of each possible end state. Each possible end state, given our construction, corresponds to the demon producing a particular program of opening and closing the door. And as far as the gas is concerned, the only effect that the demon has is via the door: the rest of the operation of the demon is irrelevant to how the gas evolves. But now we reach an absurdity.

The sort of demon we are considering, if possible, should be consistent with Newtonian mechanics, and Newtonian mechanics is time-reversible. So consider running the demon + target system *in reverse,* from the time-reverse of a particular final microstate to the time-reverse of the initial microstate. And focus, in this reverse running, only on the gas, the box, and the door: don't worry about the demon itself. In one respect, the reverse running of this sequence will look thermodynamically normal: the entropy of the target system, the box of gas, will go from having a temperature difference to equilibrium, from low entropy to high. But in another respect, the reverse running will look fantastically *atypical* and strange. For not only does the temperature of the gas equilibrate, it equilibrates *in the most efficient possible way.* Whenever the door is open, hotter particles *only* pass from the hotter side to the cooler, and cooler particles *only* pass from the cooler to the hotter. Such a perfectly efficient equilibration would be fantastically improbable in any gas: if one randomly opens and closes such a door when the temperature on the two sides is nearly the same, particles ought to pass the "bad" way almost as often as they pass the "good" way. So although some equilibration of temperatures is to be expected just because the door is being opened and closed, such a perfectly efficient equilibration would be fantastically improbable.

Here, then, is the problem. Consider tracking the trajectories of all the phase points in the initial macrostate together. They spread out to fill the phase volume of each possible end state, and each end state is associated with a particular sequence of opening and closing the door. When played in reverse, these same microtrajectories retrace their routes, flowing back into the time-reverse of the initial macrostate, while on each route the door opens and closes in reverse sequence. But if the phase space associated with each possible final macrostate has been *filled*, then the reverse sequence will be a *thermodynamically typical* evolution for a gas in the time-reverse of the final macrostate that is subject to the time-reverse of the sequence of opening and closing the door. However, if one were to prepare a box in the final macrostate, with the temperature difference between the sides, and open and close the door in any fixed sequence at all, such perfectly efficient equilibration of the temperatures will essentially never occur. Far from being thermodynamically typical, the microscopic efficiency of the reverse evolution of the gas is massively atypical.

The upshot of all this is that *no Albertian demon can be perfectly efficient in both senses*. And since there is no plausible barrier to building one that is perfectly efficient in the second sense (only allowing particles to flow the right way between the sides of the box), it must be impossible to build one that is perfectly efficient in the first sense. That is, no Albertian demon, constructed to act like Maxwell's, can produce a flow on phase space that fills the phase volumes of the possible macroscopic end states.

Indeed, no Albertian demon that is perfectly efficient in the second sense can come anywhere near being perfectly efficient in the first sense. The perfectly efficient equilibration of temperature displayed in the time-reversed sequence is so fantastically unlikely (given any fixed sequence of opening and closing the door) that the volume points in the phase space of any final macrostate that generate such time-reversed behavior must be unimaginably tiny. So in order for the Albertian demon to reduce the entropy of the target system by $k_B \ln(N)$, there must be fantastically more than N possible end states, contrary to what Figure 8.1 suggests.

A Back-of-the-Envelope Calculation

Let's review the bidding. Thermodynamics arose from practical questions: how can mechanical work be extracted from heat energy? Given a gas in a box, for example, can work be extracted from the system and paid for by having the temperature go down?

Clearly, work can always be extracted in one way: let the box expand and do work. Then we have a piston, and as the piston expands, the temperature of the gas goes down. The work done by the gas can be put to whatever mechanical use we like.

But from a practical point of view, this observation is underwhelming. It is not much use to advise extracting work from the gas by putting it in a piston, letting it expand, *and then throwing the piston away.* Thus arose the desire to design a heat engine that could work on a cycle. But such a heat engine requires two thermal reservoirs at different temperatures, and can cyclically produce mechanical work only by extracting energy from the higher temperature and expelling some of it into the lower. It would be convenient to be able to convert the energy in the gas into work without the need for a cooler reservoir.

Now in some cases it is easy to extract energy from a gas and use it to run machines without a cooler external reservoir. For example, if the *pressure* of the gas is different in different places, then we can set up a windmill. And if the *temperature* of the gas is different in different places, then we can run a heat engine between the locations. But if our gas-in-a-box has come to equilibrium, so the temperature and pressure are everywhere uniform, and if we are not allowed to expand the volume of the box, and make it into a piston, and we do not allow heat to flow out of the box, using it as a reservoir for a heat engine, then we seem to be done. There is no longer any obvious way to convert the heat energy into mechanical work.

This is what Maxwell took to be implied by the second law: "It is impossible in a system enclosed in an envelope which permits neither change of volume nor passage of heat, and in which both the temperature and the pressure are everywhere the same, to produce any inequality of temperature or of pressure without the expenditure of work." If such an inequality could be produced *without doing work on the gas,* then heat energy of the gas could be converted into mechanical energy without any investment in work. And of course, Maxwell's demon does just this: he never so much as touches *the gas,* he only opens and closes the door. So if we take the second law to imply what Maxwell took it to imply, his demon shows that it cannot follow from fundamental physics. Considerations about the entropy of the *demon* are neither here nor there.

Once we switch from Maxwell's understanding of the second law to the MEV, the situation changes entirely. The demon is part of the closed system, so its entropy does become relevant to the validity of the MEV. And we have seen that one can derive the impossibility of a macropredictable demon + gas system, and hence the impossibility of a cyclic demon, from Liouville's theorem and the Boltzmannian definition of entropy. But, as Albert points out, if one eliminates the requirement of macropredictability, this argument does not go through. So the question arises whether a noncyclical demon is possible.

From a purely practical point of view, a noncyclical demon is of no more use than the throw-away piston. True, it can extract mechanical work from heat energy in a gas in equilibrium, without changing the volume of the gas or doing work on it, but so what? If all we wanted to do was extract mechanical work from the gas, put it in a piston and *voilà*. We are not satisfied with this solution because we don't want to use the piston once and then throw it away. A throw-away demon is of no more practical use than this.

Albert can claim that his concerns were never practical in the first place: all he wants to know is whether one can design a closed system that will reliably result in the entropy of the universe going down. And by his divide-and-conquer strategy, he has shown that Liouville's theorem does not rule this out. I have not challenged this conclusion. But at the end of his chapter, Albert poses a rather practical question. He asks whether it might be possible to parlay the fact that the demon cannot operate on a cycle

> into a demonstration that the construction of a Maxwellian demon system, or the *operation* of a Maxwellian demon system, or the *extraction of mechanical energy* from a Maxwellian demon system by means of heat engines (once its operations are done), or the *exploitation* of that energy (once its been extracted) will necessarily—as a statistical matter—somehow prove prohibitive or self-defeating or uncircumventably pointless. (Albert 2000, p. 112)

I'm not sure what "prohibitive" and "pointless" exactly mean here. One might take the view—a designer of heat engines would—that a use-it-once-and-throw-it-away demon is already pointless: any mechanical energy extracted from the system by its use is not worth the effort. But the last observation I would like to make is that there is another obvious sense in which the operation of an Albertian demon is prohibitive and practically pointless. That arises simply from the scale of magnitudes involved.

As a rough example, suppose we have 10^{20} gas particles in a box. If the gas is at equilibrium, and we can't change the volume of the box, and there is no heat exchange with the world, then it is not obvious how to get mechanical work out of the heat energy. But if all of the gas particles happened to be on the right side of the box, we could use windmills or internal pistons to extract mechanical energy without changing the overall size of the box. So getting an Albertian demon to collect all the particles on one side, without doing any work on them, would pay off in terms of the mechanical energy available.

Such a demon would work like Maxwell's, but with a simpler job: instead of letting fast molecules go one way and slow molecules the other, producing a temperature difference, he merely allows molecules to pass in only one direction. By opening and closing the door in the right way, he will eventually capture all the molecules on one side.

Supposing the gas starts out in equilibrium, by how much will its entropy be reduced in this process? That's easy to calculate: the velocity distribution of the molecules is completely unchanged, and the spatial distribution goes from being uniform over the whole box to being uniform over half the box. So the phase space available to each molecule is cut in half. The total volume of the phase space of the macrostate is reduced by a factor of $2^{10^{20}}$. So in order for an Albertian demon to achieve this much of a reduction in it's target system (the box of gas at equilibrium) by the divide-and-conquer strategy, a perfectly efficient demon must have $2^{10^{20}}$ possible macroscopically distinct final states that it can end up in.

We have furthermore seen that a perfectly efficient demon is *not* possible. The time-reversed behavior of a gas separated by this demon will be far from typical: whenever the door is open, molecules only go one way. Of course, at the beginning (in the time-reversed sense of "beginning"), when the molecules are all on one side, they only can go one way. But when 40 percent of the molecules have come to occupy the initially empty side, we would expect the chance of a molecule passing in the right way when to door is opened to be only 60 percent. Since the demon only allows particles to go one way, the time-reversal of the process only allows particle to go the other. Just for the final 10^{19} particles, the chance of this time-reversed behavior is less than $(.6)^{10^{19}}$. So the volume of the phase space of each final macrostate that is filled by the operation of the demon must be much less than $(.6)^{10^{19}}$. In order not to violate Liouville's theorem, the realistic Albertian demon must have more than $2^{10^{14}}$ possible end states for each state that the perfectly efficient demon needs. In other words, a realistic demon must operate so as to choose among more than $2^{(10^{20}+10^{14})}$ macroscopically distinct end states— states that differ in the macroscopic placement of billiards balls, or planets, or buildings—in order to confine the gas on one side. Note that the necessary inefficiency, which makes most of the phase volume of the end states inaccessible, creates less than a rounding error on this number.

What would the payoff be? Well, if the gas were at room temperature, initially confined to a two-liter box, and after the operation of the demon were confined to only 1 liter, we could let the gas expand back to its original volume and extract mechanical work from it. The energy in work thus recovered would be less than one tenth of a joule: not enough to lift a small apple three inches straight up. And after the demon has done its work, we would throw it away.

And, of course, we here assume that the demon manages to move its billiard balls, or buildings, or planets without any *friction*. If, in placing the billiard balls in the correct one of the $2^{(10^{20}+10^{14})}$ possible macroscopic states, the demon loses more than a tenth of a joule of mechanical energy to friction, then the whole thing is a dead loss.

I said that I'm not sure what Albert means by "prohibitive or self-defeating or uncircumventably pointless," but as a means of recovering mechanical energy from the heat energy of the gas, this seems to qualify. We could have gotten about half of our tenth of a joule just by letting the original equilibrium gas in the two-liter box expand like a piston, and then throwing *it* away. Furthermore, the extra inefficiency of the realistic demon over the impossible perfectly efficient demon does not really matter: even the perfectly efficient demon would be just as pointless.

But still we have learned something in our investigation. We have seen the important gap between the second law as Maxwell conceived it and the MEV. We have seen the importance of picking the right macrotaxonomy if entropy, as defined by that taxonomy, is to make any connection to issues of heat and work and energy. A bad choice of macrotaxonomy allows us to create the dust-mote-in-a-box demon, which seems beside the point. We have seen that "designing a demon" cannot usefully be understood as "picking whatever Hamiltonian I like." And we have seen that modern concerns about the entropy increase of the demon have no bearing on Maxwell's original point.

References

Albert, David. (2000). *Time and Chance*. Cambridge, MA: Harvard University Press.
Maxwell, J. Clerk. (1902). *Theory of Heat*. London: Longmans, Green, and Co.

PART III

UNDERWRITING *the* ASYMMETRIES *of* KNOWLEDGE *and* INTERVENTION

Chapter Nine

Reading the Past in the Present

▶ NICK HUGGETT

1. Introduction

Why is our knowledge of the past so much more "expansive" (to pick a suitably vague term) than our knowledge of the future? And intimately related, how can we capture the difference(s): i.e., in what sense is knowledge of the past more "expansive"? As a first stab, one might be convinced by the first four chapters of David Albert's *Time and Chance* (2000)[1] that the "Newtonian statistical mechanical contraption for making inferences about the world" (96) captures everything that can be inferred by statistical mechanics at any time. Indeed, in this paper I will assume that it does. One might then reasonably wonder whether it would even be possible to know anything that didn't follow from the contraption—so doing would apparently take more knowledge of the past than the past hypothesis (PH), or more knowledge of the present state than its current macrocondition, or a more informative probability distribution, and how could one obtain any of those?[2] So one might suggest that:

---

[1] Henceforth *T&C*—all quotations from Albert are from this work unless otherwise stated.
[2] Here and elsewhere in this paper, "knowledge" refers to "empirical knowledge"—knowledge that might be obtained from experience either directly, indirectly, or inductively. Granted, that is not

> *A proposition can only be known if it is made likely by taking a uniform probability over all states compatible with the current macrocondition of the universe, and with its initial macrocondition, given the laws of mechanics.*

Then, since the initial state is one of considerable "order," given by the PH,[3] many more interesting propositions about the past than about the future will satisfy the necessary condition.

Such a necessary condition can't be the end of the story: it is asymmetric, permitting more knowledge of the past than of the future, but it doesn't make clear how we come to have any such knowledge. For instance, it would be manifestly wrong to suggest that the condition describes how we reason about other times, because clearly we have knowledge of other times without knowing anything much about the nature of the early universe, and so presumably has humanity since before the earliest creation myths. Our ability to reconstruct the past is not evidence of our *knowledge* of the PH! But the condition is indeed necessary, then it could be a great help: providing a decent statement of what the asymmetry is supposed to be, before attempting to explain it.

And in fact, the treatment of the knowledge asymmetry in *T&C* (chapter 6), does proceed in much this way. It places giving a clear description of the asymmetry forefront in the discussion, as a prerequisite to an explanation, and so presents and defends a necessary condition along the lines just proposed. However, David's condition is significantly different in an important way that is not explored in *T&C*. My main goal in the first part of this paper (§2) is to explain and clarify the difference and say why it is important (§2.1). Also different is *T&C*'s argument for the condition, as I'll explain in §2.2. That is, the first part of this paper is concerned with exposition and fleshing out of David's position.

In §3, with a clear account of the proposal before us, I will argue that it doesn't succeed as presented. First (§3.1) I will describe and explore a concrete mnemonic system—an "IGUS"—that fits the general conception of memory given in *T&C*, but manifestly does not condition on the PH. Such a model of memory is needed if we are to make progress on the question of whether or not the condition of *T&C*

---

a completely precise definition, but I hope it's clear enough for present purposes. Ultimately the kind of conclusion discussed here should itself shed light on the concept; probably there has to be some sort of reflective equilibrium between the idea of empirical knowledge and the specific cognitive mechanisms by which it can be had.

[3] Two points: First, the PH is not merely that the universe started with "low entropy" (a claim that some say is "not even false") but that it started in some macrostate, which the full PH specifies. Second, for brevity, throughout this essay, and generally without further comment, I will make the usual assumption of an even probability function over microstates.

is necessary for knowledge, and it constitutes my substantive positive contribution to this project. In §3.2, I will analyze the functioning of the IGUS in more detail, exploring its physics and machine table, to provide a model of mnemonic knowledge suitable for investigation from a physical point of view. In §3.3 I argue on the basis of these considerations that the IGUS is capable of knowledge that exceeds that which is probable according to the Newtonian contraption. In §3.4 I reply to David's response to this argument, in a somewhat concessive mood.

Thus, I have three goals: first to unpack and develop the account of the knowledge asymmetry given in *T&C*; second to present a more complete model of memory than that found in *T&C*, though based on its essential insight; and third to use the model to analyze *T&C*'s claims about the knowledge asymmetry. I conclude in §4 with some comments on the knowledge asymmetry in the light of this investigation.

2. The Knowledge Asymmetry in *Time and Chance*

2.1. The "Presently Surveyable Condition"

Suppose the "ice pachinko" of *T&C* (83) is run with a single ice cube in a sealed room with a human observer watching (but making no notes or other record of events). The cube falls through the device and randomly ends up in one of the beakers—the leftmost one, say. Eventually, after the ice cube has first melted and then evaporated, there will be no trace of which beaker it fell into remaining in the macrostate of the universe: the water molecules will be randomly distributed in the air of the room, whatever beaker the ice cube fell into. The macrocondition of the universe will be the same whatever happened, and the initial macrostate of the universe surely won't determine the outcome (certainly it isn't supposed to do that kind of work in *T&C*), so the contraption should assign equal conditional probabilities for the ice having falling left or right—that it fell left is not knowable according to our necessary condition.

That's not to say that the *micro*state doesn't contain the necessary information: if the evolution is (backward) deterministic, then the earlier state can be recovered from the present microstate. But the whole game here is to work with macro-facts—idealizing, we take those to be what's knowable. What about the fact that we can discover aspects of the microstate? Perhaps there are insensible mineral traces in the beaker that could be detected by chemical analysis. Such discoveries amount to magnifying micro-differences to produce macro-differences: paradigmatically, by focusing a microscope on something otherwise too small to see. But the proposed formula for the knowable only takes into account the *present* macrocondition; that some aspect of the microstate could or will be magnified into the macrostate at a later time is irrelevant. The necessary condition says that the outcome is knowable now only if it can be inferred using the macrostate now, and it is beside the point to note that it could be satisfied if the macrostate were different, or even that it will be satisfied by some future macrostate. It isn't satisfied at the time in question.

But of course, that the ice cube fell in the leftmost beaker is perfectly knowable at that time—it is known by the observer! So the condition is not necessary, and it cannot help explain the knowledge asymmetry. And neither does this example seem especially recherché. Macro-records are being erased all the time, and the observer is relying on no strange procedures for obtain knowledge—she's just watching!

As I said, the account just refuted is not the one offered in *T&C*. Instead of conditioning on the current "macrocondition," *T&C* conditions on (what it terms elsewhere) the "presently surveyable condition" of the world: "Everything we know of the past and present and future history of the world can be deduced ... from the following four elements: what we know of the world's present macrocondition—and of our own brains, perhaps; the standard microstatistical rule; the dynamical equations of motion; the past hypothesis" (119). In fact, the "Newtonian contraption," or "Mentaculus" in more recent discussions, uses exactly these elements, not those in the condition I gave. (Apologies for the earlier misdirection, which I found useful for setting things up.) We can say for short that we can know something of the past or future only if it is given a high probability by the contraption / Mentaculus.[4]

The difference with our first condition is that we are also to condition somehow on the physical state of our brains. Before I elaborate this crucial idea, I do want to emphasize that it is rather easy to overlook in *T&C*. The passage quoted downplays its relevance; and it first appears with almost no comment or explanation (96). The fullest discussion appears in a footnote: the surveyable condition includes "whatever (perhaps [microscopic][5]) features of the present condition of the *brain* of the observer in question may be accessible to her by means of direct introspection" (114). But I agree with *T&C* that conditioning on the brain state in some way

---

[4] The passage quoted says "deduced," though generally what can be deduced from the contraption are probabilities. Perhaps the claim is supposed to be that our ideal credences align with these probabilities. If so, we will idealize by supposing that some high degree of belief is necessary and sufficient for knowledge. We won't be dealing with any problematic cases that make this simplification problematic.

[5] The text reads "macroscopic," which seems to be a typo; compare with page 96.

is essential in this project, in order to address mnemonic knowledge; the disagreement is in the details.

The idea is that the pachinko observer's knowledge satisfies this new condition: that her memories entail something about the microstate of her brain that, with the PH (and the rest of the macrocondition), make it likely that the ice fell in the leftmost beaker. The previous example should make clear why some such conditioning is absolutely necessary if we are to include memories of the past in our treatment of the knowledge asymmetry—in particular any in explanation of why we remember the past and not the future.

T&C may intend "direct introspection" to be more expansive, but in the case in hand it seems that what we need to add to the surveyable condition are our memories; or rather, since we will have to worry about their veracity, our *putative* memories.[6] By "putative" I mean that all that's relevant is the fact that the observer's brain is in a state *compatible* with her truly remembering the beaker; of course, just being in that state is no guarantee that the memory is veridical. There are many normal psychological processes that produce false memories, of course, but in the context of T&C the relevant concern is the possibility that instead of being veridical, putative memories are the result of extremely improbable fluctuations in the microstate of the world. If one conditions on *veridical* memories, then, by logic rather than physics, it follows that the remembered events really happened: that the pachinko observer veridically remembers that the ice cube fell in the leftmost beaker entails that it did. But if we want to play the game of assuming about the present only that which could (in a very broad sense) be learned by inspection of the universe now, we can't condition on the truth of memories, only their existence. Put another way, to know that a memory is veridical is to know something about the state of the past that doesn't follow from the present state and the laws, and including such information undermines T&C's whole project of understanding time asymmetries in terms of statistical mechanics. (Thus, a note on terminology: the unmodified term "memory" will now mean "putative memory.")

Even assuming (against some critics) that the Mentaculus generally works as advertised, it might seem prima facie implausible that just conditioning on an appropriate brain state could be enough to significantly raise the probability that

---

[6] Here's another reservation about the treatment of the brain state in T&C. Talk of "direct introspection" carries a suggestion of armchair psychology: it sounds as if what counts is "second order" knowledge of what our putative memories are. But the discussion should make it clear that what actually matters is just the memories themselves—not whether in some sense that we also know that we remember. I'm pretty sure David is aware of this, but for a while I was confused by the language.

the ice cube fell in a particular beaker. After all, there are lots of ways that such belief formation might go wrong—optical illusions, tricky magicians, evil scientists, wishful thinking, forgetfulness, and so on. But this objection is not a good one, because nonveridical memories formed by such processes will not make their objects likely, if such processes leave other traces in the presently surveyable condition. To give a fanciful example, consider the group of GIs "brainwashed" to hold false memories about the heroic actions of Raymond Shaw, in *The Manchurian Candidate*. Applying the laws to those memories (consistent with the PH), we assume for now, makes the true past unlikely; conditioning on the states of their brains (and the PH) makes it likely that Shaw was a hero. But whether or not he really was makes a difference to the present surveyable condition beyond those memories: in the memories of the Soviet observers, in records kept by the North Koreans, in messages sent to conspirators in the United States, and so on. If those traces are present, and if the Mentaculus is working, once we condition on the whole surveyable condition, by the same logic, it ought to make the GIs' beliefs unlikely to be true after all. More commonly, it's easy to imagine cases, like the pachinko observer, in which a veridical memory is the only effect of an event on the presently surveyable condition: watching how a wave breaks, or the milk spills, or where the first raindrop lands, and so on. But it is much harder to imagine the formation of a nonveridical memory (given the PH) leaving no traces: conspiracies require work, and work leaves traces. So it's plausible that the Mentaculus is generally good at assigning low probability to the objects of nonveridical memories—and in fact, it should be fairly obvious that the kinds of remarks just made apply, not only to memories, but to any kind of traces of the past.[7] Thus, such cases will not concern us further.

2.2. Records and the PH

Now that we understand the necessary condition more clearly, we can turn to *T&C*'s argument for it. The strategy is to give an idealized account of the methods by which knowledge of other times can be had, thus obtaining (asymmetric) upper bounds on what can be known of past and future. One method for obtaining such knowledge involves inferences from the laws and surveyable condition (and

---

[7] And what about a case in which the Newtonian contraption makes the object of a false memory highly probable, because traces of the deception have been erased? Would that be a counterexample to the proposals of *T&C*? No, because ultimately the game is to balance our beliefs and theory so that our beliefs turn out to be likely true according to the theory—and any program like that, simply because it is probabilistic, will end up implying that some false things are true.

standard measure) alone; that generally gets the future right, but of course is radically wrong about the past.[8]

The second method is that of record reading: comparing the current state of a system with its state at another time (the "ready state"), to infer (using the dynamical laws and uniform probability) the occurrence of an interaction in between the two times—at an earlier time than the present, if the ready state occurred earlier. To literally, *explicitly* make such an inference requires knowledge, not only of the current state but of the ready state as well. Whether there is some other nonliteral, implicit procedure for obtaining knowledge from a record is the subject of §3. The argument of *T&C*, however, does invoke a procedure of literal inference (without prejudice concerning the existence of other methods):

> And the puzzle is about how it is that we ever manage to *come* by [information about the ready state of a record bearing system]. It can't be by means of retrodiction/prediction. ... It must be because we have a *record* of that other condition! But how is it that the ready condition of this *second* device is established? And so on (obviously) ad infinitum. There must be ... something we can be in a position to *assume* about some other time ... the mother (as it were) of all ready conditions. And this mother must be *prior in time* to everything of which we can potentially ever *have* a record, which is to say it can be nothing other then the initial macrocondition of the universe as a whole.
>
> And so it turns out that *precisely* the thing that makes it the case that the second law of thermodynamics is (statistically) true ... is *also* the thing that makes it the case that we can have epistemic access to the past which is not of a predictive/retrodictive sort [i.e., the PH]. (*T&C*, 117–118)

That is, *T&C* gives a regress argument to conclude that knowledge of the literal record-reading kind requires us to assume the PH. If so, the basis of that kind of knowledge of other times is the presently surveyable condition, the PH, and the laws: what we can know of other times is given by the consequences of the Mentaculus, as *T&C*'s necessary condition says.

But because we do have nonretrodictive knowledge of the past—knowledge presumably garnered from records and in almost complete ignorance of the PH (i.e., of the specific macrostate of the early universe)—the regress just shows that we aren't just literal record readers. (Or perhaps we subconsciously know the PH,

---

[8] *T&C* has a nice discussion of why this method doesn't get the future quite right either, and how to fix the problem (119–122), but we don't need to take account of that here.

and are literal record readers—in which case we need psychology, not cosmology, to discover the PH!)

I hope David will forgive me having a little fun here. Of course that's not what's intended—as he's pointed out repeatedly, he does not take the argument that way! Instead we have to bear in mind that we are considering an idealized procedure, capable of capturing anything knowable, not an account of how *we* know.[9] That said, the passage perhaps invites misinterpretation by speaking in the first person; perhaps it would be clearer if it asked how "an idealized system" that employed the method of literal record reading could come by information about ready states.

A procedure that works that way is logically consistent, but there are a few important points that will feature prominently in §3. First, because the regress argument shows that literal record reading is not the procedure by which we obtain knowledge of the past, how do we do it? Until we have an answer to that question, the argument makes knowledge of the past a bit of a paradox—the resolution is not difficult, but seeing clearly how we actually do use records will advance the discussion. Second, taking the argument as I have just suggested reveals a logical gap: Since there must be other procedures, why think that literal record reading is maximal? In other words, why think that *T&C*'s necessary condition holds?

Now, approaches to the knowledge asymmetry that appeal to the second law are familiar, and *T&C* grounds the second law on the PH, so one might suspect that we have here an appeal to entropy in understanding the knowledge asymmetry. But the argument is supposed to be a novel one, breaking from views which assume that records are inevitably entropy-increasing. First, such an assumption is not made in the argument. But more importantly, as *T&C* points out, there is no way to connect its conception of a record with the second law: whether the final state is of higher or lower entropy than the ready state, or before or after it, knowledge of the two states will generally allow one to infer something about the intervening time. (Even knowledge that both states are equilibria for a system may allow one to infer that nothing happened to the system in between!)

So things are set up for §§3–4 now. First, we saw that the kind of account of the knowledge asymmetry developed in *T&C* must include our (putative) memories in the "currently surveyable state"—for there can readily be things we remember about the past that don't leave any traces in the present macrostate. Now we have seen how *T&C* characterizes knowledge of other times that is not of the prediction/retrodiction sort—not in terms of the actual processes by which we

---

[9] It's true that *T&C* doesn't say this explicitly, but the footnote on page 116, describing retrodiction in such terms, shows that this is what's intended.

know the past, but instead by reference to a hypothetical process of literal record reading. But how do we actually know things about the past? In particular, how should we understand *memory* in this framework? It seems to me that until we have thought about the actual processes, it is not at all obvious that being knowable through hypothetical explicit record reading is a necessary condition.

3. Memories Are Made of This

In this section I will develop a physical model of mnemonic knowledge, suitable for advancing the kind of analysis found in *T&C* and understanding the role—if any—of the Mentaculus in the knowledge asymmetry. As such, the following is intended as a contribution to David's project, specifically by clarifying the concept of the "presently surveyable condition" of the world. I will offer two main points: First, a computationally efficient mnemonic system will read its internal records *implicitly* (§3.1). Second, such a system will be such that its initial and final computational states are compatible with a unique series of intervening inputs (§3.2). I argue that these points suffice for mnemonic knowledge of the past, and so characterize what having a memory means for the presently surveyable condition, thereby providing the basis for an analysis of the knowledge asymmetry in physical terms. With this model in hand I will suggest that the system can have mnemonic knowledge that does not satisfy the necessary condition of *T&C* (§3.3); but I will also consider David's response, and argue that if it works, it establishes the necessary condition on very different grounds requiring the kind of model I propose (§3.4).

3.1. An Information Gathering and Utilizing System

One thing that may strike you in thinking about record reading is that when we remember, there's no consciousness of record reading going on at all: you don't generally note that you have a memory, and *infer* from that that something occurred. Generally, part of a memory just is the belief that its object occurred.[10] So if we can get a handle on such mnemonic knowledge, perhaps that will give us an "inference method" quite different from literal record reading. Anyway, memories are both central to the knowledge asymmetry and in some way surely critical to our having knowledge of the past. So they are a good topic to investigate. We'll

---

[10] Of course, that's not always true: it makes perfect sense to doubt a putative memory, and to think about what might have caused it instead of its apparent object. But we'll stick to the straightforward cases here.

approach things with a simplified model of the physical basis of our mnemonic knowledge, both as a test case for *T&C* in its own right, but also under the assumption that it provides a reasonable model of our mnemonic knowledge. I have in mind some version of Gell-Mann's (1994) "information gathering and utilizing system," or *IGUS* (see also Hartle, 2008).

In general terms, such a device has sensors capable of responding to its environment and devices capable of affecting its environment: inputs and outputs broadly construed. The system takes inputs from its environment (visual inputs, for example), processes the received information, and stores it in its registers; that's gathering. Then it has the capacity to operate algorithmically on the contents of its registers and its current inputs to determine a course of action implemented by its output devices (emit a beep or move away, for instance); that's utilizing.

The one simplifying assumption I will make is that the IGUS's registers are in some special "empty" ready state initially, before any data is gathered. That assumption is no restriction at all on the computational power of IGUSs, and anyway I believe that the points I wish to make could be made in a more complicated way without it. Clearly an IGUS can be realized as a digital computer, though the intent is that an IGUS has greater autonomy than the typical personal computer. A little more specifically, I want to consider an IGUS rather like us: especially responding to the same kinds of features of the world that we do, drawing inductive inferences from them, and drawing on those inferences for action in the pursuit of specific goals.[11] Below I shall say more about the operation of the IGUS, specifically its algorithm.

Insofar as the system has reliable procedures for forming veridical memories, insofar as it draws reliable inductive inferences, and insofar as its actions realize its goals on the basis of those inferences, I say that the IGUS models an important aspect of our knowledge. Computationally it is relevantly like us, and an IGUS could even be physically implemented like us. I trust it is clear enough and uncontroversial enough why I say that. For now I want to focus specifically on its memories. ("Memory" can be ambiguous between mnemonic knowledge and the registers of a Turing machine; I will always use it in the former sense.)

An IGUS "remembers" by reading the records stored in its registers, so let's think about how that might go. One might initially wonder whether an IGUS could remember by literally reading records: explicitly deducing an intermediate event from knowledge of ready and later states of its registers. But it's easy to see that that would be a poor procedure to employ. Such an algorithm could form representations of the current state of a register (perhaps the register is its own representation

---

[11] For work on algorithmic inductive reasoning, see Waltz and Buchanan (2009).

TABLE 9.1. A (partial) IGUS machine table, specifying what is to be inscribed in register C given the contents of P_0 and P_1

P_0	P_1	C
?	!	Yes
?		No

in that regard), and *also* of the register's ready state, and then compute a representation of the cause of the difference. That third, derived representation would then serve as a computational input for any decision-making processes related to the memory (perhaps to report that the IGUS witnessed a particular event an hour ago). But what would the point of the computation be? A more efficient algorithm would *simply assign to the contents of the register whatever computational role the derived representation plays in the algorithm discussed.*

For instance, suppose that among its registers, P_0 contains the current input and P_1 records the input from one cycle ago. Then Table 9.1 implements the kind of mnemonic algorithm that I have in mind.

Suppose that if the IGUS "saw" an exclamation mark, then P_1 will now contain a "!" and otherwise be left empty. And suppose that if the IGUS is asked whether it just witnessed an exclamation mark, then P_0 will contain a "?". After the IGUS applies the machine table, its output depends on the contents of C; for instance, the IGUS will say "yes" if C contains yes. Thus, this IGUS answers veridically. The point is that this IGUS "remembers" whether or not it saw a "!" without computation involving explicit representations of the P_1's ready and final states; the machine table depends directly on the present state.

That is, the states of the registers themselves are perfectly capable of playing the computational role directly, without any interpretational process—the mnemonic knowledge is "immediate." But then, no *explicit* record reading of the kind considered in *T&C* has occurred in remembering, because no role has been played by knowing or assuming the ready state. Since the algorithm employs no routine to deduce anything from representations of the present and ready states of its registers, the record reading is *implicit*. This is the first main principle of a mnemonic system, the central lesson of this section. Note that it follows from considerations of computational efficiency.

Previously I emphasized that we shouldn't take *T&C* as giving a literal account of how we read records of the past. The observation just made shows how we (or at least the IGUS that models us) do it instead, and is central to the account of memory developed here. In particular, the IGUS remembers without being subject to the record regress; there is no question of the IGUS having to compute the ready state of its registers by reading some record of an earlier time, since the meaning of the registers' contents is implicit in its algorithm. The IGUS shows that there is no paradox about memories, because it remembers without knowing anything about the PH! Now, resolving the paradox doesn't refute the position of T&C, which is that the PH plays a novel role, separate from the second law, in explaining the knowledge asymmetry—a conclusion argued for by a record regress argument. But now that we're absolutely clear that the regress does not apply literally to memories, we should ask seriously whether it applies at all.

Before moving on, it's probably worth justifying the claim that the IGUS "knows" anything. In what sense can we call the contents of the registers "knowledge"? It will suffice here to reiterate that the IGUS can satisfy a wide range of behavioral criteria for knowing: actions depending on the memories, production of representations of the memories, and communicating the memories to other IGUSs, for example. I do not mean to rely on a behavioristic account of knowledge—these are just evidence, not constitutive.

3.2. An IGUS Remembers

Now that we have described the basic mechanism by which an IGUS—as a physical model of our knowledge—remembers, we can start to investigate whether the IGUS can have mnemonic knowledge that does not satisfy T&C's necessary condition for knowledge of other times—whether it can know things of the past that are not highly probable according to the Mentaculus. Because the IGUS does not function by conditioning on the PH in remembering, it does not satisfy the assumptions of the regress argument, and can perhaps recall things that do not follow from conditioning on the PH. To address this question we will consider a specific memory of the IGUS, and unpack what it implies physically for the presently surveyable condition of the world. In so doing, we will draw a second lesson about the principles governing a mnemonic system.

So, let the IGUS be the witness of the pachinko experiment discussed at the start, so that there are no macro-traces of the outcome, only the properly formed inscriptions in the IGUS's registers, which by construction suffice for its veridical memory of the event: the ice falling left, say. By stipulation, the current macrostate of the universe, the statistical hypothesis, dynamics, and PH do not make it likely that the ice fell left; it might just as easily have fallen right. (And once again, the circumstances in which the only records of an event are in the brain are perfectly common: add to earlier examples of this kind of thing, remembering the result of a coin toss, or the exact spot the ball bounced, or who ate the last pretzel, and so on.) Our question is whether also including the "presently surveyable condition" of the IGUS will make left likely. But what is this condition? A first guess might be that it amounts to the states of all the registers.[12]

But these will not suffice to assign a high probability to the remembered outcome of the experiment Any collection of inscribed registers is compatible with many machine tables and initial states: the computational state does not entail the computational role. Hence, the process by which the records were made cannot be determined by the registers alone.[13] Nor would it help to expand the surveyable condition to include the *full* microstates of the registers; many different Turing machines could be realized with registers in identical microstates—the differences would lie in the physics of the rest of the device. In the present case then, any physical realization of any computational state is equally compatible with the ice having fallen left or right (or any other possible event), so presumably conditioning on the state of the registers makes each outcome equally likely. And this even though the IGUS has a machine table along the lines of Table 9.1, and the appropriate inscriptions, so that it does know in which beaker the ice fell. But that is to say that we have not yet accounted for everything presently surveyable by the IGUS, since it can survey its possession of a memory of the ice falling left, which entails something more than the register contents, presumably something of their computational roles: at least what behaviors they will produce.[14]

So what if we include the full microstate of the IGUS (together with the macrostate of the universe) in the present surveyable condition? A real, material IGUS is, computationally, nothing but a physical realization of a Turing machine, with no magical processing powers. Therefore, its machine table should supervene on its physical state and the laws of physics, and by including the microstate we include the full computational roles of the registers. Of course, the current microstate of one's brain is considerably more than one can actually survey, as are the exact contents of ones memory registers, or their full computational role in the machine

---

[12] If the IGUS constructs a "schema," or general theory of the world, its encoding in the IGUS should also be included.

[13] If the IGUS is a universal Turing machine, then its algorithm will also be encoded in the registers. Include those registers too; just remember that mere marks on a tape (or whatever) aren't an algorithm, *except in relation to some specific machine* that acts on them in some specified ways. (I.e., one can't infer a machine table from marks alone.) In other words, even in a universal Turing machine the algorithm fails to supervene on the physical states of its registers.

[14] I'd like to thank some very bright students at Oberlin College for pushing me on this point.

table of one's mind. But if we include the IGUS microstate, then we surely include whatever it can survey of its own mind in remembering that the ice fell left: the record corresponding to that memory, and its computational significance, are determined, because the computational function supervenes on the physical constitution. But together, the record and its computational role determine the content of the memory, as we showed in §3.1. Thus, including the full IGUS microstate yields the most expansive possible conception of the presently surveyable condition, and is surely too expansive. But if we adopt it to avoid leaving anything out, we can at least fix an upper bound on what is knowable according to the Mentaculus.

With this account, we can return to the proposal of *T&C* that "everything we [IGUSs included] know of the past and present and future history of the world can be deduced from" the Newtonian contraption: in particular, that the surveyable condition so conceived—the statistical hypothesis, dynamics, and PH—make it likely that the ice fell left. Let us see why it does not—even though by construction the IGUS veridically remembers that the ice fell left.

Take the current IGUS microstate. Can we turn the nomic handle in reverse, and evolve things backward to earlier times, through changes to the registers, back to the states of the IGUS's input devices when they were triggered? And thence to the stimuli they received, from which the causes of those stimuli—the recalled events—are (let's suppose) determinable?[15] The backward determinism of the laws means that a closed system can be uniquely wound backward from a given microstate. But that won't help in this case because the IGUS is not a closed system, but an open one: its state is the result of an initial (or final) state *and* a sequence of inputs (and outputs), which are nothing but interactions with the rest of the world (even if the IGUS is otherwise perfectly screened off). Therefore, without those boundary conditions, the laws will not determine the IGUS's earlier microstates from its present state. But those boundary conditions are things such as "light arrived from the left side of the experiment," and so determine what the outcome was; so by stipulation they are not part of the current macrostate of the universe at the later time, when the IGUS is remembering. Computationally, the point is that for a typical IGUS, despite being deterministic, a given state of its registers is compatible with many initial states and many sequences of

---

[15] If the IGUS works as we do to build up a coherent "schema," then it will have routines to infer more from its inputs than is logically entailed by the raw physical stimuli by means of the laws. These are empirically revealed (and studied in cognitive science) by susceptibility to illusions and sleights of hand, though they are generally reliable. Let's shelve this complication.

inputs: a machine table and current state do not in general determine the earlier states or computational history.[16]

Given this new underdetermination, one might wonder how the IGUS manages to read the records in its registers, apparently beating physics! But there's no real puzzle: although the IGUS doesn't explicitly condition on the registers' ready state, it does do so implicitly. Consider, for instance, Table 9.1: there we simply supposed that the correct inscriptions would end up in P_1. But suppose, for instance, that the mechanism for inscribing the contents of P_1 assumes that it is initially empty; if instead P_1 starts with an inscribed "!" then the IGUS will end up saying "yes" even if it did not see an exclamation mark. In other words, the IGUS's algorithm implicitly assumes an initial state for its registers for its proper functioning, and so is not merely conditioning on its current condition.

Now, there are IGUSs for which given initial and final states are compatible with many sequences of inputs, but a system with a machine table like that could not reliably recall the events that led to such underdetermined inscriptions. Suppose, for instance, that there are two possible inputs compatible with a given inscription and initial state. The computational role of the inscription could be, as in Table 9.1, that of a memory of one possible input, so that the inscription functioned just like an IGUS's memory of a specific event. But the underdetermination of the inscription's cause means that the system's algorithm cannot reliably determine which input produced it, so which possible input the inscription represents is arbitrary. Under these conditions, even if the apparent memory of the system was, by chance, of the inscription's actual cause, we would not say that the system *knew* it, exactly because of the arbitrary nature of the putative memory. In other words, *an IGUS that models memory correctly will not have a machine table of this sort; instead, it will have an algorithm such that the ready and current states of its registers are compatible with a unique series of inputs.* This is the second principle of the functioning on a mnemonic system, and the main lesson of this section. Whereas the first principle is required for computational efficiency, this one follows if the IGUS is to remember reliably.

Before applying the two principles, I should note an objection in *T&C* (119) to the kind of model described here: that the brain's ready state can't be the "mother" of all ready conditions because memories are defeasible by external records. But I don't see that as a problem for my approach. The IGUS may simply have a set of conflicting beliefs: some memories, some concerning current experiences, and some inductive beliefs, generalizing other memories. That they have to be rationally

---

[16] Even without external inputs, it is, of course, the case that Turing machines are not in general past-deterministic.

(or algorithmically) resolved does not undermine the account of how the memories are possible.

3.3. Remembering More than the Mentaculus?

The discussion so far has revealed two principles for a mnemonic system, which I have attempted to show will make it function efficiently and reliably. No doubt it could be made more efficient and more reliable, but I argue that these are the most salient requirements, and should inform any investigations of the memory asymmetry in physical terms. That innovation is intended as the major positive contribution of this essay, and I hope furthers the project of T&C. In this section I will illustrate the significance the of these principles in an investigation of T&C's necessary condition for knowledge. We have already seen that the IGUS does not explicitly read records, contrary to the assumption of the regress argument; here I try to show that as a result it also escapes the conclusion of that argument, and may have knowledge of the past not captured by the Mentaculus. In §3.4, I will acknowledge a possible response, which may show by other considerations that IGUS knowledge is bounded by the Mentaculus, after all.

Our second principle of the IGUS's algorithm constrains its physical constitution, since it realizes such a procedure; thus, the satisfaction of the constraint is part of the presently surveyable condition, specifically it is entailed by the IGUS's microstate. So the question is, given that the IGUS physically implements such an algorithm, will its current microstate in conjunction with backward deterministic laws entail the stimuli received by the receptors? (And hence, if everything else works out, entail a high probability for their causes.) No!

There are algorithms for which the current *and* initial states of the IGUS determine the series of inputs—thus satisfying the second constraint—but for which the current IGUS state (even its current microstate) *alone* does not. For such an IGUS, although it follows from the presently surveyable condition, including the record of ice falling left and the satisfaction of our principle, that the IGUS saw the ice fall left *if* it was initially in its ready state, that condition (plus the laws) does not imply that the IGUS *was* initially in its ready state. As before, the IGUS could instead have ended up in the very same state from a different initial state and different series of inputs—for instance, one in which it saw the ice fall right; or rather, there is a nontrivial probability that it did. So nothing is changed merely by including the satisfaction of the second principle in the surveyable condition: using the laws to time-reverse the evolution from the microstate of a physical realization of such an IGUS need not entail the prior inputs, or the stimuli that caused them. Yet, in virtue of its record, the IGUS veridically remembers where the ice fell.

Essentially, the point is parallel to the first step in *T&C* regress argument: the presently surveyable condition only "remembers" (that is, makes highly probable) a past event, conditional on the correct ready state of the IGUS, but that state itself is not highly probable given the presently surveyable condition. (Maybe it's very improbable, or 50 percent, or more; the conclusion is the same.) In that case, the object of the IGUS's memory is itself not probable, hence not "remembered" by the Mentaculus. However, crucially, now we are in a different situation from the regress argument, because we are not asking whether the Mentaculus (or other literal record reader) remembers the event, but whether the IGUS does. Thus the question is whether it satisfies our two principles and in fact started in the correct state, all of which we answer in the affirmative by stipulation in the pachinko thought experiment. Thus the record is veridical, and the IGUS created and acts on it in the appropriate manner, so it is a legitimate memory of a past event—but one, we just saw, that the Mentaculus does not capture.

Some may be tempted to say that such "memories" of our IGUS aren't really knowledge *because* it has no record of its initial state, and its memories are only veridical if it was in the correct initial state. But that would be a mistake: veridical memory requires only that a record *was* properly formed, not that the IGUS also keep a record of that proper formation: it doesn't have to know that it knows, in order to know what happened. To deny that has the same logic as denying that we know anything from experience, even in the case that the experiences *are* properly produced by their objects, just because identical experiences *could* also have been caused by an evil demon. Down that path lies debilitating skepticism that must be rejected. (Of course, it will have to assign a high credence to the correct initial state, unlike the Mentaculus. But that's fine; the whole point here is that the IGUS does *not* take Mentaculus probabilities as its credences.)

In summary (putting skepticism aside), our IGUS veridically remembers in virtue of the following features: it has a machine table such that (a) the present contents of its registers represent a particular past experience in virtue of their role in future computations (especially, producing behavior reflecting that experience); (b) if the IGUS's registers were initially in their ready state, then it follows that the input corresponding to the represented experience was received; and (c) the registers were in fact that ready state (and (d) the IGUS followed its machine table, the appropriate stimuli were received from the event observed, the stimuli were properly processed by the input devices, etc.). As long as these hold, the IGUS remembers veridically. The precise issue for the Mentaculus is that (a)–(c) require nothing in the *presently* surveyable state that would guarantee, or even make it suitably probable (according to the standard statistical postulate), that the registers started in their ready state. (c), of course, is a fact about the surveyable state, but the *past* surveyable state, so is not available to the contraption of T&C. But as the preceding discussion has explained, including (a) and (b) in the contraption without (c) will not make it likely (according to the standard statistical postulate) that the putatively remembered event occurred. So we can have veridical memory, but not high probability (comparable to that we routinely assign our memories) according to the Mentaculus—showing that *one can know more of the past than the Mentaculus.*

The argument is in some ways related to the idea of a "Boltzmann brain," insofar as from the Mentaculus point of view the issue is that the process by which the IGUS obtained a putative memory might not have been a reliable one. More precisely, the argument turned on the probability of such a situation being nontrivial. But there are important differences: First, being the result of a random fluctuation is only one of many ways in which an IGUS might form a memory by an unreliable method; perhaps it malfunctioned, or was built by a malicious or incompetent AI firm, for instance. So one cannot answer my challenge simply by showing that Boltzmann brain-producing fluctuations are improbable. Second, my point is not at all a skeptical one. Quite the reverse! I say that IGUS should *not* doubt its memories just because the Mentaculus assigns them a low probability; or even lower its credence in their objects to that of the Mentaculus. No, my whole point is that the IGUS has knowledge that the Mentaculus does not, and so cannot be affected by skeptical arguments that assume the Mentaculus bounds our knowledge. (If I am right, then it would be irrational for the IGUS to set its credences to Mentaculus probabilities.)

I have not explicitly considered the PH in this discussion, but it won't be of any immediate help. For we are considering memories of events that are themselves not probable according to the full contraption, including the PH: e.g., where the ice cube fell, once it has melted and evaporated. But if the far past state of the universe does not make that event probable, why would it make the ready state of the IGUS probable? In fact, I will discuss a proposed response in the next section, which will attempt to give an indirect answer. I'm not convinced, but at least for now I hope it is clear why the PH is not directly relevant.

Foreseeing that some readers may agree with that response, against my conclusion that the IGUS outdoes the Mentaculus, let me state here some weaker conclusions that all should accept. First, the IGUS that I have described provides a physically realizable model for memory that functions in a very different way from the Mentaculus. As such it should be the starting point for statistical mechanical accounts of the knowledge asymmetry. Moreover, it provides an explicit answer to the rhetorical question of "what else" could be the mechanism of memory but the Mentaculus? Namely, an IGUS!

3.4. The Evolutionary Response—a Reply

Since I first wrote this paper several years ago, I have had the opportunity to hear David's reply to this argument. As I understand him, David would not claim a direct connection between the big-bang state of the universe 10 billion years ago and the last hour of the IGUS. Instead he proposes an indirect connection along the following lines: the Mentaculus makes it overwhelmingly likely that the kind of state we presently survey, with IGUS-like systems, is the product of natural selection; and, because selection produces fit systems, further makes it likely that any IGUS is well-functioning and thus assigns a high probability to the correct IGUS initial state, and hence ultimately to the objects of its memories—the ice falling left, say.

I would like to take this opportunity to make some brief observations regarding this proposal, although it should be noted that the following is an addendum to the original paper, and not one David has had the opportunity to hear.

First, the evolutionary response threatens to be question begging. If my argument is correct, then we have more mnemonic knowledge of the past than the Mentaculus has, and can thereby obtain even more inductive knowledge than the contraption entails. In particular, it would be on the basis of such amplified mnemonic knowledge that we judge it likely that the IGUS-containing state of the universe is the result of selection, and that IGUSs likely start in their ready state. To then simply assert that selection or the well-functioning of an IGUS is probable according to the contraption would be to simply deny what is at stake: that we have epistemic access to such knowledge through memory that exceeds the scope of the Mentaculus. (After all, the IGUS has mnemonic knowledge in a very different way from that considered in *T&C*.) On my proposal, when we reason from selection to fitness to the correct ready state, it would be on the basis of mnemonic knowledge, without any implication that it is also likely according to the contraption. The assumption that mnemonic knowledge has a high Mentaculus probability is an additional claim, tantamount to denying that memory provides knowledge beyond the scope of the contraption, so it could not be an unsupported premise in a noncircular argument against my proposal.

The argument against my claims, then, is not the question-begging one that because we know that evolution occurred, evolution was likely according to the Mentaculus; instead, the argument is that because the PH entails lower past entropy, the current state is much more likely a result of evolution than of random fluctuation, and those are the only serious alternatives. (I assume that attending to the more detailed distribution of matter shortly after the big bang would not shed any additional light on the evolution of life.) Let me grant this premise, though I worry that only a lack of imagination prevents us from seeing further alternatives. I still question the further inference from the high probability of evolution, to that of the well-functioning of a mnemonic system.

For instance, an advanced civilization, itself the product of evolution, could produce an arbitrary number of IGUSs in incorrect initial states, so that the overwhelming number of IGUSs in the universe failed to have veridical memories. In such a scenario, an IGUS picked at random would most likely *not* have veridical memories, contrary to David's fitness proposal.[17] Clearly this scenario is compatible with a high Mentaculus probability of evolution. It seems bizarre, but is it improbable given the PH, and hence evolution? I think it is really hard to say, and correspondingly hard to conclude that the well-functioning of an IGUS is highly probable according to the Mentaculus. Granted, it is hard to see how the evolutionary fitness of a civilization might lead to such behavior, but that may well be a failure of imagination.

But suppose that David's proposal works, and the Mentaculus does assign high probabilities to the objects of our memories. Then I want to emphasize an important aspect of that fact. Recall from the start of this paper that the high probability of the physical event of the ice cube having fallen left, in our example, is not at all due to the PH and current macrostate, since by construction all macroscopic traces of that event are erased. The high probability is due to the physical state of the IGUS's brain because it contains a microscopic record of the event (and because the contraption makes it probable that the IGUS started in its ready state). That is, the ice cube falling left has a high Mentaculus probability *only* because the IGUS remembers it falling left.[18]

Finally—and this is the main point I want to make—if the proposal works, and memories do satisfy the necessary condition for knowledge that I quoted in §2.1, then the reasoning behind this conclusion turns out to be quite different from that given in *T&C* and discussed in §2.2 above. No regress of record reading plays an essential role in the logic. Rather, the IGUS remembers provided that it is reli-

---

[17] Let's suppose that the advanced civilization itself disappeared long ago without trace, so that nothing in the presently surveyable state makes probable their strange history of IGUS production.

[18] It seems to me that this observation has important implications for the connection between chances and credences in any attempt, such as Loewer (2001), to identify Mentaculus probabilities as "the" objective chances. In a sense, chance follows credence, rather than the other way round.

ably in its initial state; as I have emphasized, there is no need for any record reading of its earlier state, for memories rely *implicitly* on the meaning and well-formedness of records, not *explicitly* as the regress argument supposes. The IGUS is a different route to mnemonic knowledge than the Mentaculus, and so one cannot immediately conclude that they have the same epistemic bounds, as *T&C* suggests. If they do have the same bounds, the argument is quite different, based on the kinds of realistic modeling of memory that I have attempted here, plus the evolutionary argument. Indeed, the current argument for the necessary condition, is simply incommensurable with the regress argument of *T&C*—because only the former turns on a reasonable account of memory.

4. The Knowledge Asymmetry

In this paper I have not addressed the knowledge asymmetry directly; my focus has been on how we read records. Of course that's a step toward the asymmetry, for it apparently lies in the existence of readable records of the past but not of the future. None of my remarks are intended to express skepticism that the PH is relevant to the explanation of the knowledge asymmetry, perhaps even in ways separate from its role in the second law. For instance, brains and the computers we know how to build are highly time-directed systems: synapses aren't time-reversible, and neither are logic gates. The processes in these systems are governed by cause and effect, so perhaps the arrow of causation lies at the heart of the knowledge asymmetry—and the PH at the heart of the causal asymmetry? (This problem is related to the difficulty of building an IGUS with memories of the future—if future events can't influence it now, it seems the maker just has to fill the registers themselves!)

Hartle (2008) suggests that an IGUS is time-asymmetric because its operation involves erasing data to free storage space (since there are only finitely many registers), and because the permanent loss of such information is entropy-increasing; thus, the system can only work in one temporal direction in virtue of the second law. I'd like to point out that this proposal can't be right as stated: For instance, because Hartle's IGUS has five registers, the final one remains empty until the fifth cycle, so it isn't erased for that initial period. So he hasn't explained the knowledge asymmetry until that time; by his lights, nothing rules out a suitable IGUS from remembering the future *until* its last register was inscribed, and erasing had to commence! To apply Hartle's idea, some more general argument about the necessity of erasing would be required. For instance, doesn't shuffling the records down the registers involve erasing? But that's not an essential feature. Suppose, for example, that the registers are buckets and data are stored by placing given numbers of balls in them: shuffling could be carried out by tipping the balls from one bucket to the next. The sequence of records inscribed by the IGUS's detectors can easily be inferred from the collection of registers (by subtraction, of course). The problem is that this or that part of the IGUS can be constructed without erasing, so something quite general about the physical implementation of Turing machines would be needed to pursue Hartle's strategy.

Whether or not such an analysis can work, it would not fully address the knowledge asymmetry, because it does not address our ability to obtain knowledge of the past from records other than memories. Clearly we do read records in the way described in *T&C* all the time: the footprints in the snow, the dent in the bumper, the puddle in the street. We know what they signify because we infer what caused the changes from a "ready state," even if we don't remember the causes—and even when we have no memory of the ready states (perhaps I've never even been down this street before). Does a record regress still threaten here? No.

We can readily imagine an IGUS that can read the records because it possesses *generalizations* about the ready states of systems, generalizations that it obtains by induction on the contents of its memories. (It's for this reason that I gave it inductive powers when I introduced it.) Significantly, true generalizations, if informative at all, are often more informative about origins than fates: all acorns start much the same, as do all sandy beaches as the receding tide exposes them, and all rocks—at least compared to the countless ways in which acorns can develop or not into oaks, the different kinds of impressions that can be made on beaches, and the different kinds of marks, including those made by humans, that might be found on a rock. And similarly for freshly fallen snow, bumpers, and streets.[19] That is to say, there are such records of the past because one can infer an earlier ready state simply from the *kind* of system involved; but that there are no corresponding records of the future, because there are no generic later states for a given kind of system. Thus, we would have an explanation of the knowledge asymmetry for the inferred (rather than remembered past), if we could only characterize and explain this initial-final state asymmetry!

I won't address that question here, but note that we have a model of our knowledge of other times. There are memories, modeled by the IGUS, in which the meaning of records is determined by their computational role. Using these, and a capacity for induction, we can learn to read other records for what they tell us of the past.

---

[19] This point is emphasized by Lockwood, 2005, chap. 11.

References

Albert, D. Z (2000). *Time and Chance*. Harvard University Press.
Gell-Mann, M. (1994). *The Quark and the Jaguar*. W. Freeman.
Hartle, J. (2008). "The Physics of Now." *Am. J. Phys.,* 73, 101–109. Earlier published (2005): gr-qc/0403001.
Loewer, B. (2001). "Determinism and Chance." *Stud. Hist. Phil. Mod. Phys.,* 32(4):609–620.
Lockwood, M. (2005). *The Labyrinth of Time*. Oxford University Press.
Waltz, D., & Buchanan, B. G. (2009). "Automating Science." *Science,* 324(5923):43–44.

Chapter Ten

Causes, Randomness, and the Past Hypothesis

▶ MATHIAS FRISCH

1. Introduction

Why do we think of events in the present as causally responsible for future events but not for events in the past? Why are there records of the past but not of the future? David Albert (2000) and Barry Loewer (2007) have argued that these temporal asymmetries ultimately have the same origin as the second law of thermodynamics and that it is possible to derive these and other temporal asymmetries from the core assumptions of a neo-Boltzmannian account of the thermodynamic asymmetry. These assumptions are the posit of a low-entropy initial state of the universe, a probability postulate, and a time-symmetric deterministic micro-dynamics. In this paper I want to investigate the prospects for such an "entropy account" of the causal asymmetry.

I will begin in §2 by examining one important role causal reasoning plays in physics: causal inferences allow us to gain information about the state of a system in the past, even when we do not have access to the system's full present state. As we will see, an implicit assumption in such inferences is an assumption of initial randomness. A similar assumption is also part of Albert and Loewer's account, in the form of the initial equiprobability postulate. The question, thus, is whether the randomness assumption should itself be thought of as a causal assumption or as part of a purely reductive account. On the one hand, the assumption appears to be a natural consequence of representing the world causally. On the other hand, the assumption is also central to Albert and Loewer's neo-Boltzmannian account of the thermodynamic asymmetry. Thus, in §3 I will examine what role, if any, the other core assumption of the entropy account—the assumption of an extremely low-entropy initial state of the universe, which Albert calls the "past hypothesis"—can play in underwriting the asymmetries of causation and of records. I will argue that extant arguments by Albert and Loewer for the role of the past hypothesis in a reductive account are not successful. I conclude that a convincing case for the claim that the causal asymmetry is reducible to thermodynamic considerations remains to be made.

2. Causal Representations and Initial Randomness

Imagine that you are looking up into the night sky and, for some extended period, are observing the light emitted by a particular star. What licenses your inference that the light you are observing was indeed emitted by a star—that is, by a radiating source—instead of source-free radiation coming in from the infinite past?

David Lewis (1986) has argued, as part of his attempt to ground the causal asymmetry in a putative asymmetry of counterfactuals, that the present radically overdetermines the past in the sense that many past events have multiple localized traces in the present, each of which is *individually nomologically sufficient* for the occurrence of the past event of which it is a record or trace. Thus, according to Lewis's thesis, each individual packet of radiation observed at an instance is nomologically sufficient for the earlier existence of the star emitting the radiation. If Lewis were right, the question as to how we can justify our belief that the light was emitted would have a simple answer: The radiation observed by us together with the laws of electrodynamics imply the earlier existence of a star emitting the radiation. But Lewis is mistaken, and there is no overdetermination of the kind envisaged by him.[1] The existence of a localized packet of coherent radiation does not nomologically imply the presence of a source and is compatible also with the assumption that the radiation is source-free.

---

[1] See Elga (2000) and Frisch (2005, chap. 7), for detailed criticisms of Lewis's claim.

Contrary to Lewis's suggestion, the dynamical laws of physics, such as the wave equation, which governs electromagnetic radiation phenomena, allow us to make inferences from one time to another only if we "feed into" them a fully specified initial or final state. The wave equation determines a Cauchy problem, which as input requires data on an entire cross-section of the forward lightcone of the putative source to the determine the state of the world at the source's location. Given this feature of the laws, one might perhaps think that the only way we could infer the existence of the star would be from the complete state of the world on a full initial value surface.[2] But then we could never come to know whether the light was emitted by a star, since we never in fact are in possession of the data required. All we have available are the highly localized observations of the radiation field here on Earth (plus, perhaps isolated observations from space-based telescopes). Consequently, if our knowledge of physical systems were exhausted by what we can deduce from the dynamical laws together with knowledge of a system's initial state, then our knowledge of the physical world would be extremely impoverished.

How, then, do we come to know that the light we observe has been emitted by a star? What in addition to the dynamical laws and our localized observations can ground this belief? The solution to the puzzle is that the inference to the existence of a star as source of the radiation is a paradigmatic example of a causal inference. The focused packets of radiation observed by us at different times are highly correlated with one another and we explain these correlations, which otherwise would seem extremely puzzling, by positing the existence of a single localized object, the star, as their common cause.

This kind of inference is one we employ rather frequently, both within and outside of science. Here is an example from outside physics, but with the same inferential structure: In a string of bank robberies the police discover similar kinds of little plastic toys left behind at each crime scene—a fact that is not made public by the investigators. The police infer from these discoveries that the robberies were committed by one and the same gang as their common cause rather than by different groups that operate in complete independence from one another.

In both this example and in the case of the star we engage in common cause reasoning. The values of the electromagnetic field variables and of the variables characterizing the state of the star satisfy a principle of the common cause that can license our inference to the existence of the star. Let us take as causal relata

---

[2] I am using the term "initial value state" to refer to states of a system that can be used both for prediction and for retrodiction, thereby avoiding the more cumbersome formulation "initial and final value state."

values of the variable $F(t, x)$ characterizing the field at a spacetime point and of the variable $S(t, x)$ characterizing the state of the source. The fields we observe at two different moments t_1 and t_2 are highly correlated and satisfy

$$\Pr(F(t_1, x_1) \ \& \ F(t_2, x_2)) \gg \Pr(F(t_1, x_1)) \times \Pr(F(t_2, x_2)).$$

Conditionalizing on the state of a star $S(t_{ret1}, x_{ret1})$ that has intersected the backward lightcone of our observation point at t_{ret1} and x_{ret1} increases the probability of observing the field at t_1:

$$\Pr(F(t_1, x_1) \,/\, S(t_{ret1}, x_{ret1})) > \Pr(F(t_1, x_1)).$$

Since the state of the star at one time is highly correlated with its state at later times, and $\Pr(S(t_{ret2}, x_{re2}) \,/\, S(t_{ret1}, x_{ret1}))$ is approximately equal to 1, conditionalizing on $S(t_{ret1}, x_{ret1})$ also raises the probability $\Pr(F(t_2, x_2))$:

$$\Pr(F(t_2, x_2) \,/\, S(t_{ret1}, x_{ret1})) = \Pr(F(t_2, x_2) \,/\, S(t_{ret2}, x_{re2}))$$
$$\times \Pr(S(t_{ret2}, x_{re2}) \,/\, S(t_{ret1}, x_{ret1})) > \Pr(F(t_2, x_2)).$$

Finally, conditionalizing on $S(t_{ret1}, x_{ret1})$ renders the observed fields independent of each other:

$$\Pr(F(t_1, x_1) \ \& \ F(t_2, x_2) \,/\, S(t_{ret1}, x_{ret1})) = \Pr(F(t_1, x_1) \,/\, S(t_{ret1}, x_{ret1}))$$
$$\times \Pr(F(t_2, x_2) \,/\, S(t_{ret1}, x_{ret1})).$$

The overall process is, of course, deterministic: initial fields in the remote past together with the field associated with the star nomologically determine what the observed fields will be. Yet, since we have observational access neither to the free fields prior to the putative emission events nor to the present fields on a complete initial value surface, we cannot set up a full-fledged initial- or final-value problem. Instead we somehow have to infer, based on our localized observations, what *both* the initial fields *and* the fields associated with any sources might have been. Without any additional assumptions this would be impossible: our localized observations of the total field do not provide us with enough information to infer both the state of the source and the initial field and are compatible with multiple different combinations of initial fields and radiation fields. What we observe are relatively focused packets of radiation (which we interpret as light emitted by stars) within a background field that is approximately equal to zero (or at least very weak). It is consistent with this evidence not only that the radiation coming from a single direction is due to a star as its common cause, but also that there existed strong correlations among source-free initial fields in spatially distant regions at some remote time in the past that resulted in macroscopic fields converging onto the putative trajectory of the star, passing over it, and then rediverging, mimicking (for later observers) the presence of a star. The further back in time we followed such source-free fields, the weaker these fields would be as they become more and more dispersed toward the past, originating in what ultimately might have been extremely delicately coordinated microscopic correlations among very distant field regions.

Yet such highly correlated initial free fields strike us as radically improbable, and thus we invariably infer that the correlated packets of radiation we observe are emitted by a star. That is, an implicit assumption in our inference to the star as common cause of the observed field values is that the initial fields are effectively random—or rather, as random as possible, given our observational evidence. Thus, instead of having to posit a precise value for the initial fields, which we cannot know, we can nevertheless infer something about the history of the radiation field if we make the weaker assumption that the initial fields at some point in the remote past were effectively random. Without knowledge of full initial or final conditions our inferences cannot be based solely on the fully deterministic laws, but given the initial randomness assumption we can probabilistically infer the existence of the star as the most likely explanation of our observations. What is more, because the randomness assumption allows us ignore the initial fields, we can even appeal to part of the dynamical laws—the "retarded" or "causal" Green function specifying the field associated with a source in its past—to infer more detailed information concerning the state of the star without having to plug that function into the full deterministic equation of motion.

How then should we think about the probabilities in our inference to a common cause? In the first instance the probabilities in question can be thought of as epistemic probabilities. On this understanding the randomness assumption partly reflects our ignorance of the precise initial conditions: we presuppose, whatever the precise values of the initial fields are, that it is extremely unlikely that they contain strongly correlated spatially distant disturbances that are coherently focused on spacetime points in their future. Note that this is a time-asymmetric assumption: we do not assume—and would be wrong to do so—that final fields contain no strong correlations between spatially distant regions. But one might also want to assign a more objective status to the randomness assumption or even treat it as lawlike constraint, equivalently to the way Albert and Loewer propose to treat the postulate of an initial equiprobability distribution of microstates in the foundations of statistical mechanics.

My aim here is not to resurrect a principle of the common cause as a metaphysical principle that states that whenever two simultaneous events are correlated and neither of the two events is a cause of the other, then the two events must be joint effects of a common cause. Such a principle is arguably false—a point that has been much discussed in the literature.[3] Instead, all we need in order to account for our inference to the existence of a star is an epistemological version of the principle, according to which observed correlations often, but by no means always, allow us to infer a common cause in their past. While our expectation generally is that correlations can be explained by a localized common cause, there may be cases where no such explanation is possible (as appears to be the case for certain quantum mechanical phenomena) or cases where a separate-cause explanation of the correlations is superior to any putative common cause explanation. The inference to a localized common cause, thus, ought to be thought of as involving a comparison between the common cause explanation and potential separate-cause explanations (see Sober 1984). Elliott Sober proposes that such a comparison between different hypotheses takes the form of a comparison of likelihoods. In our case, two possible explanations of our observations are that the correlations among the fields observed at different times are due to the star as common cause or that the observed fields are the forward evolutions of completely independent earlier initial free fields. The first hypothesis renders the observed fields much more likely:

$$\Pr(F(t_1, x_1) \,\&\, F(t_2, x_2) \,/\, S(t_{retl}, x_{retl}) \,\&\, \textit{random weak initial fields}) \gg$$
$$\Pr(F(t_1, x_1) \,\&\, F(t_2, x_2) \,/\, \&\, \textit{random weak initial fields}).$$

Thus, we accept the first hypothesis. Note that the probabilities in question are strictly between 0 and 1, since it is nomologically compatible with the randomness assumption that there are strongly correlated but very small disturbances in the free fields in the remote past that at later times become much larger and converge on the (putative) worldline of the star, either to mask the fields associated with a star or to mimic the field of a star where there is none. Yet such correlations are extremely improbable.

Traditional formulations of the common cause principle are restricted to correlations among simultaneous events. But the principle can easily be broadened to include the kind of correlations we have been discussing. For correlations among timelike related events, such as our successive observations of starlight, there is, in principle, a third kind of hypothesis one ought to consider; namely that the first

---

[3] For an overview of such arguments, see Arntzenius (2010).

observation is a cause of the second observation. But because there is no plausible mechanism for such a direct causal link between our observations in our case, this hypothesis can be rejected as radically implausible.

Because the full classical theory governing radiation phenomena is deterministic, it predicts that the correlations between different values of the field will also be screened off from one another by events in their future. Such future screening-off events, however, will in general not be localized events, but will correspond to widely spread-out regions in phase space. Thus, it is crucial that the randomness assumption allows us temporally asymmetrically to infer the existence of a *localized* common cause in the past.

A similar kind of inference to a localized common cause also plays a role elsewhere in cosmology. I said above that the initial randomness assumption ought to be expressed as stating that initial fields are as random as possible. Yet according to modern cosmology the initial fields cannot be completely random, since there exists a weak cosmic microwave background radiation that appears to be almost isotropic. Cosmologists want to resist having to postulate that this background radiation is approximately the same throughout regions that could have had no causal connection. Here inflationary cosmology is meant to help by offering an account that provides a common cause for the background radiation: according to the theory, the early universe underwent exponential growth within the first 10^{-30} s after the Big Bang, allowing the radiation to have originated in a small, causally connected region.

I want to draw two conclusions from my discussion so far: First, causal reasoning is widespread in physics. I have only investigated a single example here, but the discussion readily generalizes. Only very rarely do we know the state of a system on an entire Cauchy surface and can rely on the dynamical laws to make inferences about the past. Instead we need to exploit correlations among different observations to infer past common causes. Second, and more important for my present purposes, a necessary condition for the reliability of causal inferences is the assumption of initial (micro-) randomness. That is, we need to assume that the values of all variables on which the value of the correlated variables depends in addition to a common cause C are randomly distributed. Without that assumption there could be correlations among these variables characterizing an earlier initial state of the system at issue that resulted in correlations among the effect variables even in the absence of a common cause or that render the effect variables probabilistically independent despite the presence of a common cause (see Arntzenius 2010).

Once we have acknowledged the central role of an initial randomness assumption in causal reasoning, two strategies present themselves concerning the place of causal representations in physics. First, one might take the initial randomness assumption to be itself an intrinsically causal assumption. On this view the fact that causal representations play a successful role in reasoning in physics need not be reducible to any noncausal features of the world (or of us as human agents investigating the world). Second, one might take the assumption to be part of a set of noncausal assumptions to which causal representations can ultimately be reduced.

That the randomness assumption is itself a causal assumption is *prima facie* quite plausible. In a world that can be successfully represented by common cause structures we would expect the values of variables associated with spatially separated regions of space to be randomly distributed unless they are linked through a common cause in their past. Thus, the initial randomness assumption and the assumption that physical systems can be represented by time-asymmetric causal structures can be seen as mutually supportive: Only under the assumption of initial randomness will correlations be a successful guide to causal structures; and in the absence of common causes in the past we expect the values of certain variables to distributed randomly. Moreover, the pair of assumptions fits well with our explanatory practices: We generally search for explanations of correlations by earlier factors. That is, certain correlations strike us as utterly mysterious unless they can be accounted for in terms of an earlier common cause, while pointing to any later (either localized or nonlocal) screening-off condition does nothing to remove our sense of puzzlement. By contrast, we do not similarly think that randomly distributed values of variables call for an explanation in terms of earlier factors.

The second strategy is the strategy that Albert and Loewer pursue in their attempts to reduce the causal asymmetry to the same principles that are at the heart of their neo-Boltzmannian account of the thermodynamic asymmetry. Unlike philosophers of physics, such as John Earman (2011), who maintain that causal notions are hopelessly vague expressions of a philosopher's metaphysics, Albert and Loewer believe that a kind of causal reasoning plays an important role in how we come to know about the world. Yet they do not take causal representations to be fundamental and argue that the asymmetry of the causal relation—the fact that we take the future but not the past to causally to depend on the present—is ultimately reducible to the asymmetry of thermodynamics.

One of the core principles of the thermodynamic account is a probability postulate—the assumption that all microstates compatible with the initial macrostate of the universe were equiprobable. This assumption—which ensures that the temporal evolution of the universe from its initial macrostate is overwhelming probability "typical" and exhibits thermodynamically normal behavior—plays a role analogous to the assumption of initial micro-randomness in my discussion above. Thus, Albert, Loewer, and I are in agreement on the importance of such an assumption to the success of causal representations. In §3 I want to examine what role the additional assumption that the universe began its life in an extremely low-entropy state, what Albert dubbed "the past hypothesis," can play in a reductive account of the causal relation (see also Frisch 2006; 2010; 2014, chap. 8).

3. Albert and Loewer's Entropy Account

The central assumptions of Albert and Loewer's reductive account are (i) the "past hypothesis," according to which the universe began its life in whatever extremely low-entropy macrostate cosmology eventually presents us with; (ii) a probability postulate, which posits an equiprobability distribution over all microstates with this initial macrostate; and (iii) a time-reversal-invariant deterministic microdynamics. These assumptions, they argue, entail several asymmetries, which they take to be intimately connected to the causal asymmetry:

1. A temporal asymmetry of records or traces, which consists in the fact that there are localized records of the past but not of the future;
2. a closely related temporal asymmetry of knowledge, which is meant to capture the fact that we can in some sense know more about the past than about the future;
3. a temporal asymmetry of influence or control, which consists in the fact that we take ourselves to have some control over the future but to have absolutely no control over the past;
4. and, finally, a temporal asymmetry of counterfactual dependence and the asymmetry of causal influence.

In §2 I have argued that we can acquire knowledge of the past state of a physical system, given merely localized data, through causal inferences. By contrast, Albert and Loewer focus in the first instance on the notion of records or traces of the past. Yet from my perspective this is merely a shift in emphasis, since inferences that appeal to records or traces can be thought of as a special case of causal inferences. Traces exhibit the same kind of fork asymmetry that is at the core of inferences to a common cause. Multiple records of the state of a system (such as the light emitted by a star recorded at multiple locations simultaneously) are all separate effects of a common cause, and even the interaction of a single recording device with a system is an example of a causal fork (see also Frisch 2014, 224–228).

In a talk, Albert described what he takes to be a crucial difference between our inferences toward the past and our inferences toward the future as follows: Present traces of the past allow us to infer facts about the state of a system at some past time t_p without any knowledge of the history of the system between t_p and the present. Since traces can be relatively isolated from the system at issue after the recording interaction took place, the reliability of a trace may be completely independent of the evolution of the system after the recording interaction. By contrast, any prediction concerning the future state of a system at a time t_f crucially depends on the evolution of the system between the present and t_f.

Albert did not make this connection, but the asymmetry he described is exactly the asymmetry associated with causal forks that are open toward the future. Any correlations between a present effect E and the occurrence of an earlier cause C are independent of what occurs causally after C along other causal routes distinct from that leading from C to E. By contrast, whether some later effect E' occurs, does not depend merely on the occurrence of a single present cause C' and the route from C' to E', but depends on the occurrence of other causes of E' as well. We can express this asymmetry in interventionist terms: Interventions in the causal future of C on causal routes other than that linking C and E cannot affect whether or not the occurrence of E is a reliable record of the occurrence of C. By contrast, interventions in the causal past of E' on causal routes distinct from the one leading from E' to C' can interfere with any correlations that might otherwise exist between E' and C'. Thus, in order to predict E' on the basis of the occurrence of C', we need to know what occurs on other causal routes leading into C', while other causal routes leading away from C are irrelevant to the reliability of the correlation between E and C.

Under what conditions can a localized event E be a reliable record of an earlier event C that caused it? Albert has convincingly argued that records are always, at least implicitly, inferences from two times to a time in between. In addition to the recording event E we also need to make an assumption about the "ready state" of the recording system prior to its interaction with the recorded system. In §2 I arrived at a similar conclusion. Inferences from correlations among different effects (playing the role of records) to a common cause as the recorded state depend on an assumption of initial randomness.

Albert (2000) has forcefully argued that not only the probability postulate but also the past hypothesis plays an important role in ensuring the reliability of records: the *past hypothesis* can at least in principle, as he puts it, play the role of the "mother of all ready conditions." Conditionalizing the occurrence of a putative record on the past hypothesis, that is, can ensure that that record is reliable. This reliability of records, according to Albert and Loewer, has the consequence that the past, unlike the future, is dynamically insensitive to small changes in the present macrostate: If we feed a state that differs only slightly macroscopically from the actual present state of the world into the dynamics, conditional on the past hypothesis and assuming the probability distribution induced by the initial equiprobability postulate, then we find that the macro-past with overwhelming probability would have been what it actually was, and this is so because conditionalizing putative traces of the past on the past hypothesis ensures that these traces will with overwhelming probability have been reliable. For Albert and Loewer it is this dynamical robustness of records, when underwritten by the past hypothesis, that underlies the asymmetry of causal influence. We do not take ourselves to have any influence over the past, because the thermodynamic account entails that the past would have been what it was anyway even if we had decided to act differently from the way we actually did. Thus, while I have suggested that the asymmetry of records or traces is a special case of the asymmetry of causation, Albert and Loewer's strategy is to reverse the order of explanation. The thermodynamic account underwrites the asymmetry of records and hence of knowledge, which in turn allow us to account for the asymmetries of influence, counterfactuals, and causation.

In what follows I want to critically examine whether the past hypothesis can play the role that Albert and Loewer assign to it. I want to begin by asking more generally what the connection might be between the existence of records or traces of the past and the fact that the entropy of a closed system is overwhelmingly likely not to decrease—a fact that Albert and Loewer take to be underwritten by the past hypothesis. Albert and Loewer emphasize that many of our inferences about the past would be radically false if these inferences were made only on the basis of the present macrostate and an equiprobability distribution over microstates compatible with that state. Because of the time-reversibility of the microlaws, we would mistakenly conclude that entropy was overwhelmingly likely to have been higher in the past. For example, if we encountered a half-melted ice cube in a glass of water, the dynamical laws plus an equiprobability postulate would lead us to infer, not that the present state of the ice cube was evidence of someone having put an unmelted ice cube into the glass earlier, but instead that it was evidence for the presence of a glass with water that anti-thermodynamically had spontaneously begun to freeze. To block this mistaken retrodiction, Albert and Loewer maintain, we need to postulate the past hypothesis. That is, without an assumption that can ensure thermodynamically normal behavior, many of our retrodictions would be radically mistaken.

But this argument can only establish that the past hypothesis is a necessary condition for the reliability of many of our records and does not yet show why we do not also have records of the future. After all, in contrast to our inferences toward the past, our inferences toward the future based on the dynamical laws face no similar "thermodynamic obstacle." In the first instance, then, postulating a past hypothesis merely ensures that our dynamical inferences toward the past are *as good* as our inferences toward the future. But what we are still looking for is a reason we can know *more* about the past—that is, why our inferences toward the past are in some sense more powerful than our inferences about the future. What else, then, can we say about the connection between records and the thermodynamic arrow?

The example of the half-melted ice cube might suggest that the connection between records and the direction of entropy increase is that the record state is one of nonmaximal entropy and that the systems ready state must have been one of even lower entropy. Albert (2000) has argued, criticizing a suggestion by Reichenbach along these lines, that this is false and that a system locally in equilibrium can also function as a record. I want to amplify Albert's discussion here. Since the measuring system cannot be a closed system during the measuring interaction, there are in fact no restrictions on the relative entropy of the record state and the measurement state, and the entropy of the record state can be higher, lower, or the same as that of the ready state. It is easy to find examples of each kind.

First, an exposed film is an example of a system that ends up in a state of higher entropy than its ready state. An unexposed film consists of an emulsion of silver bromide molecules. During the exposure silver atoms are formed. The sum of the molar entropies of atomic silver and bromine is higher than that of silver bromide, and hence the entropy of the system constituted by the exposed film goes up, despite the fact that after being exposed the film intuitively contains a much larger amount of "information" and might seem to be much more highly "ordered" than the unexposed film. Thus, one has to be careful not to confuse the thermodynamic notion of entropy with a more intuitive notion of information.

Second, an example of records that have much lower entropy than their respective ready states are tree rings, which climate scientists use as a proxy for historic temperatures. Roughly, the size and density of tree rings are a function of the yearly temperature and therefore can be used as a record of the temperature. The chemical entropy of cellulose and the other constituents of a tree trunk is much lower than that of the CO_2 and water out of which they are formed through photosynthesis.

Finally, an arrow or other sign arranged out of pebbles on a beach to signal the way is a record that has the same thermodynamic entropy as the random pattern of stones out of which it was formed.

So, the question remains: Because the ready state of a measuring system can have higher, lower, or the same entropy compared to the record state, what accounts for the fact that the only records are of the past? Now, there are several further interesting and important connections between the existence of records and the second law of thermodynamics, but none of these, I think, get at the kind of connection that Albert and Loewer are after: On the one hand, very many (and perhaps most) records depend on the presence of friction and on the availability of low-entropy energy reservoirs. This suggests that the second law of thermodynamics does play an important role in the existence of records. On the other hand, the second law of thermodynamics also is "a great destroyer of records". Some of what Albert and Loewer say might be taken to suggest that records of past macro-events are pervasive and that traces of virtually the entire past are somehow "baked" into the present macrostate of the world. But it is important not to exaggerate how prevalent records of the past really are. Precisely because of the increase of entropy and the evolution toward equilibrium, very many past events have left no macro-traces in the present. Thus, as I have argued elsewhere (Frisch 2010), Albert and Loewer's account has the consequence that we can influence the past: every time our decision to perform a certain action is correlated with the occurrence of some past event E that has left no other macroscopic traces in the present, the decision will come out as a cause of the past event E, according to their account.

There is one argument, however, for the claim that the past hypothesis entails that we can know much more about the past than the future, which parallels Albert's discussion of the claim that the past hypothesis can function as an ultimate ready condition and which therefore might be able to support the latter claim. This argument, which I want to call the "constraint argument," appeals to the fact that low-entropy states correspond to much, much smaller regions of the phase space available to a system than states of higher entropy do. Imposing a low-entropy constraint on the past of the universe (or of a finite subsystem) thus puts a severe constraint on the system's past. In explicit premise-conclusion form, the argument can be represented as follows:

1. Low-entropy states correspond to extremely small regions of the phase space available to a system.
2. Therefore, imposing a low-entropy constraint on the past of the universe puts a severe constraint on the system's past.
3. Since the universe evolves toward states of ever higher entropy, its future evolution is much, much less constrained.
4. Therefore, we can know much more about the past of the universe than about its future.

But what this argument does not take into account is the fact that the different macrostates of a system that are accessible and of interest to us do not correspond to phase-space regions of comparable size. Thus, when we know that a system is in an equilibrium state, then we know all there is to know about the system macroscopically, even though we have not been able to narrow down very much what region of its phase space the system occupies. While in some sense we know much less about a system when we know that it is in equilibrium than we know of the system that it is in a specific extremely low-entropy state—we have much less narrowed down the region of phase space that it currently occupies—in another sense we know just as much about the system: in both cases we know its exact current macrostate.

In fact, if all we know of the system's past is that it was in *some* extremely low-entropy state, but we know of the system that it will evolve into its equilibrium state, then macroscopically we know *more* about the system's future than its past. Take as an example the case of a body of gas that is expanding in a box and about which we know that it was compressed into 1/100 of the total volume in one of the eight corners of the box (but know nothing more about the past state). Then even though the system's past microstate is much, much more constrained than its future microstate—we have narrowed down the phase-space regions that were accessible to the system in the past much more than the regions accessible in the future—we have full knowledge of the system's future macrostate but do not know everything there is to know about the system's past state.

Imagine that we were playing the following game: We are each given a box with a gas in some randomly prepared macrostate, and our goal is to acquire as much information as possible about the system's macrostate. Whoever can determine more of the macroscopically available information about the gas in the box wins the game. Imagine now that I received a box in which the gas is in equilibrium and evenly spread throughout the container and I am able to determine this (along with the values of all the thermodynamic parameters characterizing the state of the gas); and that you received a box with a gas very far from equilibrium compressed into a small volume in one of the corners of the box and with the partitions constraining the gas just removed. Let us assume that you were able to determine that the gas was compressed into one of the corners, but perhaps because the box is rotating and you were not able to keep track of its rotations, you were not able to determine into which corner the gas was compressed. Then it seems clear that I won the game: I have been able to determine all the information about the gas that is macroscopically accessible, while you have not been able to do so. And this is so, even though if our goal had been to constrain the possible microstates of the system as much as possible, you would have won. And it would not even have been close: you would have won that contest by many, many orders of magnitude, and the fact that you failed by a factor of eight to maximally constrain the available region of phase space would make no difference to that other contest. Thus, even though you would have handily won the game of narrowing down the phase-space region the system occupies, you lost of the game of determining the system's macrostate as precisely as possible.

Thus, the constraint argument fails and therefore cannot successfully be adopted to support the claim that the past hypothesis can function as ultimate ready state. In fact, we can now see that analogous considerations also call into doubt the claim that the past hypothesis provides a constraint sufficient to ensure the reliability of records. Here is an explicit version of that argument—the ready-state argument:

1. Inferences based on records presuppose assumptions about an earlier ready state.
2. The past hypothesis can function as the mother of all ready states.
3. Therefore, putative records of the past are reliable conditional on the past hypothesis: If R is a record of a past state S, then $Pr(S/R\&PH) \approx 1$.
4. Therefore, small counterfactual changes to the present are associated with changes to the future but not to the past.

Both the ready-state argument and the constraint argument rely crucially on the idea that the past hypothesis puts a severe constraint on the past evolution of the universe. But like in the case of the constraint argument, it is unclear whether positing that constraint can on its own carry the burden in the argument that it would need to carry. As our discussion above has shown, positing that there is *a* low-entropy constraint is not enough to ensure a unique past evolution. It is compatible with the assumption of an extremely low-entropy initial state that a system that evolved into a macrostate slightly different from its actual macrostate began its life in a different, nonactual low-entropy state. Thus, if the macrostate of the gas in our example, when it is close to equilibrium, had been slightly different, it might have evolved from a low-entropy state that had the gas constrained into a different corner than the one in which it was actually located. Recall that the phase-space regions corresponding to the forward evolutions of macroscopically different low-entropy states are highly fibrillated—they have "fingers" spread throughout the entire phase-space volume that corresponds to the later high-entropy state. This has the consequence that if we move the system's present phase-space point away from its actual trajectory to one corresponding to a slightly different macrostate, we are as likely to land on a trajectory that has evolved from a macroscopically different low-entropy past as on a trajectory that evolved from the actual low-entropy past. (Of course, the overwhelming majority of trajectories on which we could land evolved from a high-entropy past. But these trajectories are excluded by positing the past hypothesis.)

Now Albert may respond (and in fact has responded) to this kind of argument by insisting that the past hypothesis contains much more than the claim that the universe began its life in an extremely low-entropy state: it states that the universe began its life in whatever "Big Bang-ish sort of state" cosmology eventually presents us with. Yet it is unclear how much any additional cosmological constraints might help, since any such constraint will be much more coarse-grained than the kind of small-scale differences in the states of medium-sized macroscopic objects involved in commonsense causal claims. No matter what the constraints on the initial state of the universe cosmology will eventually present to us, these constraints will be compatible with many different histories of "medium-sized dry goods." That is, even Albert's richer past hypothesis appears not be sufficient to ensure the reliability of records.

There is one further reply one can give to my criticism of the ready-state argument. Once we focus not merely on the history of a single system, like the body of gas or the entire universe taken as a whole, but consider a complex system with many different interacting subsystems, it seems plausible that we would find that otherwise reliable records or traces of the past of a system could be misleading or unreliable in a close counterfactual situation only if the microstate of the system in question exhibited correlations of the kind that the initial randomness assumption, which we discussed in §2, is intended to exclude. Imagine, for example, that we took a picture of the body of gas as it was expanding and that the picture clearly indicated that the gas was compressed into a corner marked with a red dot. Then in the actual situation this picture serves as a record of the gas's past state. If we now consider a counterfactual change to the macrostate of the gas, such that the resulting counterfactual macrostate is overwhelmingly probable to have evolved from a nonactual low-entropy past, then the resulting "world" would have to have contained strong correlations among the variables characterizing its initial microstate. For it would have to be the case *either* that the photograph of the initial state of the gas was unreliable and there were strong correlations among the incoming light waves that mimicked the light waves reflected by the gas at its actual past location and masked that the light was reflected at the gas's counterfactual past location; *or* that there were strong initial correlations among the constituent atoms of the gas that resulted in improbable fluctuations in the gas's macro-history—or perhaps both. In any case, the counterfactual "world" would have to violate the assumption of initial randomness.

Thus, Albert and Loewer's premises do after all seem to entail that traces or records of the past are overall reliable, and hence that there are certain counterfactuals—those associated with small macroscopic changes to the present—that are temporally asymmetric, and hence, perhaps, that there is a causal asymmetry. But if what I have just argued is right, then the entire burden of the argument is carried by the probability postulate or the assumption of initial microrandomness. The past hypothesis does no work—no work, that is, aside from providing a *necessary condition* for the reliability of records by preventing an overwhelmingly anti-thermodynamic past and thereby ensuring that inferences to the past do not go radically wrong.

4. Conclusion

Where does this leave the debate about the two strategies I distinguished above? The initial randomness assumption, I have argued, is quite naturally thought of as a causal assumption that fits well with our explanatory practices of trying to explain correlations between distant events in terms of earlier common causes. We would expect initial states to be randomly distributed in the absence of an even earlier common cause. Thus, I think the burden of proof lies with defenders of a reductive strategy to show that the interlocking pieces of our causal representations of phenomena—our time-asymmetric explanatory practices, the positing of time-asymmetric causal relations, and the assumption of initial randomness—can be reduced to a set of noncausal assumptions. I am skeptical that Albert and Loewer's thermodynamic account can discharge this burden. If, as I have argued, the entire work in their account is done by the equiprobability or randomness assumption, they have not succeeded in deriving the causal and explanatory asymmetries from what are clearly acausal assumptions.

References

Albert, David Z. 2000. *Time and Chance.* Cambridge, Mass.: Harvard University Press.
Arntzenius, Frank. 2010. "Reichenbach's Common Cause Principle." In *The Stanford Encyclopedia of Philosophy* (Fall 2010 ed.), edited by Edward N. Zalta. http://plato.stanford.edu/archives/fall2010/entries/physics-Rpcc/.
Earman, John. 2011. "Sharpening the Electromagnetic Arrow(s) of Time." In *The Oxford Handbook of Philosophy of Time,* edited by Craig Callender. Oxford: Oxford University Press.

Elga, Adam. 2000. "Statistical Mechanics and the Asymmetry of Counterfactual Dependence." *Philosophy of Science* suppl. 68: 313–324.
Frisch, Mathias. 2005. *Inconsistency, Asymmetry and Non-locality: A Philosophical Investigation of Classical Electrodynamics*. New York: Oxford University Press.
——. 2006. "Causal Asymmetry, Counterfactual Decisions and Entropy." *Philosophy of Science* 72 (5):739–750.
——. 2007. "Causation, Counterfactuals and the Past-Hypothesis." In *Russell's Republic: The Place of Causation in the Constitution of Reality*, edited by H. Price and R. Corry. Oxford: Oxford University Press.
——. 2010. "Counterfactuals and Entropy." In *Time, Chance, and Reduction*, edited by Andreas Huettemann and Gerhard Ernst. Cambridge: Cambridge University Press.
——. 2014. *Causal Reasoning in Physics*. Cambridge: Cambridge University Press.
Lewis, David. 1986. "Time's Arrow" In *Philosophical Papers*, vol. 2. Oxford: Oxford University Press.
Loewer, Barry. 2007. "Counterfactuals and the Second Law." In *Causality, Physics, and the Constitution of Reality: Russell's Republic Revisited*, edited by H. Price and R. Corry. Oxford: Oxford University Press.
Sober, Elliott. 1984. "Common Cause Explanation." *Philosophy of Science* 51: 212–241.

Chapter Eleven

Time, Flies, and Why We Can't Control the Past

▶ ALISON FERNANDES

1. Introduction

David Albert explains why we can typically influence the future but not the past by appealing to a low-entropy initial state of the universe, a postulate he calls the "Past Hypothesis" (2000).[1] He argues (2014) that in the rare cases where we can influence the past, we cannot use this influence effectively to knowingly gain future rewards. So, it does not constitute control. But there is an important case Albert fails to consider: a case in which our action in the present is reliably correlated with several events in the past, and can, it seems, knowingly be used to gain future rewards.[2] To deal with this case, we need to appeal to epistemic

---

[1] An earlier version of this paper is referred to in Albert (2015, 49), Loew (2017), and Loewer (2020, 19).

[2] My target here is Albert (2000, 2014). While the case is relevant for other entropy-based accounts (Loewer 2007, 2012, 2020; Kutach 2002, 2013), I will not take up those arguments here—see Fernandes (2022a). For further criticism of Albert's account, see Frisch (2007, 2010) and Price and Weslake (2009).

conditions on deliberation: being agents requires our decisions being epistemically undetermined at the time we make them. In a world with the Past Hypothesis, this implies that deliberation will typically come prior to decision. Once deliberation in this direction is established, correlations toward the past cannot then be exploited for control. Deliberation is required to explain why we cannot effectively control the past, whether as part of a defense of Albert's account or used independently to explain the asymmetry of control.

I begin with Albert's explanation of the asymmetry of influence—why the future depends on what we do now in a way the past does not (§2). I then argue against his treatment of exceptional cases. His response to a case presented by Mathias Frisch fails because it introduces an unexplained asymmetry of rewards (§3). Moreover, he cannot deal with a new, importantly different case by appealing to conditions on rewards (§§4-5). In this new case, it seems we *can* knowingly influence the past to gain future rewards. To deal with this case, we need to appeal to evidential undermining and epistemic conditions on deliberation that explain why agents deliberate before they decide (§6). By appealing to deliberation, we can explain the asymmetry of control.

If we can recover Albert's explanation, we revive his more general project of unifying and explaining a range of asymmetric phenomena. This follows a tradition, beginning with Boltzmann's work in statistical mechanics, of explaining apparently fundamental phenomena in scientific terms: why ice cubes in glasses of water tend to melt (thermodynamic asymmetry), why our knowledge of the past is so different from our knowledge of the future (epistemic asymmetry), why we can influence the past but not the future (asymmetry of influence), and why causes come before their effects (causal asymmetry). What typifies this program is attempting to account for these asymmetries by appealing merely to time-symmetric dynamical laws, assumptions, and methods, and the Past Hypothesis.[3] The program uses a form of explanation that is already well accepted in scientific explanations of the second law of thermodynamics. Moreover, by tracing a range of asymmetric phenomena back to a single boundary condition, the program explains, at a deep level, how these phenomena relate. So, this kind of explanation is worth preserving.

---

[3] For an overview of foundational work in statistical mechanics, see Sklar (1993). For work broadly in this program, see Reichenbach (1956), Horwich (1987), Price (2002), Callender (2004), Greene (2004), Penrose (2005), Loewer (2007, 2012), Carroll (2010), and references therein. For criticism, see Winsberg (2004), Earman (2006), Maudlin (2007), Price and Weslake (2009), and Frisch (2007, 2010).

2. The Asymmetry of Influence

Albert's explanation of the asymmetry of control proceeds in two parts.[4] In the first, Albert uses the Past Hypothesis to explain why the present contains many more "causal handles" on (i.e., means of influencing) macroscopic aspects of the future than of the past. In the second part (considered in §3), Albert argues that any causal handles there are on the past cannot be used effectively—we cannot know about them or use them to gain rewards. So, even if we can influence the past, we can't control it.

The first thing to note is that talk of causal handles as "ways of influencing" should be compatible with deterministic fundamental dynamical laws. Here, for simplicity, I'll take the fundamental laws to be Newtonian laws of motion. Although such laws are not exceptionless or fundamental, whatever laws replace them will have to recover Newtonian laws on everyday length and energy scales, leaving the explanation largely intact. When we claim that manipulating a local feature of the present is a way of influencing the future, we are not claiming, contrary to fact, that two different futures are compatible with the present state and the laws. What we are claiming is that, counterfactually, *if* the present were otherwise than it is, *then* the future would be different. This, in turn, means that we need some way of evaluating counterfactuals.

Albert proposes the following method.[5] Start with phase space, a continuous space containing six dimensions for every particle in the system, one for each dimension of position and momentum. Points in phase space represent possible microstates of the system at a time. Phase space can be partitioned into volumes, using macroscopic parameters such as pressure, temperature, and density. These volumes represent possible macrostates of the system. Macrostates that occupy the largest volume of phase space are high-entropy equilibrium states, while those that occupy the smallest volumes are low-entropy states. Given deterministic laws, some possible microstates of the universe at a given time are compatible with the Past Hypothesis: with the universe beginning in "whatever particular low-entropy highly condensed big-bang sort of macrocondition it is that the normal inferential procedures of cosmology will eventually present to us" (Albert 2000, 96). The Past Hypothesis claims that the universe began in a particular low-entropy

---

[4] Albert sometimes refers to a merely *apparent* asymmetry. I take it that the asymmetry is real, for any reasonable concept of control, but this choice does not significantly affect the explanation.
[5] Albert claims that any reasonable time-symmetric method of evaluating counterfactuals yields the same asymmetry (2000, 125). Loewer's account (2007) is broadly compatible. Lewis's account (1979) is symmetric, but beset by further problems (Elga 2001).

macrostate.[6] Counterfactuals are evaluated as follows. Take the location in phase space of the microstate of the actual world at the time of the antecedent. The closest world is one whose microstate at the time of the antecedent (as measured in phase space at that time) is as close as possible the actual microstate, provided that the world also satisfies the following criteria:

(1) The antecedent of the conditional is satisfied.
(2) The Past Hypothesis is true.
(3) The world's macrohistory, given its macrostate and the Past Hypothesis, is assigned a reasonable probability by the Statistical Postulate.
(4) The fundamental dynamical laws are those of the actual world.

The counterfactual is true if and only if the consequent is satisfied at this closest world.

The Statistical Postulate in the third condition takes probabilities of volumes of phase space at a time to be uniform over the standard Lebesgue measure. This implies, roughly, that for any two volumes of the same size, the system is as likely to be in one as the other. The third condition excludes cases where the closest world is an anomalous one—in an improbable microstate, i.e., with an unusual macrohistory, given its current macrostate and the Past Hypothesis. We will see some of the interesting results of this condition below (§4).

Note that any change to the present microcondition of a world will necessarily imply changes to its past as well as future microhistory. Any counterfactual change to the present is necessarily a change to the past and future. This means that, if we are to derive an asymmetry of influence, only some changes to the past, present, and future can be considered, and the possible antecedents and consequents of the counterfactuals must be restricted. Albert restricts the consequents to localized areas of the world that can be characterized macroscopically in relatively simple, everyday language (2000, 121). He restricts the antecedents by introducing what he calls a "fiction of agency." This is a "primitive and un-argued-for" conception of what lies under our "*direct* and *unproblematical* and *unmediated* control" (2000, 128). While "fiction" suggests that such control is illusory, it need not be—for the moment we can simply take it as a primitive conception of what we take to be directly under our control. This "black box" conception can be filled out in various ways—to allow us direct control of our limbs, for example, or the

---

[6] This claim will need to be amended if the concept of entropy does not apply at the start of the universe (Albert 2000, 85; Earman 2006, 412).

electrical nerve impulses in our brains. Albert claims that under any reasonable conception, direct control will be localized to a very small area of the universe.

Given these restrictions, counterfactuals determine what aspects of the present, past, and future we can influence. Albert claims that, if the antecedents are features that are under our direct control and the consequents are readily characterizable macroscopic features, then the present contains many more "causal handles" on the future than the past. The motion of my hands turning the steering wheel, for example, is a causal handle on my car beginning to turn left in the next moment. But it is not a causal handle on the car's previous motion.

Albert's reasoning is based on the claim that the Past Hypothesis accounts for an asymmetry of *records:* it explains why we can reliably read local macroscopic or directly introspectable features of the present, such as our memories, as reliable indicators of the macroscopic past but not the future. According to Albert (2000), when we reason using records, we reason to a state at a time between two known states—the "ready" state of the recording device and its "record-bearing" state. Following standard usage, I will take "record" to refer to the localized macrostate (or directly introspectable state) that is the "record-bearing" state.[7] We have epistemic access to the Past Hypothesis, which functions as "the mother (as it were) of all ready conditions" (2000, 118). But we have no access to an analogous "Future Hypothesis," or any other future state that can function as a suitable ready state—one that provides information about the future states of a system beyond what can be derived from relevant knowledge of the past, the Statistical Postulate, and the dynamical laws. States now can serve as records of the past but not the future. While there may always be *microscopic* indicators of any past event at any time, whether there are *records* is a contingent matter on Albert's account—it depends on whether there are localized macroscopic (or directly introspectable) states now that are reliably correlated with the past event (given the Past Hypothesis, the laws, and the Statistical Postulate).

Assuming that the same inferential procedure is used for reading records and evaluating counterfactuals, the asymmetry of influence can be derived from the asymmetry of records. The method for evaluating counterfactuals requires leaving most of the actual present as it is, including records. If these records are reliable, when we evolve the system backward in time, the past we infer to will contain all those recorded events, and thus will be very much like the actual past. But in

---

[7] I am skating over some of the details of what records are. To my mind they are never sufficiently defined on Albert's account, and they don't feature in his final statement of the epistemic asymmetry. But discussions of his accounts are standardly framed in terms of localized macroscopic records, and I follow that custom here.

regards to the future, there is no such guarantee. Because the present contains no records of the future, a small change in one part of the present may well imply large changes in the future—the present state of the world does not constrain these changes to the same degree.

Consider an example. In the actual world, my cousin begins to saw away at a tree, and then waits for my signal. I signal by waving my arm; she continues sawing, and the tree comes crashing down. If I had not waved my arm, could I have influenced anything significant about the future? When we consider the nearest counterfactual world in which the antecedent is satisfied, where I do not wave my arm, and evolve this state forward, it is likely the tree does not come crashing down. No signal is given, and my cousin does not continue sawing. Could I have influenced anything significant about the past? The nearest counterfactual world in which the antecedent is satisfied has me not waving my arm. But it is likely to have lots of *records* of the sawing having begun—my memories, sawdust lying on the ground, birds flying startled in the air. If these records are kept intact in the nearest counterfactual world, and if evaluating counterfactuals involves the same inferential procedures as reading records, these reliable records keep the past as it was. Given the restriction imposed by the Past Hypothesis, it is likely that my cousin has still been sawing away at the tree.

This same type of reasoning applies to any macroscopic event in the past for which there are records in the present. Therefore, it seems that if there are records in the present that are not under our direct control, we cannot influence these aspects of the past, and so our influence of the past is severely restricted in a way that our influence of the future is not. By explaining the asymmetry of records, the Past Hypothesis explains the asymmetry of influence.

3. Dealing with Exceptional Cases

However, cases presented by Adam Elga (2001), Douglas Kutach (2002), and Mathias Frisch (2010) suggest that Albert's account entails that we *can* influence the past. As a result, Albert can't explain the asymmetry of influence. Albert (2014) responds by considering *control* of the past, rather than *influence,* and introducing further conditions on control. While some of the conditions are problematic, they sufficiently deal with these exceptional cases—but not with a new case that I present.

Elga and Kutach introduce the following case.[8] There is a large-scale event in the past, such as the sinking of Atlantis, for which there are no records in the

---

[8] The case is based on observations in Elga (2001), cited in Loewer (2012) and Albert (2014).

present. This means that small changes in the present, such as the movement of my finger, may be counterfactually correlated with whether Atlantis exists. Given that we do control such small changes in the present, Albert's account implies that we can also control past macroscopic events.

Albert (2014) and Loewer (2012, 128) respond by arguing that while there may be counterfactual dependence of the past even on the present in these cases, this does not constitute *control*. This is because there are no macroscopic correlations that we could become aware of between what we control in the present and the past. If there were, the present would contain records of the past macrostate—but the Atlantis case relies on the absence of such records.

Albert and Loewer's response succeeds, but only by conceding that an asymmetry of counterfactual dependence is not enough to explain the asymmetry of control. Instead, we need further epistemic conditions on what counts as *effective* control: it is control we can know about. Such epistemic conditions are plausible. We don't have control of an event merely because it is correlated with our action. For example, I do not control what a detective records merely because he has decided to write down my shopping choices—to take a case from Anscombe (1957). I need to at least know about this correlation in order to effectively use it to bring about my ends and so for it constitute control. With this added condition, Albert's account survives Elga and Kutach's objection.

Frisch (2010) raises a separate challenge to Albert's account. He introduces a case in which the unique record of an event is under an agent's direct control. Say Bert is playing a piano piece, and comes to a point in the music where he can decide whether to play the first or second ending—the decision being under his direct control. Stipulate that Bert has no conscious memory of what music he has played, and there are no external records present. But Bert is a reliable pianist. Therefore, whatever decision he makes counts as a record of where he is up to in the piece and what music he has played. So, by having control of his decision in the present and his decision being a unique record, he has control over what music he has played in the past. Albert's account does not imply the asymmetry of influence.

Albert (2014, 167) responds to Frisch's case by adding another condition on effective control. He argues that, even if Bert knows he can influence the past, he cannot use this influence for any further benefit, beyond what he secures through influencing the present. If Bert wants to finish playing sooner, for example, he can influence the past so that he began the piece earlier. But he can just as easily decide to play the second ending and finish playing. The past event only influences the future via a feature that Bert already has direct control over. Bert's influence of the past gives him no additional benefits. Hence, the Frisch case is not a case of effective control.

But Albert's response fails. He only considers additional benefits in the future, which introduces an unexplained asymmetry. If we consider benefits in the past as well, influencing the past in the Frisch case may well give Bert additional rewards. For example, if he started playing earlier, he may have spent less time bored at the piano. Influencing the past is essential for securing this gain. For Albert's response to work, the asymmetry in which benefits we consider (those of the past or future) must be explained in a noncircular fashion. Moreover, even if this can be done, there is an important new case for which Albert's response is ineffective.

Before we move on to the new case, we should note that there is a better response to Frisch available. To effectively control the past, it is not enough to know that in some cases one's decision or action is correlated with the past event. To know that one controls the past, one must have some way of checking whether one has controlled it on occasions when one has *attempted* to do so. Bert cannot do this in the Frisch case. Because his decision in the present is the only record of the event, there is no time at which he knows both his decision and his past playing by independent means. So, he cannot determine whether he has controlled the past, and this is not control he can know about. Frisch's objection fails. Albert's explanation of the asymmetry of control survives. But there is a new case that evades this response.

4. The Fly Case

The Elga-Kutach and Frisch cases can be dealt with. However, there is another case that threatens Albert's explanation of the asymmetry of control. Here it is: In the actual world, a fly flies in front of Daphne's face at t_1, and at t_2 she swats it away. Consider the counterfactual: If she had not swatted at t_2, the fly would have been somewhere else at t_1. If this counterfactual is true, and she can knowingly exploit this counterfactual dependence, she can control the past. So, Albert's account does not explain the asymmetry of control. In this section, I will argue that the counterfactual is true, and in §5, that Daphne can knowingly exploit this counterfactual dependence.

To evaluate the counterfactual, we consider nearby worlds in which Daphne does not swat, the Past Hypothesis is true, the macrohistory of the world is statistically normal, and the fundamental dynamical laws are those of the actual world. Perhaps she doesn't swat because a stray thought or itch distracts her. But say she is not the type to get distracted by stray thoughts or itches—since she was a young child, many varied experiences have led her to treat flies as small flying vermin, to be efficiently swatted away whenever they appear. Given these experiences, her swatting will itself count as a record of the fly's location in the past. Moreover, say there are lots of records of these harrowing experiences and her reliability at swatting. If these records are to remain reliable, if she hadn't swatted at t_2 it would be because the fly was out of range or sight at t_1. The counterfactual looks true, and set to give Daphne influence on the past.

The counterfactual would be false if the present contained many large-scale records of the fly's actual position at t_1. Keeping these records intact and reliable would prevent the location in the past from changing—as we saw in the tree-sawing case above. But no entomologists are filming the fly's activities. The fly's location at t_1 is a small event, perhaps recorded only in its location at t_2, a slight pressure gradient in the air, and Daphne's memory. These records are small enough that they may change in a nearby world where she doesn't swat. Some such changes are required if her tendency to swat and records of this tendency are to remain reliable. While intuitively we might not think the size of the record should be relevant to whether it remains reliable in counterfactual scenarios (and I don't subscribe to this idea myself), it is a standard feature in the tradition following Lewis (1979). Given this feature, one will always be able to engineer scenarios in which the records of the reliability of the agent are larger than the records of the event she attempts to influence—and so less susceptible to change.

The important difference between the fly case and the Frisch case is that in the Frisch case there are no macroscopic changes to the present beyond those specified in the antecedent. Bert's influence in the present is limited to what he controls directly. Not so in the fly case. By directly controlling a record of the fly's past location, Daphne influences other records in the present. This allows her more wide-ranging control over the past and future, as we will see in §5.

What is doing much of the work in making the counterfactual true is the third condition: the world's macrohistory, given its macrostate and the Past Hypothesis, is assigned a reasonable probability by the Statistical Postulate. There are worlds in which there are lots of records of Daphne's reliability as a flyswatter and yet her swatting behavior is uncorrelated with the location of the fly in the past. But these are all unusual worlds compared to a given standard, and assigned a low probability by the Statistical Postulate. We should not infer to such worlds. Similarly, there are worlds in which the fly's location at t_2 changes, but not its location at t_1: the fly flies incredibly fast. But again, these are unusual worlds. We should not infer to them.

There is a nearby world in which all the relevant records remain reliable. It is a world in which Daphne does not swat and the fly's locations at t_1 and at t_2 both change. This is the world we should infer to. These changes are required to satisfy the antecedent and make the world's history sufficiently probable, given its macrostate. While Albert doesn't provide a measure for how the improbability of a world's macrohistory should be traded off against changes to its present state, I'll provided reasons below (Objections 1–3) for why some changes should be allowed for. Daphne's swatting remains a reliable record of the fly's location. By directly controlling her swatting, a record of the fly's location, Daphne influences the fly's location in the present and past. Albert's account does not yet deliver the asymmetry of influence.

I will now defend this counterexample from a number of preliminary objections.

Objection 1: The relevant counterfactual is false. The third condition only requires us to pick a microstate that is probable *with respect to other microstates in the same macrostate*. Given Frisch's interpretation (2007, 2010), no macrostates are ruled out by this condition. In the nearest world (as measured in phase space) in which the antecedent is satisfied, nothing macroscopic outside the antecedent changes. That is the macrostate we use to evaluate the counterfactual. So, the fly's location in the past does not change.

Response: Albert's account is consistent with this reading of the third condition, but it favors my own. My reading is more in keeping with Albert's appeal to "normal procedures of inference" and his defense of these conditions as capturing how we ordinarily infer (2001, 130). It is normal to infer from a record of one event to other records of the same event at that time. My morning newspaper allows me to infer the news of yesterday as well as the contents of your newspaper today. If counterfactuals are to capture such inferences, those like the following should come out true: If my paper had been different, yours would have been as well. If this counterfactual is to come out true, we can't rule out making changes in the macroscopic state outside the antecedent. In addition, Albert (2014, 163) more explicitly requires us to "find the possible world which is closest to the actual one, as measured by distance in phase-space, at the time of the antecedent." If there are to be no macroscopic changes outside the antecedent, this requirement is difficult to make sense of. Once the antecedent is specified, there would only be one macrostate to consider. Why should closeness *within* this same macrostate matter? The condition makes better sense if its work is to *minimize* macroscopic changes, not rule them out altogether.

Objection 2: Albert's intentions aside, we should follow Kutach (2002) and Loewer (2007) and not allow macroscopic changes outside the antecedent when evaluating counterfactuals.

Response: Satisfying antecedents nearly always involve making changes elsewhere in the system, under any method of evaluating counterfactuals. For example, if my hand is to be located in region x instead of y, air molecules will have to be moved out of x and into y. This is not an unusual feature of Albert's method. Moreover, the method gives us a useful guide to what other changes are required to satisfy the antecedent. Because worlds with vacuum pockets around me are very unusual, we should not infer that were my hand at x, there would be a vacuum pocket at y.

In addition, allowing the macrostate outside the antecedent to change is in keeping with our normal inferential reasoning. We reason from states of one system to states of another system at the very same time. If the counterfactual method is to preserve this reasoning, it must allow for changes in the antecedent to be correlated with macroscopic changes outside the antecedent. While, ultimately, we may want a method that doesn't allow for changes outside the antecedent, the more informative project is to consider how we get to such a method without first simply stipulating against such changes. Price and Weslake make a similar point (2009, 426).

Objection 3: Condition 3 can only rule out exceedingly unlikely worlds. A world where Daphne doesn't swat and the fly was still in view is not sufficiently unlikely.

Response: If Condition 3 is to ensure good inferential reasoning, it should not be too weak. We should not infer that measuring devices always break down when we consider changes to their records. While unusual events do happen in our world, and reliable devices do sometimes break down, we shouldn't assume such breakdowns will *always* happen if there are other nearby worlds in which recording devices remain reliable. Otherwise the method for evaluating counterfactuals will often lead to unhelpful results.

Objection 4: If we're requiring that records remain reliable, why should the size of the record matter? Why not also consider *large* macroscopic changes outside the antecedent?

Response: While such a method is appealing, my target is Albert (2000, 2014, 2015). Albert follows Lewis (1979) in thinking that changes from the actual world should be minimized. What's important is that, even with this requirement, counterexamples can be generated.

Objection 5: The fly's location is a small event. The influence on the past is only minimal, and so Albert's explanation succeeds for the most part.

Response: While the fly's location is a small event, a small change in the near past can be parlayed into a much larger change in the more distant past. For example, say the event I control in the past is the last remaining newspaper record of a nineteenth-century fire. In this case, a small event I control (the newspaper existing) is a unique record of a larger, more distant event (the fire), and I control the larger event as well. Nothing so far in the account rules out such large-scale changes—or us training ourselves and others to develop responses that can be exploited in this way.

Objection 6: The setup for the case itself is too fragile—add some large records in the present, and the influence on the past is destroyed (Albert 2015, 50–51). Fragile cases are not sufficiently common and robust to challenge Albert's explanation.

Response: While the case is fragile, this is not enough to dismiss it. We should consider why we can't make our influence on the past more robust, as we can our influence on the future. The response I recommend below (§6) can account for this fact.

Objection 7: Agents aren't reliable responders. Their neurological makeup is fragile and easily changes in counterfactual worlds. Their decisions and actions aren't records, and so don't allow for influence of the past under Albert's account.

Response: It is essential for the fly case that Daphne's reliability is recorded in large-scale macrofeatures of the world, such as in eyewitness reports and her four-volume treatise on her dedication to fly swatting. These macrofeatures can't easily change and are reliable evidence for her behavior and neurology.

Objection 8: Agents aren't reliable responders unless they observe the events they respond to. Having observed them, these events should be held fixed when evaluating counterfactuals.[9]

Response: Daphne's swatting may be an unconscious, instinctive action and yet reliable. But more importantly, we should ask why we *don't* need to observe future states in order for them to be correlated with our decisions now, whereas we *do* need to "observe" (consciously or not) past states for them to be correlated with our decisions now. Calling actions correlated with the past 'responses' hides this asymmetry. Once this asymmetry is explained, Objection 8 turns out to be compatible my account below (§6).

5. Can the Fly Case Be Used for Effective Control?

Having considered some preliminary objections, how else might we respond to the fly case? Albert responded to the Elga-Kutach and Frisch cases by invoking conditions on effective control. While these cases involve counterfactual dependence and influencing the past, the agent cannot knowingly use this influence to gain rewards. Hence, they do not threaten the asymmetry of control. I will show

---

[9] Albert (2014, 167) can be interpreted as offering a version of this response.

how Daphne can knowingly gain rewards in the fly case, and why it remains a threat to Albert's account.

The fly case can be enriched by adding a sensor that records the location of the fly at t_1 and displays this location on a screen at t_2. By stipulation, the screen is the only record from the sensor of the fly's location that remains at t_2—the rest of the device resets. Also stipulate that this record is small in size and can change in the counterfactual world. This means that both the record and Daphne's hand swatting are reliably correlated with the presence of the fly in a localized area at t_1. Say Daphne is offered a reward (anytime before t_2) for keeping this area free of flies at t_1. She will attempt this feat by not swatting at t_2. The sensor will tell whether she has succeeded, and she will be rewarded when the screen is checked after t_2. From what has been said so far, she can influence the fly's location to gain the reward. So, she can control the past, and Albert's account doesn't imply the asymmetry of control.

My response to the Frisch case will also not work here. In the Frisch case, there is no independent means of checking whether Bert has controlled the past. The only record of the past event is a record he controls directly, and there is no way to confirm that this record remains reliable. What is importantly different about the fly case is that there are records of the consequent *outside* the direct control of the agent—the fly's location in the present and the recording on the screen. This means that there is an independent means of checking whether she has controlled the past.

Albert's (2014, 167) response to the Frisch case will not work here either. His response was that any rewards Bert gains in the future are due only to Bert's direct control of the present, not his influence on the past. Not so in the fly case. Here, Daphne's mere ability to not swat, uncorrelated with the absence of the fly, is not enough to secure the reward. She has to influence the fly's location in the past. Otherwise, the sensor's record in the present will not be changed, and she will not receive the reward. Again, this is because there are records of the consequent outside her direct control. Daphne can influence these records, but only by influencing the past.

Nor is Albert's response to the Elga-Kutach case effective here. In the Elga-Kutach case, there are no records of the past event in the present, and so no macroscopic correlations between the action and events the agent could become aware of. But, in the fly case, there are such records. There are macroscopic correlations between Daphne's swatting now and the fly's location in the past, correlations she can know about. So it seems that she *can* control the past to gain the reward.

Albert (2015) suggests a new response to the Frisch case. Say agents' psychological states are reliable responses to past events and also under their free and direct control. If so, Albert argues, there must be correlations between the decisions of distinct agents on how they *exercise* that control; and "the existence of a correlation like that seems very obviously at odds with the idea of such decisions as free and spontaneous and autonomous acts of the sort we are thinking of when we entertain the fiction of agency" (2015, 45). But this is too fast. Consider a case. A lecturer asks their students to come to class early tomorrow. Every cooperative student freely decides to do so and shows up early. Every contrary student freely decides to flout the request and shows up 10 minutes late. The time of each cooperative student's arrival and each contrary student's arrival may count as a record of the lecturer's request. But there are no mysterious correlations. It is no surprise that, having heard the same request, students with similar motivations respond in similar ways. While these may not be *paradigmatic* cases of free agency (for reasons I explore below), they can't be dismissed as not being cases of agency at all.

There is also a more promising response: Daphne cannot use her influence on the fly's location for gain because any attempt to exploit that influence destroys the correlation on which it is based. If she commits to gaining the reward, and, for that reason, decides not to swat, the stillness of her hand is no longer a reliable record of the fly's absence. Instead, it is a reliable record of her commitment to gaining the reward. In the nearest counterfactual world, the macrohistory will then not be one in which the fly is somewhere else, but merely one in which she wishes to make it so. More generally, if the antecedent is brought about in order to bring about the consequent, the antecedent is no longer a reliable record of the consequent, but only of a desire or reason to bring it about. Thus, the counterfactual is false, and she does not control the past. I call this "evidential undermining."[10]

Evidential undermining works equally well against the Frisch case. If Bert decides to play the second ending in order for it to be the case that he started playing earlier, and so will finish sooner, his playing the second ending is no longer a reliable record of his having begun earlier—just of his desire to finish sooner. While Frisch is right to claim that we can learn inductively of correlations between our decisions and past events (2010, 31), this is not the same as learning of correlations we can make use of; these need to be correlations which are robust when we attempt to use them to control the past.

This evidential undermining is analogous to responses made by evidential decision theorists to medical Newcomb cases (Price 1986, 1991). The general idea is

---

[10] For further discussion of evidential undermining, see Price and Weslake (2009), Blanchard (2014), and Fernandes (2017).

that the evidential correlations an agent makes use of in decision must survive her knowledge of her deliberation—a deliberation that includes a belief in the correlation. In medical Newcomb cases, the agent has no reason to believe that correlations between her actions now and past events will be preserved if she decides to act on the basis of a belief in the correlation (in contrast to traditional Newcomb cases). For this reason, evidential decision theory gives her no reason to decide on past events. The corresponding response in the fly case is that evidence for a correlation between the fly's presence and Daphne's swatting is no longer evidence for that correlation whenever she decides not to swat in order to exploit the correlation.

Albert (2015, 49–50) comes to adopt the evidential undermining response (as a second response)—partly as a result being presented with the fly case (49). But while the response is largely correct, it is inadequate in its current form. It relies on an asymmetry—that agents in our world deliberate *before* they decide.[11] It is because deliberative reasoning comes temporally prior to decision that it can disrupt correlations between decision and past states, but not correlations between decision and future states. This disrupting may be due to general features of influence—such as the fact that influence propagates through time. Or it may be due to features peculiar to deliberation. But it is straightforwardly true. If an agent's deliberative structure were reversed, such that she acted, then decided, then deliberated, correlations between her decision and past states would be preserved when she attempted to make use of them.

Here's another way to press the concern. When Albert first presents the undermining response, he puts it in terms of the correlations in the Frisch case not being sufficiently *robust*: they only hold in certain circumstances, while it is part of our concept of free choice that one can act "in any number of different hypothetical deliberational contexts, and for any number of different hypothetical reasons, and in the service of any number of different hypothetical ends" (2015, 50). What prevents the correlations being robust in the fly case, however, is not events that happen in the present or future of the act—such as Daphne's current hopes to impress her friends, or her later donating money to charity—these may occur or not, and still the correlation can obtain. It is events that happen *in the past*, such as her being offered the reward while having the desire for it, that undermine the correlation.[12]

---

[11] Loew (2017, 448) raises a similar concern.

[12] It is partly for this reason that I take issue with Albert's earlier suggestion that we are never reliable responders—for what explains evidential undermining is precisely that we do sometimes respond to past events in reliable ways.

Altogether, if evidential undermining is to succeed as a response, we must explain why deliberation comes before decision, and so undermines correlations in that direction only. In §6, I will explain this fact by appealing to epistemic conditions on deliberation and an epistemic asymmetry. What will emerge is that while we can appeal to the above response to defend Albert's account of the asymmetry of control against cases like the fly case, another option is available. We can explain the asymmetry of control more directly by making use of the fact that agents deliberate before they decide, and their deliberation undermines correlations in the same temporal direction.

6. Deliberative Agency

I will now explain why deliberation comes temporally prior to decision. I will argue for two epistemic conditions on deliberation (*ignorance* and *efficacy*), that allow us to explain a key feature of deliberation: its seriality. *Seriality*, combined with an epistemic asymmetry, explains deliberative asymmetry and the asymmetry of control.

Ignorance

I take it to be a condition of deliberation that an agent cannot be certain of her decision as she is deliberating and as she decides.[13] If an agent is deliberating on whether to *A*, she cannot be certain that her deliberation terminates in her deciding to *A*: decisions are *epistemically undetermined*. Why think that decisions satisfy this requirement? While many theorists simply note the condition as necessary, a promising response appeals to the function of deliberation: there is no point deliberating if an agent is already certain of her decision. Holton (2006), for example, argues that choice scenarios arise in cases where we are unsure of what to do—if we were sure, there would be no need for deliberation. Ginet (1962) makes a similar point.

Accepting this requirement places an *ignorance condition* on decision. There cannot be connections between an agent's decisions and states of the world (including her beliefs and desires) that allow her to infer her decision while deliberating and deciding. In particular, if she is certain of other states of the world, she

---

[13] Defenders of this condition include Hampshire and Hart (1958), Ginet (1962), Taylor (1964), Schick (1979), Levi (1986), Holton (2006), and Ismael (2007, 2012). The general shape of this account of agency draws much from Ismael's work. For further discussion of different kinds of ignorance conditions, see Fernandes (2016a).

can't also be certain of the correlations between those states and her decision. *Ignorance* may be satisfied if the correlations between states of the world and her decisions are complicated and difficult to track. This will be the case if our internal makeup is complex, constantly changing, and hidden from view. Think of a fireman and whether he can use the heat of the floor as evidence that he will decide to leave a burning building (Holton 2006). In order to make an accurate inference, he must take into account his values and desires, the way he balances the risks involved against the possibility of saving the building, and the significance of all this evidence. In addition, his inference must take into account what effect the inference will have on his decision—making a prediction can change his evidence about what he will decide to do. Because of this, his decision is very unlikely to be susceptible to simple modeling. *Ignorance* can be satisfied by these kinds of epistemic limitations.

Efficacy

There is a second epistemic condition on decision: agents must take their decisions to be *efficacious* in producing their results. An agent, while deliberating whether or not to A, must believe that her decision to A will result in her A-ing: her decision must be good evidence for her A-ing. This does not imply a causal connection between the decision and outcome. It is a minimal epistemic condition, which many defend (including those mentioned in footnote 13).

These two conditions explain how features of the world, particularly bodily movements, appear to be under our direct control. We can do better than Albert and not take direct control as a primitive. As far as our decisions are epistemically undetermined, other states of the world we take to be reliably correlated with them will appear under our direct control. Further conditions may be needed to ensure direct control. But as necessary conditions, *ignorance* and *efficacy* partly determine the appropriate scope of our conception of direct control: it will include only features that we take to be reliably correlated with our decisions.

Seriality

These two conditions on decision explain a key feature of deliberation: its *seriality*. Seriality is the ordering of deliberation, decision, and action, as schematized in Figure 11.1. This is not yet a temporally asymmetric ordering—we're not presuming that decision takes place *before* action in time. Decision might come after. But *seriality* does imply that decision falls temporally between action and deliberation.

Here is how *seriality* results. *Ignorance* implies that an agent cannot be certain of her decision while deliberating and deciding. *Efficacy* implies she must be

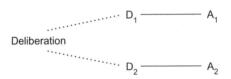

FIGURE 11.1. The seriality of agency. While deliberating, the agent is uncertain of her decisions (represented by dotted lines). Decisions are epistemic possibilities (D_1 or D_2). The agent takes there to be a correlation between her decisions and actions (A_1 and A_2) (represented by solid lines). Credit: © Alison Fernandes

certain of the correlation between her decision and her action while deliberating. Together they imply than an agent cannot be certain of her action while deliberating—given *efficacy*, this would violate *ignorance*. But to satisfy *efficacy*, the agent also needs some way of becoming aware of correlations between her decisions and actions in general. There must be a time at which she knows both her decision and her action. Unless an agent becomes ignorant of her decision between acting and deciding (a very unusual epistemic structure indeed), she cannot typically deliberate at times between deciding and acting—this would violate *ignorance*. She also cannot act between deciding and deliberating—this would prevent there being a time at which both decision and action are known. So agency must take this serial form. It is a form that allows there to be a time at which both decision and action are known, but which does not prevent the agent deliberating at another time.

It does not yet follow that agency is oriented in a particular temporal direction—only that decision comes between deliberation and action. However, given the epistemic asymmetry that Albert explains (or even a more limited epistemic asymmetry), *seriality* implies that agency will typically lie in a temporal direction that runs past–future. *Ignorance* and *efficacy* can be satisfied in a forward temporal direction. Our decision-making is sufficiently complex to satisfy *ignorance*, as we saw in the fireman case. Records in the form of memories also allow agents to pick up on general correlations between their decisions and actions, allowing *efficacy* to be satisfied. Not so in the backward temporal direction. Here, the reliability of records prevents *ignorance* from being satisfied. If our macroscopic acts occur in the recent past, we will typically have records of them in the present in the form of memories. Given *ignorance*, this prevents us deliberating about these acts: the decisions are not epistemically undetermined. An agent could still take the mental state to be epistemically undetermined, by ignoring or forgetting the correlation with action. But then the mental state would not satisfy *efficacy* and would no longer count as a decision. Note also that if the temporal distance between the decision and the action becomes too great, the reliability of their correlation is likely to be lost, and so *efficacy* will not be satisfied.

It is because there are readily available records of macroscopic events in the recent past, but not future, that deliberation is typically directed toward the future, and we deliberate before we decide. Records of the past prevent decision from being directed toward the past. But they don't prevent decision from being directed toward the future, and they allow agents to track correlations between decisions and actions. This argument does require an epistemic asymmetry. But we have independent grounds to accept such an asymmetry, whether it is a general asymmetry of records (Albert 2000) or something limited to records like memories. If a suitable epistemic asymmetry can be explained in terms of the Past Hypothesis, as Albert suggests, we also have an argument for why the Past Hypothesis is relevant to the direction of deliberation. It is relevant because it explains features of our epistemic access to the world (*contra* Price and Weslake 2009).[14]

The seriality of agency and an asymmetry of records explain why agents deliberate before they decide and why their decisions are typically directed toward the future. But we haven't yet explained why agents *never* deliberate about the past. There are two ways that agents could generalize from the typical case and adopt a conception of agency that rules out deciding on past states of affairs. One is by overgeneralizing and believing that there is no counterfactual dependence between decisions now and macroscopic states that are accessible at the time of deliberation. Given that the entire past seems 'in principle' accessible, we cannot influence the past. So-called "bilking" arguments against influencing the past exploit this idea (Dummett 1964). A second way is by overgeneralizing and believing that there is no counterfactual dependence of past macroscopic states on present decisions (Frisch 2007, 365). Both of these ideas can explain what seems odd about the fly and Frisch cases and why we take influence of the past to be *impossible*.

But we need not go so far. We can combine the deliberative asymmetry with the previous response to the fly case to finally explain the asymmetry of control. I argued in §5 that agents who deliberate prior to decision will be unable to exploit correlations between their decisions and states further in the past. If they attempt to use such correlations to control the past, they will undermine them. Their decisions will only be evidence for their desires or reasons, rather than for

---

[14] While I agree with Loew (2017, 446) that Albert misconstrues the role of the Past Hypothesis, this does not imply that the Past Hypothesis has no role to play in explaining the asymmetry of control. See Fernandes (2016b, 2022b) for discussion of the relation between statistical-mechanical accounts and agency approaches.

the past states they hope to control. So, as far as an agent's evidential map encodes correlations she can use, no correlations between decisions and previous states will appear. While she can believe that there are evidential correlations between her decision and previous states, they are not ones she can exploit to achieve her ends.

Consequently, agents cannot freely establish correlations between their decisions and past events and use these correlations to control the past. In the fly case, for example, if I decide to develop a reliable correlation between the state of the fly and my swatting, I cannot then exploit this correlation to control the past. Deciding to swat upon observing the fly already assumes that deliberation comes prior to decision. Given this asymmetry, I will undermine any correlations to the past I attempt to exploit.

7. Conclusion

An asymmetry of records together with epistemic conditions on deliberation explain why deliberation comes prior to decision. Given this asymmetry of deliberation, evidential undermining (§5) explains why we can't typically control the past. Agents with very unusual epistemic capacities may be able to deliberate after they act, and so effectively control the past. There may also be unusual circumstances (such as traditional Newcomb cases) in which we do control the past—cases in which correlations toward the past are not undermined by our deliberating on the basis of them. But in ordinary circumstances, agents like us, able to form stable memories of past macroscopic events, won't be able to deliberate toward the past. We can't pick up on correlations between past actions and present decisions in ways that allow decisions to be epistemically undetermined. Given that we deliberate before we decide, we cannot, in ordinary cases, exploit past directed correlations without undermining them.

By using epistemic conditions on deliberation, we have a good explanation for why we cannot control the past. Our epistemic capacities prevent past-directed deliberation, and, given future-directed deliberation, we can no longer exploit correlations between our decisions and events in the past. This explanation can be used to defend Albert's original account of the asymmetry of control. *Contra* Price and Weslake (2009), appealing to features of agency is compatible with explaining the asymmetry of control in terms of entropy—particularly if explaining agential features requires appealing to the asymmetry of entropy. But the explanation I have offered here also stands alone. By appealing to plausible conditions on deliberation and an epistemic asymmetry that is more limited than Albert's version, we can explain why deliberation comes prior to decision and why we cannot control the past. Such an account may then be the foundation for explaining the more general temporal asymmetry of causation (Fernandes 2017, 2022b).

Acknowledgments

My warm thanks to David Albert for many inspiring and encouraging discussions on these issues. For their comments, criticisms, and helpful suggestions I would also like to thank John Maier, Jenann Ismael, Mat Simpson, Huw Price, Michael Hicks, Thomas Blanchard, Brad Weslake, Heather Demarest, Barry Loewer, Matthias Frisch, Sidney Felder, and audiences at the University of Sydney, the University of Pittsburgh, and the University of South Florida.

References

Albert, David Z. 2000. *Time and Chance.* Cambridge, Mass.: Harvard University Press.
———. 2014. The Sharpness of the Distinction between Past and Future. In *Asymmetries of Chance and Time,* ed. Alistair Wilson. Oxford: Oxford University Press.
———. 2015. *After Physics.* Cambridge, Mass.: Harvard University Press.
Anscombe, G. E. M. 1957. *Intention.* Basil Blackwell.
Blanchard, Thomas. 2014. Causation in a Physical World. PhD diss. Rutgers University.
Carroll, Sean. 2010. *From Eternity to Here.* New York: Dutton.
Callender, Craig. 2004. Measures, Explanation and the Past: Should "Special" Initial Conditions Be Explained? *British Journal for the Philosophy of Science* 55:195–217.
Dummett, Michael. 1964. Bringing about the Past. *Philosophical Review* 73 (3): 338–359.
Earman, John. 2006. The "Past Hypothesis": Not Even False. *Studies in History and Philosophy of Modern Physics* 37:399–430.
Elga, Adam. 2001. Statistical Mechanics and the Asymmetry of Counterfactual Dependence. *Philosophy of Science* 68 (3): S313–S324.
Fernandes, Alison. 2016a. Varieties of Epistemic Freedom. *Australasian Journal of Philosophy* 94 (4): 736–751.
———. 2016b. *A Deliberative Account of Causation: How the Evidence of Deliberating Agents Accounts for Causation and Its Temporal Direction.* PhD diss. Columbia University.
———. 2017. A Deliberative Approach to Causation. *Philosophy and Phenomenological Research* 95 (3): 686–708.
———. 2022a. Back to the Present: How Not to Use Counterfactuals to Explain Causal Asymmetry. *Philosophies* 7 (2): 1–12.

———. 2022b. The Temporal Asymmetry of Causation. Unpublished manuscript.
Frisch, Mathias. 2007. Causation, Counterfactuals, and the Past Hypothesis. In *Causation, Physics, and the Constitution of Reality*, ed. Huw Price and Richard Corry, 293–326. Oxford: Oxford University Press.
———. 2010. Does a Low-Entropy Constraint Prevent Us from Influencing the Past? In *Time, Chance and Reduction*, ed. Gerhard Ernst and Andreas Hüttemann, 13–33. Cambridge: Cambridge University Press.
Ginet, Carl. 1962. Can the Will Be Caused? *Philosophical Review* 71 (1): 49–55.
Greene, Brian. 2004. *The Fabric of the Cosmos*. New York: Alfred A. Knopf.
Hampshire, Stuart, and H. L. A. Hart. 1958. Decision, Intention and Certainty. *Mind* 67 (265): 1–12.
Holton, Richard. 2006. The Act of Choice. *Philosophers' Imprint* 6 (3): 1–15.
Horwich, Paul. 1987. *Asymmetries in Time*. Cambridge, Mass.: MIT Press.
Ismael, Jenann. 2007. Freedom, Compulsion and Causation. *Psyche* 13 (1): 1–11.
———. 2012. Decision and the Open Future. In *The Future of the Philosophy of Time*, ed. Adrian Bardon, 149–168. London: Routledge.
Kutach, Douglas. 2002. The Entropy Theory of Counterfactuals. *Philosophy of Science* 69 (1): 82–104.
———. 2013. *Causation and Its Basis in Fundamental Physics*. New York: Oxford University Press.
Levi, Isaac. 1986. *Hard Choices*. Cambridge: Cambridge University Press.
Lewis, David. 1979. Counterfactual Dependence and Time's Arrow. *Nous* 13: 455–476.
Loew, Christian. 2017. The Asymmetry of Counterfactual Dependence. *Philosophy of Science* 84 (3): 436–455.
Loewer, Barry. 2007. Counterfactuals and the Second Law. In *Causation, Physics, and the Constitution of Reality*, eds., Huw Price and Richard Corry, 293–326. Oxford: Oxford University Press.
———. 2012. Two accounts of laws and time. *Philosophical Studies* 160 (1): 115–137.
———. 2020. The Mentaculus Vision. In *Statistical Mechanics and Scientific Explanation: Determinism, Indeterminism And Laws Of Nature*, Valia Allori (ed.), Singapore: World Scientific.
Maudlin, Tim. 2007. *The Metaphysics within Physics*. New York: Oxford University Press.
Penrose, Roger. 2005. *The Road to Reality*. New York: Alfred A. Knopf.
Price, Huw. 1986. Against Causal Decision Theory. *Synthese* 67:195–212.
———. 1991. Agency and Probabilistic Causality. *British Journal for the Philosophy of Science* 42 (2): 157–176.
———. 2002. Boltzmann's Time Bomb. *British Journal for the Philosophy of Science* 53:83–119.
Price, Huw and Weslake, Brad. 2009. The Time-Asymmetry of Causation. In *The Oxford Handbook of Causation*, ed. Helen Beebee, Christopher Hitchcock and Peter Menzies, 414–443. Oxford: Oxford University Press.

Reichenbach, Hans. 1956. *The Direction of Time*. Mineola: Dover Publications.
Schick, Frederic. 1979. Self-knowledge, Uncertainty, and Choice. *British Journal for the Philosophy of Science* 30:235–252.
Sklar, Lawrence. 1993. *Physics and Chance*. Cambridge: Cambridge University Press.
Taylor, Richard. 1964. Deliberation and Foreknowledge. *American Philosophical Quarterly* 1:73–80.
Winsberg. Eric. 2004. Can Conditioning on the Past Hypothesis Militate Against the Reversibility Objections? *Philosophy of Science* 71 (4): 489–504.

Chapter Twelve

The Concept of Intervention in *Time and Chance*

▸ SIDNEY FELDER

In the course of his profoundly illuminating explanation in *Time and Chance* of the incomparably greater influence we are able to exercise over the future than over the past, David Albert provides brief informal characterizations of both (1) the class of conditions that should be taken as being directly accessible (or "given") to an agent, and (2) the class of conditions that are most reasonably taken as falling within a human being's direct, unmediated control. In this contribution, I will outline an alternative approach to the interrelated concepts of direct accessibility and control, and say something about how well this approach fits into the fundamental explanatory framework developed in *Time and Chance*.

A way to understand one central thread of argument in *Time and Chance* is to see it as establishing that, in order to obtain the totality of dependencies characteristic of our world and of our condition as agents in it, the system of fundamental dynamical laws (interpreted expansively to include the whole underlying ontology associated with the theory—and hence defining, most notably, what counts as a completely determinate state of affairs) must be augmented by (1) the hypothesis—whose introduction constitutes an essential extension of the primary system (which, to simplify exposition, is assumed deterministic)—that the (unconditional) probability of each set of conditions is proportional to its standard measure in phase space (the Statistical Postulate, SP), and (2) the hypothesis that the universe, or at least the present cosmic epoch, began in a particular very low-entropy state (the Past Hypothesis, PH). The best way to interpret the above system (obtained by conditionalizing SP's functional representation Pr on PH) is as specifying the world's characteristic probabilistic structure Pr_{PH}, understood as yielding—without the necessity for the introduction of auxiliary hypotheses—the entire class of dependencies relevant to determining which conditions are subject to an agent's control.

Invariably, what obtains, in the final analysis, is some absolutely definite state of affairs (variously designated "the world," "the entirety of past and future history," "the totality of what is the case," and so on). Thus, when considering what follows from the stipulated existence of a certain set of conditions S, we cannot be evaluating what follows from the supposition *that only S holds*; rather, we are evaluating what follows from *supposing only* that S holds. Although the authority we assume in framing such suppositions, and in evaluating their consequences under freely specified constraints, is manifestly implicit in the plainly not coherently deniable legitimacy of hypothetical reasoning itself, the situation is somewhat different when we set out to employ theoretically established dependencies with the object of determining what occurrent conditions *in fact* obtain. Specifically, we then confront the following circumstance. On the one hand, any non-vacuous utilization of these dependencies presupposes that some less than completely comprehensive set of actual conditions function as categorically *given*. On the other, nothing less comprehensive than the totality of conditions (whatever they happen to be) can possibly obtain. The implications of the latter are not universally understood. Thus, some imagine that the order in which events occur is also the order in which they are "fixed."[1] We assume here, however, that whether the universe is deterministic or indeterministic, "that which exists," to whatever degree it is known or unknown, simply (i.e., "timelessly") *is*. Thus, events occurring at t_1 are not "already given" prior to events occurring at t_2 in any sense beyond that explicit in the fact that the former occupy earlier positions than the latter in the serial arrangement of events in time.

---

[1] By those who subscribe to the particular version of this picture according to which the laws of nature, conceived as generating future conditions out of past conditions, embody this asymmetrical creating or determining, the hypothesis that the laws of nature are indeterministic tends to be interpreted as implying that each instant t marks a fundamental division between those conditions (subsequent to t) to which, at or prior to t, it is possible to meaningfully ascribe non-extremal objective chances, and those unknown conditions (at or prior to t) to which, at or subsequent to t, such ascriptions are inadmissible.

In any of its ordinary employments in explanation and prediction, the given corresponds to some expression of the *known*. There is no such thing as *the* known (except in derivative, specially contrived senses); there are only things known to particular individuals at particular times. Note that although each particular representation of the condition of the world is given through some epistemic state, it is not in the first instance the epistemic state itself, but the state of affairs represented in it, that occupies the position of argument in the functional expressions of both deterministic and probabilistic dependency.[2] Thus, for each nomically admissible history w_i and each K—the latter designating that component of the total epistemic state of a given individual at a given moment t that is not implicit in Pr_{PH}—the function Pr_{PH} specifies the probability that the state of affairs represented by K (that is, s_K) is instantiated in w_i.[3]

Among the dependency relations that can be extracted from a probabilistic function, the standard relation of probabilistic dependence, most appropriately formulated in the following (perhaps unconventional) way, is the salient one here: A (causal) variable v is *probabilistically dependent* upon the pair of events s_i, s_j in relation to an epistemic state K if the probabilistic distribution among the values of v, given the hypothesized realization of s_i, fails to exactly coincide with the probabilistic distribution among the values of v given the hypothesized realization of s_j. Thus, when this relationship holds, there exists some value of v such that the agent in epistemic state K is able to ascribe one probability to that value of v under the hypothesis that s_i will be (i.e., is) realized, and to ascribe a *distinct* probability to that same value of v under the hypothesis that s_j will be realized.[4]

The reason Pr_{PH} and not Pr is properly seen as defining the world's "effective" nomic structure is formulated and demonstrated with consummate clarity and force in *Time and Chance*, and nothing in the way of reinforcement needs to be provided here. Accordingly, I'll assume Pr_{PH} as a fixture in what follows, and will move directly to the situation of a representative early twenty-first-century human agent X at time t. The features of the world that X's present state of knowledge K

---

[2] Nothing here is to be read as any commitment to what should count as "given" at a fundamental epistemological level.

[3] The appropriate delimitation of the contributions of PrPH and K raises some exceedingly interesting questions. Here it must suffice to say that there are a number of critical epistemological lessons to be taken from the dialectic of Time and Chance, concerning, most saliently, the particular roles that the various components of theoretical structure and experience (SP, PH, the agent's sensations and memories, etc.) play in making possible the synthesis of any kind of meaningful picture of the past, the future, or the present.

[4] We are forced to omit any proper consideration of the implications of agents' limited capacities to trace the consequences of even theoretically well-defined representations of occurrent conditions.

can realistically be assumed to "fix" comprise a highly heterogeneous lot in both category and scale. Among the infinite number of questions about the past, present, or future plausibly left open by K are (1) whether a certain pair of fair dice thrown in a high-stakes game come to rest at t displaying a certain numerical sum; (2) the average velocity of atmospheric movement within a certain comparatively calm cubic-kilometer region near the center of the highest-intensity very-large-scale storm raging on Jupiter in 1900; (3) whether line 7 on page 32 of a particular copy of a reliable standard work on early modern philosophy that X is about to consult to obtain Descartes's date of birth contains the impression *1596* or *1598*; and (4) whether the prime suspect Y in a criminal investigation who just entered an apparently vacant building presses a certain button (D) at t.[5]

Assuming the standpoint of X at t, and conventionally identifying the set of all things "that can happen now" with the set of all conditions that are compatible with s_K, we interpret the answer to the question of whether the value assumed by a certain variable v *counterfactually depends* upon a certain set of hypothetical variations in the condition of the world as the answer to the question of whether (given Pr_{PH}/s_K) the probability ascribable to any specified value v_i of v is dependent upon which variation in the set is assumed to be realized. This proposal is, of course, to be understood, not as a peculiar piece of linguistic legislation, but as an analysis of what it means for one set of conditions to depend upon another. It is a fundamental feature of the approach being advanced here that a hypothesized variation is not a hypothesized alteration in (supposedly) existing conditions, but instead one hypothesis among a number of alternative hypotheses about what conditions obtain.[6]

Representative cases (1) and (2) above require little remark. A spectator X may easily possess knowledge of the events surrounding the toss of the dice that permits him to assign one set of probabilities to a great variety of concretely specifiable events occurring seconds, minutes, hours, days, years, decades, and centuries after t, given the supposition of one numerical outcome of the dice throw and a very different set of probabilities to these same events given the supposition of another.[7] On the other hand, among the totality of possible events occurring

---

[5] X's tendency to be concerned with contemporaneous events reflects the particular concerns of this essay, and not the existence of any general constraint on the temporal relationship among v, s_i, and K.

[6] Thus, think of typical annotations of a chess game such as "... but the variation K-f1, Q-f3ch; R-f2, Q×Rch; K×Q, B-d3 leads to a forced mate for Black in 5 moves ..."

[7] Under some far from fantastic scenarios, there are events of the gravest import in the arbitrarily far future that are probabilistically dependent upon the outcome of the dice throw. (Suppose that the dice are tossed as part of a desperate expedient to resolve a deadlock between equally powerful

more than a few seconds prior to t that are specifiable by X in terms that make no essential reference to the outcome of the dice throw,[8] there exist none whose probability depends upon which numerical outcome X supposes to obtain. And although the set of plausible average atmospheric velocities within the indicated cubic-kilometer region of Jupiter's atmosphere in 1900 is a direct expression of a range of variation far more extensive than that *supposed* in (1), there is nothing outside the outer boundaries of the larger storm at any instant whose probability (in relation to K) depends upon what remotely plausible conditions of atmospheric movement are hypothesized to obtain within this cubic-kilometer region at that time.

We now consider (3). It will be uncontroversial that the deductions X will justifiably make about the probabilities of a tremendous variety of events prior to and simultaneous with t, given the observation of the configuration of ink *1596* will differ substantially from those he would make about the probabilities of these same events given the observation of the configuration of ink *1598*. It is another thing to say—though it is, nevertheless, what we are asserting—that whether Descartes was born in 1596 or 1598, and whether the overwhelming majority of books recording Descartes's date of birth contain impressions representing the year 1596 or the year 1598, counterfactually depend upon whether a certain piece of paper—consulted or not—contains the impression *1596* or contains the very slightly different impression *1598* at t.[9]

In asserting that the year of Descartes's birth is counterfactually dependent upon the character of a certain numerical token on page 32 at time t, is not one committed to the claim that this past event depends upon *what just happens to appear* on a particular leaf of a recently printed book? And, in particular, isn't one then committed to the absurd claim that if it could somehow have been arranged for *1598* rather than *1596* to be present on the page, Descartes would have been born in 1598 and not in 1596?

---

blocs supporting opposing plans to deflect a giant asteroid that is projected to strike the earth and obliterate all life on it within the month.)

[8] The proviso "in terms that make no essential reference ..." reflects the fact that it is always possible to define events such as 'those conditions exactly one week prior to t that are perfectly correlated with numerical outcome 11' that are indeed probabilistically dependent upon the outcome of the subsequent dice throw. However, even supposing that X can characterize any such precursors in definite conceptually independent terms, the prospect that X can correlate these inevitably highly abstract characterizations with any mode of presentation that could be of practical relevance is extremely remote.

[9] Once and for all: The probabilistic dependency we are considering here does not (at least in the first instance) concern what depends either upon which of a number of possible things might be learned or observed at t or upon whether X decides to consult page 32 of this book at t.

"What just happens to appear..." is a highly loaded phrase. If by dependence upon "what happens to appear," one means a dependence that holds regardless of *how* what obtains obtains, then the answer to both of the above questions is no— or perhaps better, the answer is that both of these questions are simply not defined. By dependence upon "what happens to obtain," one can only reasonably mean a dependence upon which state of affairs obtains under conditions (such as Pr_{PH}/s_K) that are believed by X to characterize the actual world, conditions that (so to speak) determine the manner in which the specified state of affairs is *likely* to be realized. Consequently, the same conditions (Pr_{PH}/s_K) that permit X to infer that the inscription *1596* on page 32 is extremely unlikely to be the result of a random, highly localized statistical fluctuation, and that permit (more specifically) the token *1596* to function as an indication of the year of Descartes's birth as well as of the contents of tens of millions of texts and scores of thousands of human memories at and prior to t, simultaneously permit these spatially and temporally widespread past and present events to depend, in a robust and *a priori* transparent way, upon certain quite minutely differing hypotheses about the configuration of ink on a particular leaf of paper.

Similarly, assuming it is known by X (now a detective on stake-out) that the criminal suspect Y introduced in (4) is extremely unlikely to press button D unless in possession of certain information to which only the perpetrator could possibly have had access, various conditions at and prior to t (what certain confederates were and are doing, where certain bodies are buried, where certain sums of money are hidden) will be dependent upon whether or not the suspect presses button D at t. However, if it is assumed that the depression of button D occurs in the midst of a freak involuntary seizure that propels Y's hand onto D, we obviously obtain a situation in which the conditions presupposed in the existence of the dependency described in the last sentence do not hold. By a variation of the same reasoning, if the resting configuration of the dice referred to in (1) had been supposed to be deliberately arranged, the existence of a probabilistic dependence of highly consequential events prior to t upon the displayed numerical sum would not at all be out of the question. (Note that the manners in which the conditions supposed realizable at t are likely to obtain were at least implicit in the initial descriptions of all four of the above cases.) Of particular significance for understanding the implications of Pr_{PH} for an agent's (in)capacity to influence past events is the fact that the probabilistic dependence of the year of Descartes's birth upon which of two configurations of ink appears on page 32 *fails* to hold within any situation in which X himself at or just prior to t is the determiner of which of these configurations appears.

Let $\mathbf{m}$, $\mathbf{M}$, and s_K denote, respectively, the very far from thermodynamic equilibrium microstate (i.e., the precise condition) of the actual world w at the present moment t, the complete present macrocondition (macrostate) that $\mathbf{m}$ instantiates (where, for present purposes, a complete macrocondition is to be understood broadly, as including, in addition to all macroscopic physical conditions at a given t, the experiential contents of all mental states that the *micro*condition at t realizes), and the conditions that X at t knows to obtain; let m_i, w_i denote, respectively, an arbitrary *set* of microconditions at t and the set of complete trajectories in phase space that intersect it.

I will begin with the following picture of intervention, a highly idealized adaptation of the one vividly described in *Time and Chance*. Associated with an agent X at time t is a particular range of variability, which is given by an abstraction from $\mathbf{m}$ that leaves fixed some unspecified *proper superset* of the totality of conditions at t falling outside the region of space encompassing X's body (or perhaps brain). This range of variability consists of a family O of mutually exclusive states of affairs, each of which is conceived as being realizable by X at will. In other words, it is assumed that X, through an act of volition, "freely" selects which of these alternative "admissible" determinations (sets of microconditions) $m_{o(1)}$, $m_{o(2)}$, $m_{o(3)}$, ... $\mathbf{m}$ instantiates. This defines the class of conditions under X's direct, unmediated control. Note that in order for this (unauthorized) idealization to work, O must be defined in such a way that each element $m_{o(i)}$ of O contains some microcondition whose associated trajectory is consistent with all components of Pr_{PH}/s_K—a constraint that could not in general be satisfied if it were assumed that the agent could realize at will any set of instantaneous conditions within even the minutest connected open region of physical space.[10] X's influence over all other conditions is propagated through the medium of the fundamental dynamical laws.

The standard formulation (or "axiomatization") of the primary physical theory—in particular, its presentation of the world's history as a product of fundamental dynamical laws and "initial conditions"—together with the assumption that mental states supervene on more or less instantaneous, spatially highly

---

[10] This is because within any dynamically naturally specifiable subset of phase space of non-zero measure, there will exist an infinite number of microconditions that occupy trajectories not conforming to the conjunction of PH, SP, and s_K. (There are a number of reasonable ways to obtain certain of the constraints (notably those associated with the standard general approach to the evaluation of counterfactual conditionals) that we have elected to directly incorporate into the specification of O. Albert (unpublished) depicts some as entering in at a secondary stage, as restrictions on the indirect expression at t of the conditions directly realizable by the agent. (Barry Loewer (2007) utilizes a framework in which the macrocondition at t—where "macrocondition" is understood in a manner distinct from its characterization above—is held fixed.)

localized, physical states, yields a theoretically natural and conceptually stable demarcation between those conditions under an agent's direct influence and those conditions under his indirect influence. Though the existence of a definitely situated distinction of this kind is not an essential element of the above picture, it will be most convenient to proceed by first saying something about how this distinction operates within the framework of a conception of intervention and control (whose groundwork was prepared above) according to which a condition is under X's control (or influence) at t if and only if it is probabilistically dependent (of course, in relation to K) upon what volition X exercises at t.

Of course, any serious consideration of the difficult question of how to conceive the relationship between the "determining" and the "determined" in a volition's exercise is way beyond the scope of anything that can be embarked upon here. What must suffice for this occasion is the statement that whatever transparently in the very act of willing constitutes one *something* as opposed to *something else*, that "whatever" should count as a condition that is *immediately* realizable by the agent. (This is to be read as implying that a necessary and sufficient condition for the nonrealization of whatever is immediately intended in the exercise of a particular volition is the equally free exercise of an alternative volition.) Note the consequence that it is a necessary condition for the classification of any condition s_j as being under X's *indirect* control in a specific situation that there must be some distinguishable condition s_i—however difficult that condition may be to reproducibly intuitively summon or to categorize in ordinary terms—whose probability is (very roughly speaking) at least as sharply dependent upon X's volition as is the probability of s_j.

The set of conditions under an agent's immediate control in the present sense do not, as a matter of fact, fall under any system of descriptively or phenomenologically uniform determinations. The initiation of a particular line of thought and the exertion of tension in a finger that is in immediate contact with a button expected to offer a certain measure of resistance, for example, are not ultimately traceable to selections of specific canonical settings of a volitionally yet more anterior "central control." And in any particular situation, there will be a multiplicity of distinguishable external conditions whose degree of probabilistic dependence upon the agent's volition approximates that of the kinds of conditions (actions) just mentioned. However, the character of these external conditions will vary greatly from situation to situation—a circumstance that makes it natural to ascribe the agent's influence over these conditions (and, hence, over the more remote conditions linked to them by the probabilistic dependencies peculiar to that situation) to the specific manner in which variations on the surface and in the interior of the agent's body that are consistently volitionally realizable by the agent are coupled with the conditions external to it in each particular situation.

There is, nevertheless—both in the case of the picture presented in *Time and Chance* and in the case of the one being expounded here—a certain artificiality attending the distinction between the conditions that are under an agent's immediate, as opposed to mediate, control. In both cases, the isolation of the conditions that are categorized as being under the immediate control of the agent is coordinated with a certain abstraction—most saliently from the specified character of the physical laws in the former picture, most saliently from the probabilistic interdependencies characteristic of the agent's particular situation in the latter. The former abstraction arises naturally when the fundamental theory is considered in a "meta-theoretic" or "architectonic" setting—that is, a setting in which the precise character of the dynamics, as well as the conditions at t, appear as variables—and so one in which, correspondingly, a certain "contingency" attaches to the character of these laws and hence to the connection between conditions at distinct moments. Instead conceiving the stated character of the physical laws as a given, there is no commanding basis for X at time t to consider events at times other than the present to be metaphysically or epistemically more contingent or remote than are *any* present conditions either not actually presently known by X, or not identical to those upon which X's mental state at t supervenes. And since, in the present theoretical setting, the existence of any condition m_i is inseparable from the existence of all conditions occupying w_i, it is not unnatural to see all conditions intersected by any trajectory in $\{w_{o(i)}\}$ as being under X's direct control at t. Analogously (turning now to control understood as probabilistic dependence), since each of the agent's possible actions in the given situation, defined by Pr_{PH}/s_K, is associated with a definite probabilistic distribution among all conditions, once the agent reaches cognitive equilibrium in a particular situation (a state of affairs that itself involves a not inconsiderable degree of idealization), the operative division between means and ends vanishes, and the agent in effect directly selects which among a certain set of probabilistic distributions among worlds holds. Moreover, it is evident that a significant proportion of the "internal" determinations of the agent whose probabilities depend with maximum sharpness upon the agent's volition (determinations which, simultaneously, are the immediate expressions of its exercise) are only distinguishable as such—i.e., are only such—in conjunction with (or as a kind of "feedback" from) features of the agent's environment that (because of their situation-specific character) would conventionally be classified as falling only under the agent's indirect influence.

For what follows, some observations pertaining to the present conception of control bear noting. First, although the supposed character of the supervenience of mental states upon physical states implies that there is zero probability that the world containing the hypothesized exercise of volition (i.e., the directly willed condition) s_A at t generating the movement of X's right thumb in one direction instantiates the same precise brain state as does the world containing the hypothesized exercise of volition s_B at t generating its movement in another, this does not *imply* that X will be in a position at t to identify any concrete physical description of any brain state to which he can assign a higher probability supposing the exercise of one of these two volitions than he can supposing the exercise of the other. (This is true whether the brain states instantiating s_A and s_B differ microscopically or macroscopically.) Second, while (given that the deterministic setting in which we are operating excludes the existence, at the fundamental physical level, of any branch-points) the agent's brain state at any instant is not the site of the incipient divergence of lines of causal influence, there are certain adducible senses in which X's voluntary movement of his right thumb has its source in some determination of X's brain, one being that t (modulo the obvious assumptions and conventions) is the only moment at which worlds instantiating s_A, s_B respectively *necessarily* differ at the causal variable "X's brain state." (Neither the agent's mind nor the agent's brain form causally closed systems, and hence, although the instantaneous sections of sets w_A and w_B necessarily never overlap at any microstate, it does not follow that these sets cannot overlap at the variable "X's brain state" even at moments quite shortly after t.) Third, while X at time t can plainly influence the conditions of innumerable microscopic objects and processes (such as the locations of individual carbon atoms in X's right thumb), and while means presently exist to permit X at t to influence the relative probabilities of microscopically distinct values of selected variables, there does not exist, on any time slice, two sets of complete microconditions differing only "microscopically" whose relative probabilities can be influenced by X's choice of action at t.

Perhaps the structurally most salient contrast between the picture of intervention presented in *Time and Chance* and the one associated with the particular conception of counterfactual dependence propounded here is connected with the distinction—a variation of that drawn early on between hypothesizing *that only* condition *S* holds and *hypothesizing only* that *S* holds—between sets consisting of *conditions* that differ only slightly and sets consisting of *specifications* of conditions that differ only slightly. Expressed another way, in the former picture, **m** is taken as "base point," in the latter, s_K. Thus, given any abstraction from **m** that leaves everything outside X's brain state fixed at the actual condition of the world at t (i.e., at the conditions actually obtaining at t), no two conditions $m_{o(i)}$, $m_{o(j)}$ instantiating alternative volitions of X at t can differ by anything more than a minute degree. On the other hand, worlds instantiating distinct conditions at t whose *specifications* differ by only a minute degree consistent with s_K (e.g., various hypotheses about configurations of ink on a particular piece of paper), which, individually, correspond to conditions occupying (however diffusely) tremendously extensive regions of phase space, can (via Pr_{PH}/s_K) be associated with vastly different conditions at time t. Everything else follows from the observation that the selection of an action presents a precise parallel to the specification of a hypothesis.[11] If we identify the set of conditions under an agent's direct control with those conditions that are transparently realized in the exercise of the agent's volition—i.e., with *that which* is specified in the volition's exercise—then the range of conditions over which a human agent has direct, unmediated control turns out to be appropriately negligible.

An essential observation is that under a range of conditions wide enough to encompass any that are at all likely to have been encountered by human beings thus far, the extent as well as the degree of an agent's indirect control over either past, present, or future events are in all practical respects identical whichever of the two conceptions of intervention is employed. At the same time, the account of agents' inability to exercise any influence over past or (any but the most localized) present conditions is conceived, or at least presented, somewhat differently in the two approaches. In *Time and Chance* (wherein the existence of a *general* counterfactual dependence of past events upon present occurrences is quite pointedly noted), a human agent's incapacity to influence past events is interpreted as a particular instance of the generic counterfactual independence, in all humanly salient respects, of past events upon small variations in total present conditions that is a consequence of PH.

However, since past and present conditions are in general *not* probabilistically independent of minutely varying *specifications* of present conditions, it follows that if the alternative understanding of counterfactual dependence and intervention that I have been sketching is employed, the explanation of the inability of human agents to influence past or (any but the most localized) present conditions has to proceed differently. (From this point of view, dependencies upon states of the world that differ only minutely correspond to a special class of counterfactual dependencies, which, though useful in certain contexts of inquiry, do not answer present concerns.) There are (as illustrated, for instance, in the case of the criminal

---

[11] In this context, an action functions as a hypothesis whose truth *or* falsity the hypothesizer realizes at will.

investigation described above) innumerable plausible second-person perspectives in relation to which a great number of concretely specifiable humanly significant events prior to and simultaneous with t are probabilistically dependent upon small hypothesized variations in an agent's motions at t. However, in consequence of the special character of the agent's access to *her own* memories and intentions, it will be impossible for the agent *herself* to relate any two of her available actions to numerically distinct probabilities of any independently specifiable condition in the past or present, and hence the agent has no *control* over events either prior to or (modulo the understood exception) at t.[12] Of course, the fact that—under anything like familiar circumstances—the epistemic state of the agent at the moment of action t always screens off any dependencies of past and present conditions upon her choice of action that may hold from the standpoint of a second person at t (or from the standpoint of the agent herself at a later time) is, ultimately, most distinctively attributable to the Past Hypothesis.

These observations are by no means inconsistent with the realizability of specially composed situations in which a human agent at t is capable of influencing highly consequential past and present conditions. Newcomb's Problem,[13] under one natural interpretation, presents such a situation. Thus, assume (1) that X justifiably believes that the Predictor, days before t, has access to the value of a ("precursor") variable P that it can correlate with X's definitive selection,[14] and (2) that X, prior to his final (i.e., unappealable) decision, is unable to identify any variable

---

[12] Juxtaposing the relation of probabilistic dependency with another relation (already employed without ceremony) should cast light on what the claim that the actions of human beings have no influence over the past *does not* mean. The latter relation—given by a function that specifies, for each pair of hypothesized conditions s_i, s_j, the probability that their respective completely determinate instantiations *differ* in a specified respect—is designed to capture a distinct sense in which a variable v is "sensitive" to specified classes of variations in conditions, where—and this is the whole point of introducing this relation—the judgment that this kind of sensitivity holds carries no implication that one who is in a position to make *this* judgment is also in a position to link any particular one of these conditions to any particular value of the variable v. (Although the probabilities that two tosses of a fair die will produce any particular outcome are identical, the probability that two successive or alternative tosses of the die will yield the same outcome is only 1/6.) It is vanishingly unlikely, for example, that worlds whose conditions at t differ even to a minute degree will instantiate the same macrocondition at every instant prior to t. And, in relation to Pr_{PH}/s_K, there will be countless "ordinary" macroscopic variables at and prior to t, not excluding many that are transparently of great interest to human beings, that are sensitive in the present sense to the class of volitions exercisable by X at t. (See Felder (2005) for elaboration.)

[13] Two boxes, one transparent and containing \$1,000 (A) and one opaque (B), are placed before an agent at t. The contents of A are supposed a given, those of B supposed determined by the prior action of a highly accurate Predictor that places \$1,000,000 (nothing) in box B if it predicts that the agent will select only box B (both boxes).

[14] The character of this variable, and whether or not it has a humanly comprehensible description in other terms, is contingent upon hypotheses about the nature of the Predictor and its methods.

that he can correlate with P whose actual value he can specify independently of his selection. In this situation, the prior action of the Predictor and the contents of box B are probabilistically dependent upon X's actions at the moment of decision, and hence are under X's control.

There will, consequently, be tangible macroscopic records (traces) of X's eventual choice at t present in X's environment at and prior to t—conditions potentially both widespread and widely observable—that (see footnote 12) are consistent with *only one* of X's available choices at t. (I use the term "available" quite deliberately.) Indeed, one may suppose, as in one classic variation, that the back of box B is transparent, and that its interior has been under another person's continuous observation since before the Predictor acted. An amusing and revealing further twist can be given by supposing that X (prior to making his selection) discovers that the observer—of whose smiling but unfamiliar countenance X was already able to catch a clear glimpse—is just as likely to be a person (Z) who wishes him well as he is to be a person (Y) who wishes him ill, it being additionally supposed that X has good reason to believe that the physiognomy of each is likely to register accordingly. (Z's (Y's) face thus apt to exhibit a glad (discomfited) expression in the event of the presence of \$1,000,000 in B and a discomfited (glad) expression otherwise.) In this case, macroscopic anticipatory traces of X's choice in the form of the observer's smile are directly observed and recognized as such by X, though these are, by assumption, momentarily indecipherable—i.e., incapable of being translated by X into "actionable" knowledge of the contents of box B.[15] This latter fact implies that whether the face—whose expression and general appearance are *not* under X's control—belongs to a person (whether otherwise known or unknown to X is immaterial) who wishes X well or to a person who wishes X ill *is* a condition that is under X's control.

Such situations suggest a number of fascinating directions of inquiry (which are explored extensively in Felder (2005)), but the main question I wish to address in these final remarks is whether the kind of influence over all these present and

---

(A number of models of beings systematically capable of producing highly (if not perfectly) reliable forecasts under the relevant range of conditions are plausibly consistent with PrPH.)

[15] A quite different situation (*not just the same situation* about which X happens to possess greater knowledge) is created if it is assumed that X penetrates the significance of the observer's expression to the point of being able to deduce box B's contents prior to his selection. It is characteristic of the general category of situations presently being discussed that the existence and shape of the dependencies associated with them are highly sensitive to subtle variations in the manner in which externally observable activities of agents are mentally conditioned. However, there are strong stabilizing factors in place (some in the form of incentives to X) that, by making it highly improbable that certain epistemic states will be occupied, strongly separate X's actual situation from far less favorable ones whose conventional descriptive specifications differ from it very little.

past conditions we are attributing to X is best regarded as a mere play of perspectives. The answer is that there is nothing of the subjective or perspectival in the present conception of dependence and influence that is not implicit in any analysis of reality in terms of objective interdependencies (deterministic or probabilistic). In utilizing these dependencies in the paradigmatic way to establish what conditions ("ultimately") actually obtain, an agent is simply evaluating what follows from a certain incomplete specification of events,[16] a specification picked out by (i.e., consisting of) what is "given" to him, together with those conditions that, among those realizable at will by the agent in the situation he occupies, he chooses to bring into being. The claim that all this is objective is nothing other than the claim that conclusions deduced from these partial representations of the world can be verified—that is, that they correspond to conditions that can be relied upon to cohere, a claim that is simply an expression of the initial supposition that the pattern of events instantiates the theoretically specified regularities (i.e., satisfies a description consistent with Pr_{PH}).

The objectivity of such conclusions (i.e., the consistent satisfiability of the expectations intrinsic to them) is not undermined by the fact that what details of the world are accessible in different situations, as well as what volitions are exercised in them, are not independently specifiable—i.e., are interdependent. (There will, *ideally* (see footnote 4), be no disagreement about what follows from any particular specification of conditions.) The contrary impression derives, I believe, from the delusion that there can be an answer to the question of, of which of the relata in a dependency is the "real" source of constraint in anything other than a conventional sense.[17] Furthermore, although the present conception of counterfactual dependence at first glance might suggest the insertion of an incongruous subjective parameter into the evaluation of the "real effects" of certain objective variations of conditions, and hence the dependence of something objective upon something merely subjective, this is a misunderstanding of the basic situation. These epistemic states are not free-floating objects, but instead conditions whose systematic relationship to conditions interconnected by Pr_{PH} is presupposed in any employment of Pr_{PH} to obtain significant knowledge of the actual state of the world. The reason the year of Descartes's birth can be counterfactually dependent upon the set of impressions {1596, 1598} in relation to K, but not in relation to some epistemic state K' assumed to include awareness of Descartes's date of

---

[16] By "follows," I of course do not mean to imply any order in which events are *determined*, but only the existence of certain facts about the character of the complete extension **w** of these specifications.

[17] I assume it obvious that the identification of any event as *the* cause of another possesses meaning only relative to some particular epistemic or practical context.

birth, is that in one case the properties of the "worlds" Pr_{PH} associates, respectively, with the conditions $s_K + 1596$ and $s_K + 1598$ are being compared, whereas in the other, the properties of the "worlds" Pr_{PH} associates, respectively, with the conditions $s_{K'} + 1596$ and $s_{K'} + 1598$ are being compared. (Thus, K and K' are at bottom associated, not with competing answers to a given question, but with distinct questions.) Finally, there is nothing less comprehensive than the totality of conditions that ever "happens." From a trans-subjective perspective, there is no sense in which anything about the world is open beyond that implicit in the absolute logical contingency of its character, and no sense in which anything about the world is closed beyond that implicit in the world's self-identity.

I believe that enough has been said here to make the following plausible: (1) The conception of counterfactual dependence and intervention sketched above fits cleanly into the framework yielded by what is most essential and distinctive to the account constructed in *Time and Chance*. And (2) the faint gestures in the general direction of the standard semantics of counterfactual conditionals discernible in that work are best seen as vestiges of a picture by and large already superseded by its fundamental approach. I would also argue that the combination of Pr_{PH} and the present analysis of intervention, by situating nonstandard patterns of dependency within the same domain of discourse as the one occupied by the pattern whose predominance is so beautifully accounted for in *Time and Chance*, adds significantly to the explanatory power of the scheme constructed in the latter.[18]

Acknowledgments

I thank my great and longtime teacher David Albert—this is just the extremest tip of a colossal iceberg of gratitude owed—for two exceedingly productive conversations at the (very!) early stages of the paper's composition. Many thanks also to Porter Williams, who read the paper on short notice, and (as always) provided wise advice and highly insightful comment, and to Sebastian Watzl who (also responding on short notice) wrote a detailed and extraordinarily valuable running commentary on the draft (permitting me to substantially improve a number of passages in the text.)

---

[18] This scheme, for example, determines the structural configuration of the conditions that make Newcomb-like situations possible, as well as the general spatiotemporal pattern of the dependencies associated with them.

References

Albert, D. *Time and Chance.* Cambridge, MA: Harvard University Press, 2000.
Albert, D. "The Sharpness of the Distinction between the Past and the Future" (unpublished manuscript).
Felder, S. "Causality and Directive Agency from the Perspective of Fundamental Physical Theory." PhD dissertation, Columbia University, 2005. ProQuest/UMI.
Loewer, B. "Counterfactuals and the Second Law." In *Causation, Physics, and the Constitution of Reality: Russell's Republic Revisited*, ed. Huw Price and Richard Corry. New York: Clarendon Press of Oxford University, 2007.

Conclusion

▸ DAVID Z ALBERT

Classical statistical mechanics—the statistical mechanics (more particularly) of systems of Newtonian point particles—begins with a postulate to the effect that a certain very natural-looking measure on the set of possible exact microconditions of any such system is to be treated or regarded or understood or put to work—of this hesitation more later—as a probability distribution over those microconditions. The measure in question here is (as a matter of fact) the simplest imaginable measure on the set of possible exact microconditions of whatever system one happens to be dealing with—that is, the standard Lebesgue measure on the phase space of the possible exact positions and momenta of the Newtonian particles that make up that system. And the thrust of all of the beautiful and ingenious arguments of Boltzmann and Gibbs, and of their various followers and collaborators, was to make it plausible that something like the following is true:

> Consider a true thermodynamical law, any true thermodynamical law, to the effect that macrocondition A evolves—under such-and-such external circumstances and over such-and-such a temporal interval—into macrocondition B. Whenever such a law holds, the overwhelming majority of the volume of the region of phase space associated with macrocondition A—on the above measure, the simple measure, the standard measure, of

volume in phase space—is taken up by microconditions that are sitting on deterministic Newtonian trajectories that pass, under the allotted circumstances, at the end of the allotted interval, through the region of the phase space associated with the macrocondition B.

And if these arguments succeed, and if we allow ourselves to imagine, just for the sake of this discussion, that Newtonian mechanics is true,[1] then the above-mentioned probability distribution over microconditions will underwrite great swaths of our empirical experience of the world: It will entail (for example) that a half-melted block of ice alone in the middle of an average sealed terrestrial room is overwhelmingly likely to be still more melted toward the future, and that a half-dispersed puff of smoke alone in an average sealed terrestrial room is overwhelmingly likely to be still more dispersed toward the future, and so on.

But there is, of course, a famous trouble with all this, which is that all of the above-mentioned arguments work just as well in reverse, that all of the above-mentioned arguments work just as well (that is) at making it plausible that (for example) the half-melted block of ice I just mentioned was more melted toward the past as well. And we are as sure as we are of anything that that's not right.

And the canonical method of patching that trouble up—the method that was elaborated and defended, in some detail, in *Time and Chance*—is to supplement the dynamical equations of motion and the Statistical Postulate with a new and explicitly *non*-time-reversal-symmetric fundamental law of nature, a (so-called) Past Hypothesis, to the effect that the universe had some particular, simple, compact, symmetric, cosmologically sensible, very low-entropy initial macrocondition. The patched-up picture, then, consists of the complete deterministic microdynamical laws and a postulate to the effect that the distribution of probabilities over all of the possible exact initial microconditions of the world is uniform, with respect to the Lebesgue measure, over those possible microconditions of the universe which are compatible with the initial macrocondition specified in the Past Hypothesis, and zero elsewhere. And with that amended picture in place, the arguments of Boltzmann and Gibbs will make it plausible not only that paper will be yellower and ice cubes more melted and people more aged and smoke more dispersed in the future, but that they were all less so (just as our experience tells us) in the past. With that additional stipulation in place (to put it another way),

---

[1] The thought here, as usual in such discussions, is that imagining that Newtonian mechanics is true is not likely to result in any particularly devastating sort of distortion or loss of generality. The thought (more particularly) is that whatever we learn about the business of coming to terms with these issues in the context of Newtonian mechanics will likely be helpful in the contexts of more up-to-date ontologies and dynamical laws as well.

the arguments of Boltzmann and Gibbs will make it plausible that the second law of thermodynamics remains in force all the way from the end of the world back to its beginning.

What we have from Boltzmann and Gibbs, then, is a probability distribution over possible initial microconditions of the world, which—when combined with the exact deterministic microscopic equations of motion—apparently makes good empirical predictions about the values of the thermodynamic parameters of macroscopic systems. And there is a question about what to make of that success: We might take that success merely as evidence of the utility of that probability distribution as an instrument for the particular purpose of predicting the values of those particular parameters, or we might take that success as evidence that the probability distribution in question is literally true—we might (that is) take that success as evidence that everything that can be derived from that probability distribution turns out to be empirically correct.

And note that if the probability distribution in question were, in the above sense, literally true, and if the exact deterministic microscopic equations of motion were also, in the above sense, literally true, then that probability distribution, combined with those equations of motion, would necessarily amount to, not merely an account of the behaviors of the thermodynamic parameters of macroscopic systems, but the complete scientific theory of the universe—because the two of them together assign a unique and determinate probability value to every formulable proposition about the exact microscopic physical condition of whatever physical things there may happen to be. If the probability distribution and the equations of motion in question here are regarded not merely as narrowly focused predictive instruments but as claims about the world, then there turns out not to be any physical question whatsoever on which they are jointly agnostic. If the probability distribution and the equations of motion in question here are regarded not merely as narrowly focused predictive instruments but as claims about the world, then they are either false or they are in some principled sense all the science there can ever be.

And precisely the same thing will manifestly apply to *any* probability-distribution over the possible exact microscopic initial conditions of the world, combined with *any* complete set of laws of the time-evolutions of those macroconditions. And this sort of structure will consequently be worth making up a name for. Start (then) with the initial macrocondition of the universe. Find the probability distribution over all of the possible exact microconditions of the universe that is uniform, with respect to the standard statistical-mechanical measure, over the subset of those microconditions that is compatible with that initial macrocondition, and zero elsewhere. Evolve that distribution forward in time, by means of the exact microscopic dynamical equations of motion, so as to obtain a definite numerical assignment of probability to every formulable proposition about the physical history of the world. And call that latter assignment of probabilities the Mentaculus.

And *Time and Chance* was largely taken up with an attempt to look into the possibility that the probability distribution we have from Boltzmann and Gibbs, or something like it, something more up-to-date, something adjusted to the laws and the ontology of quantum theory, or quantum field theory or quantum string theory or quantum brane theory, is, in the sense we have just been discussing, true—and (moreover) to make it plausible that explanations, not only of the time-directedness of thermodynamic phenomena, but of the asymmetry of our epistemic access to the past and the future, and of the fact that it seems to us that by acting now we can affect the future but not the past, and a number of other things besides, can be dug out of that distribution, not only as a matter of fundamental principle but (indeed) in a surprisingly clean and simple and tractable and perspicuous way.

And I am happy to note that the project outlined in *Time and Chance* is still very much alive—and that its promise seems to have grown bigger, and that its appeal seems to have grown wider, with the intervening years. But it will be fitting to acknowledge—and particularly so on an occasion like this—that the way I put things in that book was sometimes sloppy, and sometimes wrong, and almost always woefully incomplete. And I want to begin my remarks by pointing to some of the more prominent of these defects.

I repeatedly referred to the Mentaculus, in *Time and Chance,* as "the complete contraption for making inferences." And there is a certain ideal and principled sense in which this is perfectly and unassailably true. If the probabilities that the Mentaculus assigns to every formulable proposition about the physical history of the world are the right ones, and if I am aware of those assignments, and if I am logically omniscient, then what I learn about the physical history of the world by learning any set of contingent facts {F} is precisely the difference between the Mentaculus and the Mentaculus *conditionalized on {F}*. And the fact that I have a very different kind of epistemic access to the past than I do to the future has to do—on this way of thinking—with the fact that the probability-distribution on which we are conditionalizing here is itself dramatically *asymmetrical in time*. The fact that I have a very different kind of epistemic access to the past than I do to the future has to do—that is—with the fact that the probability-distribution on which we are conditionalizing here satisfies a past-hypothesis at one of its temporal ends but not at the other. Period. Case closed. End of story.

But there has nevertheless been a lot of puzzlement, in the years since *Time and Chance* was first published, about the relationship between the ideal and principled and obviously impossible sort of inference sketched about above and the inferences toward the past and the future that we actually make in the everyday business of going about our lives. Many readers of that book have quite rightly pointed out (for example) that the business of making inferences about the past and the future had been up and running for any number of millennia before anybody had ever heard of the Past-Hypothesis, or of the Statistical Postulate, or of the microscopic equations of motion. And so whatever it is that I may have had in mind in referring to the Mentaculus as the "complete contraption for making inferences," I certainly cannot have meant, or (at any rate) I cannot have been right to mean, that the Mentaculus is something that we routinely and deliberately and self-consciously bring to bear on the business of actually making inferences.

I have sometimes said—by way of responding to these sorts of worries, by way of attempting to ameliorate these sorts of worries—that if anything along the lines of the Mentaculus is *true*, then some crude, foggy, reflexive, largely unconscious but perfectly serviceable acquaintance with the probabilities it generates will certainly have been hard-wired into us as far back as when we were fish, as far back (indeed) as when we were slime, by natural selection—and lies buried at the very heart of the deep instinctive primordial unarticulated feel of the world. And that sounds nice, and it is surely true as far as it goes—but it no longer sems to me to go nearly far enough, and I want to try to do a little better here.

Suppose (then) that a regular guy, untutored in statistical mechanics, and innocent of the Past-Hypothesis, is confronted with a ½-melted block of ice, sitting in a puddle on the floor of a warm room. And suppose that he is confident that the contents of this room have not been disturbed over the past twenty minutes or so. And suppose that he is interested in figuring out what the room looked like ten minutes ago. He might be expected to reason more or less as follows: He knows, from his empirical experience of the world, that blocks of ice that are left alone in otherwise unremarkable warm rooms are *extremely* likely to get more melted as time goes by. And he knows, from his empirical experience of the world, that the prior probability of running into a ¼-melted block of ice in an otherwise unremarkable warm room is more or less on a par with the prior probability of running into a ¾-melted block of ice in an otherwise unremarkable warm room. And he straightforwardly inserts those probabilities into their appropriate Bayesian slots, and performs the necessary multiplications and divisions, and concludes that the ice is enormously more likely to have been ¼-melted ten minutes ago than it is to have been ¾-melted ten minutes ago.

And this is an eminently sensible way of reasoning, and it is an eminently reliable way of reasoning, and it seems to me to provide a crude but plausible model of how our reasoning in such cases actually proceeds—and note (and this is the point) that it makes no reference whatever, explicit or otherwise, to the initial microstate of the universe, or to the statistical postulate, or even (come to think of it) to the microscopic equations of motion.

What's going on here? How does this guy manage to get away without appealing to any of the fundamental elements of the Mentaculus? Simple. He gets away without appealing to the Mentaculus *itself* by appealing to certain particular *logical consequences* of the Metaculus—consequences which he happens to have learned not by deriving them from the fundamental laws of nature, but through his own empirical experience of the world—and those consequences turn out to be all he needs for the particular inference that he wants to make.

What the Mentaculus has to offer us, in connection with an inference like this one, is an explanation of *why* the two crucial premises of that inference—that blocks of ice in otherwise unremarkable warm rooms are extremely likely to get more melted as time goes on, and that the prior probability of running into a ¼-melted block of ice and the prior probability of running into a ¾-melted block of ice, in otherwise unremarkable warm rooms, are more or less on a par with one another—both turn out to be *true*.

And maybe it will be worth rehearsing, yet again, very briefly, how that explanation goes:

The claim that blocks of ice in otherwise unremarkable warm rooms are extremely likely to get more melted as time goes on is widely presumed to follow—as we have already remarked—from the microscopic equations of motion and the statistical postulate.

But the statistical postulate is of course very much at odds, as it stands, with the claim that the prior probability of running into a ¼-melted block of ice and the prior probability of running into a ¾-melted block of ice, in an otherwise unremarkable warm room, are more or less on a par with one another—because the latter of those two macrostates is associated with a vastly larger region of phase space than the former. And one way of saying why it is that the Mentaculus requires a Past-Hypothesis is in order to correct precisely *that*—in order to make it the case (that is) that the prior probability of running into a ¼-melted block of ice and the prior probability of running into a ¾-melted block of ice, in an otherwise unremarkable warm room, are more or less, notwithstanding that they take up such vastly different volumes of phase space, on a par with one another.

And it goes without saying that if the probabilities that the Mentaculus assigns are correct, then the principled calculation—the one that proceeds from the past-hypothesis and the statistical postulate and the microscopic equations of motion—will yield the same qualitative conclusion about the condition of the block of ice 10 minutes ago as the regular-guy calculation I outlined above. And if the past-hypothesis and the statistical postulate and the microscopic equations of motion together constitute the true and complete and fundamental laws of physics, then Mentaculus calculations can always be used to refine and to improve and to access the accuracy of regular-guy calculations, and to explain why regular-guy inferences succeed exactly as well or as poorly as they do, and to bring all of the different ways of inferring certain pieces of the history of the world from other pieces of the history of the world together into a single, simple, universal, implicative structure.

And so there *is* a sense, notwithstanding all of the confusion that it has caused, in which the Mentaculus earns the name of "the complete contraption for making inferences"—and there turns out, on a little reflection, to be no great mystery about the relationship between this contraption and the sorts of inferences that we make in the course of the everyday business of making our way in the world.[2]

---

[2] Wayne Myrvold has recently weighed in on these matters, in his book *Beyond Chance and Creedence*, as follows: "My complaint about Albert's approach is not so much about where he ends up, as the route by which he gets there. He begins by introducing a disastrous statistical postulate for which there is no motivation except the lingering influence of the Principle of Indifference, a statistical postulate that, if accepted, would lead you to the conclusion that you are a Boltzmann brain and that the author of this book does not exist. We are to be rescued from this disaster by postulating something that we never had reason to doubt, namely, that we may rely on the perfectly ordinary inferential procedures of science, when applied within one particular branch of science, cosmology. Instead of taking a leap into a hole of one's own digging and then clawing one's way out, it would, it seems, be better not to take the leap in the first place." But this is all wrong. The motivation for introducing what Wayne refers to as the "disastrous" statistical postulate has nothing to do with any lingering influence of the principle of indifference—it has to do (rather) with the business of explaining the phenomenon of the approach to equilibrium, in the way we were taught to do by Boltzmann. It is a "hole" (as Wayne puts it) that we more or less *had* to fall into, being human, reasoning in the way that we do, in order to eventually find our way to a fully satisfactory theory.

Imagine (by way of analogy) that someone were to say: "My complaint about quantum mechanics is not so much about where it ends up, as the route by which it gets there. It begins by introducing a disastrous Schrodinger equation for which there is no motivation except an irrational longing for first-order differential equations, an equation that, if accepted, would lead you to the conclusion that measurements don't have outcomes. We are to be rescued from this disaster by modifying the equation in such a way as to bring it into conformity with a proposition that we

The discussion of the asymmetry of our epistemic access to the past and the future in Chapter Six of *Time and Chance* is built around an analysis of what it is to be a measuring device. I begin by pointing out that

> measuring devices are not the sorts of systems whose states become reliably correlated with the states of the systems they are designed to measure merely in the event that they interact with those systems in the appropriate way. Indeed, insofar as the basic laws of nature are exclusively dynamical ones (of which more in a minute), there simply *can't be* any systems like that.
>
> Here's what measuring devices *are*: measuring devices are the sorts of systems which reliably undergo some particular *transition*, when they interact in the appropriate way with the system they are designed to measure, only in the event that the measured system is (at the time of the interaction) in one or another of some particular collection of physical situations. The "record" which emerges from a measuring process is a relation between the conditions of the measuring device at the two opposite temporal ends of the interaction; the "record-bearing" conditions of measuring devices which obtain at one temporal end of such an interaction are reliable indicators of the situation of the measured system—at the time of the interaction—only in the event that the measuring device is in its ready condition (the condition, that is, in which the device is calibrated and plugged in and facing in the right direction and in every other respect all set to do its job) at the interaction's other temporal end. The sort of inference one makes from a recording is not from one time to a second in its future or past, but rather from two times to a third which lies in between them.

---

never had any reason to doubt, namely, that measurements *do* have outcomes. Instead of taking a leap into a hole of one's own digging and then clawing one's way out, it would, it seems, be better not to take the leap in the first place." This—again—is all wrong. The motivation for introducing the "disastrous" Schrodinger equation has nothing to do with any irrational longing for first-order differential equations—it has to do (rather) with the business of explaining the phenomenon of the stability of matter. It is a "hole" that we more or less *had* to fall into, being human, reasoning in the way that we do, in order to eventually find our way to a fully satisfactory theory.

That's what happens in science. You solve a problem. And the solution creates a new problem. And then you solve the new one. And you go on from there. Everybody always knew (after all) that matter was stable. The point was never to *predict* that—the point was to *explain* it. And the search for explanations—even for explanations of the most familiar and universally acknowledged truths, especially for explanations of the most familiar and universally acknowledged truths—can lead you through strange and unfamiliar places.

I'm thankful to Sidney Felder for a helpful conversation about this.

And note that inferences of this latter sort can be immensely more powerful, that they can be immensely more informative, than inferences of the predictive/retrodictive variety.[3] Think (for example) of an isolated collection of billiard balls moving around on a frictionless table. And suppose that billiard ball number 5 (say) is currently at rest; and consider the question of whether or not, over the past ten seconds, billiard ball number 5 happens to have collided with any of the other billiard balls. The business of answering that question by means of retrodiction will of course require as input a great deal more information about the present—it will require (in particular) a complete catalogue of the present positions and velocities of all the other billiard balls in the collection. But note that the question can also be settled, definitively, in the affirmative, merely by means of a single binary bit of information about the past; a bit of information to the effect that billiard ball number 5 was moving ten seconds ago.

The question, of course, is how one ever manages to come by information like that—and the answer that *Time and Chance* proposes is the *Past-Hypothesis*. The fact that our world obeys a past-hypothesis—and that it does not obey any corresponding future one—is what's supposed to lie at the bottom of the fact that we can make reliable inferences to the past, but not to the future, by means of measurement. And that latter fact is what's supposed—in turn—to lie at the bottom of the general asymmetry of our epistemic access to the past and the future.

And all of this still seems to me to be very much on the right track—but I am very much indebted to Frank Arntzenius for bringing it to my attention, some years after *Time and Chance* was first published, that there is a good deal that still needs doing by way of satisfactorily nailing it down. What Frank pointed out (in particular) is that there is a certain literal and sensible and quantifiable sense of "informative" according to which the fact that billiard-ball 5 is currently at rest, and that it was moving ten seconds ago, cannot *possibly* be any more informative about the past than it is about the future. For if the equations of the motions of the fundamental constituents of the world are Newtonian, if (that is) the equations of the motions of the fundamental constituents of the world are deterministic, and time-reversal-symmetric, and satisfy Louiville's Theorem, then it simply cannot be that any conditionalization on facts about the present can eliminate

---

[3] What I mean by "predictions" and "retrodictions"—by the way—are inference procedures to other times which operate by plugging any available macro-information about the present + the standard microstatistical rule into the microscopic equations of motion.

more exact microscopic trajectories, or a larger measure of exact microscopic trajectories, toward the past than toward the future.

And this leaves us with a pressing and interesting problem. There can certainly not be any doubt (think of the billiard balls, or of an audio tape, or of a photograph, or what have you) that there is some vivid and important sense in which measurements are enormously more "informative" about times in between the time when the ready condition obtains and the time when the record-bearing condition obtains than they are about times either before the time when the ready condition obtains or after the record-bearing condition obtains—but the measure of information that we are alluding to here cannot possibly be the straightforward and trivial one that we considered in the previous paragraph, and business saying exactly and explicitly what the measure is turns out not to be easy.

Time and Chance's discussion of the asymmetry of our capacity to influence or to control or to intervene in the past and the future was also built around the idea of measurement—and (more particularly) around the example of the billiard balls—but viewed (you might say) from the opposite angle.

Here—more or less in its entirety—is the heart of that discussion:

> Think (to begin with) of the collection of billiard balls we were talking about before. And suppose that some particular one of those balls (ball number 5, say) is currently stationary. And suppose (and this is what's going to stand in—in the context of this extremely simple example—for a past-hypothesis) that that same ball is somehow known to have been moving ten seconds ago.
>
> What we learned about that sort of a collection of balls in the previous section (you will remember) was that whereas (on the one hand) whether or not ball number 5 will be involved in a collision over the next ten seconds is determined by the present condition of the entire collection of balls, whether or not ball number 5 has been involved in a collision over the past ten seconds is (on the other) unambiguously determined—under these circumstances—by ball number 5's present condition alone.
>
> And this is something it will be worth taking the trouble to put in one or two slightly different ways.
>
> One of the things this means is that whereas (on the one hand) there are patently any number of hypothetical alterations of the present condition of the balls in this set—whatever that condition might happen to be—which would alter the facts about whether or not ball number 5 is to be involved in a collision over the next ten seconds, there can (on the other) be no hypothetical alterations in the present condition of this set of balls,

CONCLUSION   361

unless they involve hypothetical alterations in the present velocity of ball number 5 itself, which would alter the facts about whether or not ball number 5 had been involved in a collision over the past ten seconds.

And another of the things it means (and this is the one that's going to be the most useful for our purposes here) is that whereas (on the one hand) there are perfectly imaginable present conditions of this collection of balls in which certain small hypothetical alterations of (say) the present velocity of ball number 12 would alter the facts about whether or not ball number 5 is to be involved in a collision over the next ten seconds, and there are perfectly imaginable present conditions of this collection of balls in which certain small hypothetical alterations of the present position of ball number 2 would alter the facts about whether or not ball number 5 is to be involved in a collision over the next ten seconds, and there are (in short) perfectly imaginable present conditions of this collection of balls in which certain small hypothetical alterations of any physical feature you choose of any particular one of these balls you like would alter the facts about whether or not ball number 5 is to be involved in a collision over the next ten seconds, there are (on the other hand) no imaginable present conditions of this collection of balls at all (so long as those conditions are compatible with the proposition that ball number 5 is currently stationary, and so long as it is taken for granted that ball number 5 was moving ten seconds ago) in which any hypothetical alterations whatsoever in the present conditions of any of these balls other than ball number 5 itself would alter the facts about whether or not ball number 5 was involved in a collision over the past ten seconds.

And so there are (as it were) a far wider variety of potentially available routes to influence over the future of the ball in question here, there are a far wider variety of what we might call causal handles on the future of the ball in question here, under these circumstances, than there are on its past.

And the trouble with *this* discussion is not merely that it is wildly incomplete, but that it is also—in a way that seems not to have been noticed over the intervening decades—wrong.

Let's start with that.

It is perfectly true that if it is somehow given to us that ball number 5 was moving 10 seconds ago, and if ball number 5 is stationary now, then we don't need to know anything at all about any of the balls other than ball 5 in order to know that ball 5 was involved in a collision sometime over the past 10 seconds. But it is also true—and this is what was completely overlooked in *Time and Chance*—that the fact that ball number 5 was in a collision over the past 10 seconds (which is to say: the fact that ball number 5 was moving 10 seconds ago, combined with the fact that it is stationary now) has implications about the present states of the other balls. And so, if the condition of ball number 5 10 seconds ago is held fixed, there is no guarantee at all that we can freely vary the condition of (say) only ball number 17 in the present. Varying the condition of ball 17 in the present may well entail (given the classical equations of motion, and given the fact that ball number 5 was stationary 10 seconds ago) variations of the present conditions of other balls—including, of course, the condition of ball 5. And so (if the ball-5 past-hypothesis is in force), it is simply *not true* (as I had implied it was both in *Time and Chance* and in a subsequent book of mine called *After Physics*) that counterfactual variations in the present state of ball 17 cannot, as a general matter, produce counterfactual variations in the facts about whether or not ball number 5 was involved in a collision over the past 10 seconds. And indeed, in a case like this one—in a case (that is) like the billiard-balls, where we are taking it for granted that all of the relevant degrees of freedom are macroscopic ones, where all inferences to other times can invariably be cast in the form of predictions or retrodictions—it is not at all clear why counterfactual variations in the present state of ball number 17 should be any less capable of producing counterfactual variations in the facts about whether or not ball number 5 *was* in a collision over the *past* ten seconds than it is of producing counterfactual variations in the facts about whether or not ball number 5 *will be* in a collision over the *next* ten seconds. Here's a way to put it: In a case like this, we can look at the past-hypothesis as simply imposing a restriction on the set of possible macro-states of the entire set of balls *at present*—and there is no obvious way in which those restrictions make the facts about past-collisions any less sensitive than the facts about future-collisions are to the present state of ball 17.

That's the bad news. The good news is that under conditions that more closely resemble the sort of world we actually live and act in—conditions (that is) in which there is a dynamically significant distinction between the macro and the micro, conditions (that is) where statistical-mechanical considerations can come into play—things come out much more congenially. Suppose (for example) that we replace the collection of billiard balls with the universe that we live in, and that we replace the stipulation that ball number 5 was stationary 10 seconds ago with a past-hypothesis *appropriate* to the sort of universe that we live in, and that we limit ourselves to counterfactual antecedents representing *agential decisions,* and that we can take it for granted that the physical difference between an agential decision to do *this* and an agential decision to do *that* is a local and microscopic difference.

In that case, it seems very plausible that the sorts of counterfactual antecedents that we are interested in are going to be thoroughly compatible (unlike in the billiard-ball case) with whatever we might take ourselves to know of the actual physical situation of the world, at the time of the antecedent, in spatial regions that are disjoint from (say) the head of the agent. And this, in turn, will imply that such decisions—although they can make all of the familiar sorts of counterfactual differences towards the future—can *not* make the relevant kind of difference to past events of which it itself is not a record. And so the *practical* upshot of this second billiard-ball example—the upshot (that is) for the sorts of decisions that we actually make in the sort of universe that we actually live in—turns out to be right after all.

But these are small potatoes. The real trouble (as things stand now) is not merely that the question of the time-asymmetry of intervention has not yet been fully *answered*—but that we are still in the dark about exactly how, and in exactly what language, and at exactly what level of generality, the question is even to be *posed*.

What is it, exactly, that we are trying to get to the bottom of here?

It might be thought that we want to understand why the future, but not the past, counterfactually depends on the present. But if we hold the laws fixed when we evaluate counterfactuals, and if the dynamical laws in question are deterministic and time-reversal symmetric—as is the case (for example) in Newtonian Mechanics—then the past cannot *fail* to counterfactually depend on the present. Maybe (then) what we want to understand is why the future counterfactually depends on the present so much more *sensitively* than the past does. But Adam Elga has pointed to cases (the so-called "Atlantis" cases) in which that the counterfactual dependence of the past on the present is enormous. And Barry Loewer has responded to Adam's examples by suggesting that what we want to understand is not why the future but not the past can be *influenced* by the present, but why the future but not the past can be *controlled* from the present—where the business of controlling X requires not only that one is influencing X, but (in addition) that one is *aware* that one is influencing X, and that one is aware of *how* one is influencing X. And Barry has shown how the cases that Adam points to necessarily lack this additional feature of control. But since then Mathias Frisch has pointed to cases in which we seem to have what Barry would count as control over macroscopic features of the past. And Frisch's examples have in turn prompted yet another proposed refinement of the time-asymmetry of intervention: not that the past cannot be *influenced* by the present, and not that the past cannot be *substantially* influenced by the present, and not that the past cannot be *controlled* from the present, but that the past cannot be *robustly* controlled—that it cannot be controlled (that is) in a way that persists across a reasonable range of what you might call *deliberational contexts*—from the present.[4] And so on.

There have also been lively and interesting discussions about what particular conception of *agency* should be built in to the best and most general and most illuminating way of formulating the question of the time-asymmetry of intervention. The discussions in *Time and Chance* relied on a "primitive and un-argued-for and not-to-be-further-analyzed conception of which particular features of the present condition of the world it is that are to be thought of as falling under our (as it were) direct and unproblematical and unmediated control"— something that I have since been referring since to as "the fiction of agency." But various investigators (Jenann Ismael, for example, and Alison Fernandes) have suggested that we might do better, without any important loss in generality, by starting out with a more realistic and more detailed and more specific conception of what it is to be an agent—something (for example) that includes distinctions, and temporal relations, between "deliberating" and "deciding" and "acting."

My sense is that there is fairly widespread confidence that as soon as we are in a position to say exactly what the time-asymmetry of intervention *is*, it is going to be a relatively straightforward matter to see that the *explanation* of that asymmetry is the *Mentaculus*—the subtle part, and the difficult part, and the interesting part, as often happens in fundamental scientific work, is the business of *posing the question*.

Let me now turn to the penetrating and wonderful essays in this book. And let me begin by saying that I am thrilled and honored and humbled that so many of the people that I most admire have been willing to take the time and the effort to react, in these pages, to a decades-old book of mine—and that I am deeply grateful to Eric Winsberg and to Brad Weslake for their work in editing these essays. I am especially and particularly grateful, grateful beyond what I have any idea of how to put into words, for the steadfast and perennial and unconditional love of my friend Barry Loewer, whose idea it was to put this book together.

Seven of the papers in this collection (the ones by Barry Loewer, Eric Winsberg, David Wallace, Sean Carroll, Chris Meacham, Dustin Lazarovici, and Eddy

---

[4] This dialectic is discussed in some detail on pages 46–51 of my *After Physics* and the argument from Frisch can be found in a paper of his called "Does a Low-Entropy Constraint Prevent Us from Influencing the Past?" which appears in *Time, Chance, and Reduction*, ed. Gerhard Ernst and Andreas Huttermann, 13–33 (Cambridge: Cambridge University Press, 2010).

Keming Chen) confront the sort of project I have just sketched out very directly, and in its entirety, and at its deepest logical foundations.

Barry (to begin with) has been my collaborator in thinking all this through from its very earliest beginnings. What I owe to him in this matter—and in many others besides—is altogether beyond reckoning. And he and I are very much on the same page about most of this stuff. And the eloquent and panoramic and magisterial account of our project that he presents here feels almost exactly right to me. And the same applies to his beautiful essay on the freedom of the will—which I think amounts to a completely decisive dismantling of the Consequence Argument.

If there is any daylight between us at all, it has to do (I suppose) with the business of evaluating those particular counterfactuals—the central and exemplary and paradigmatic ones, on our view—whose antecedents are agential decisions. What Barry has to say about that business here more or less rehearses the position that he and I put together some years ago—it stipulates (in particular) that everything about the world, at the time of the antecedent, and outside of (say) the agent's head, is to be held fixed. A stipulation like that has an immediate and undeniable intuitive appeal, and it turns out to afford a quick and easy response to some worries that Mathias Frisch has raised about the account of the asymmetry of influence proposed in *Time and Chance*[5]—but I have since been persuaded that it privileges the moment of the antecedent in a way that, on reflection, feels hard to justify. The truth is that Sidney Felder has been warning me about a worry like that for something on the order of ten years now—but it was Alison Fernandes (in her contribution to this volume) who first put it into the sort of concrete and explicit and pedestrian language that I simply could not help but understand, and by which I finally felt compelled to change my mind.

Some of the things that Eric Winsberg worries about (on the other hand) just don't seem like worries to me—or not (at any rate) insofar as I understand them at the moment. I don't see why it should count as a particularly "unappealing" consequence of Humean conceptions of lawhood (for example) that such conceptions are going to end up counting everything about certain radically simple and wildly imaginary worlds as nomically necessary. And I don't see what it is that makes it obvious that the business of writing down the Past Hypothesis in the language of the microphysics (which is to say: I don't see what makes it obvious that the business of associating the initial macrocondition of the universe with this or

---

[5] I have sketched out a response like that, for example, in "The Sharpness of the Distinction between the Past and the Future," in *Chance and Temporal Asymmetry*, ed. Alistair Wilson (Oxford: Oxford University Press, 2014).

that particular region of the universe's phase space) is going to be so terribly complicated. And while Eric is certainly right to point out that P (the proposition that adding PH and PROB to the dynamical P-laws results in a system that is only a little less simple but vastly more informative than the system consisting only of the dynamical laws) does not logically entail C (the proposition that the P-laws, PROB, and PH make up the best system), he is wrong (it seems to me) to conclude that the truth of P is (therefore) simply "irrelevant" to the business of accessing the truth of C. If P is true, and if the P-laws are indeed the best system of the world on the first of Eric's three conceptions of informativeness, then—all other things being equal, and absent any particular evidence to the contrary—C seems like a damn good bet on the second conception of informativeness.

But Eric does seem to me to put his finger on something really deep, and genuinely troubling, in his discussion of Boltzmann Brains. If the universe eventually settles in to an eternal finite-temperature equilibrium, as some of our best current cosmological guesses seem to suggest, then the business of predicting what I am going to see when (say) I walk outside, or look in a closet, or turn my head, seems to require something over and above the Past Hypothesis and the Statistical Postulate and the microscopic equations of motion, something (in particular) about which particular one of the infinity of brain-stages in the complete history of the universe whose phenomenological experience is exactly the one that I am having now is actually, currently, mine. And it seems right to say—as Eric does—that stipulations about which one of the above-mentioned brain-stages is actually currently mine seem like poor and unfamiliar sorts of candidates for fundamental physical laws.

My suspicion is that there is a way of dismissing every such stipulation—except (of course) for the usual, intuitive, commonsensical one, the one, that is, that Eric refers to as the Near-Past Hypothesis—in much the way that we dismiss familiar skeptical hypotheses about (say) the world's having been created, along with our records and memories, five seconds ago. But the business of satisfactorily articulating and defending a suspicion like that is still very much in need of doing, and strikes me as a worthwhile and important project.

David Wallace's paper is (as I said) focused on the analysis of basic logical structure as well—and the upshot of that analysis, which is presented in the map of logical relations at the end of David's paper, seems exactly and very helpfully right to me. My only quibble is that David seems to have misconstrued what it was that I was trying to do in *Time and Chance*. The point was not—as David seems to suppose—merely to explain what he calls the "Predictive Accuracy of Macrophysics" (PAM), but something much larger and more ambitious: to spell out what was referred to at the end of chapter 4 of *Time and Chance* as "the complete contraption for making inferences," and what is nowadays more commonly referred to as the Mentaculus. David's PAM, for example, will be compatible with the skeptical hypothesis that the world was created five seconds ago, in the macrostate that we normally take it to have occupied five seconds ago, and didn't exist at all before that. The business of eliminating hypotheses like that—the business (more generally) of capturing and underwriting our normal scientific procedures of inference from the present to times other than the present—requires much more than the conjunction of the Simple Past Hypothesis, the Simple Dynamical Conjecture, and the Predictive Accuracy of Microphysics (SPH + SDC + PAμ). It requires something about the initial macrocondition of the universe.

What particularly excites and intrigues me about David's paper, on the other hand, is his Simple Dynamical Conjecture itself—because if anything like that should turn out to be true, and if there should turn out to be individual microstates of the world that satisfy the relevant criterion of "simplicity," then the microscopic underpinnings of thermodynamics might turn out not to require anything along the lines of a probability distribution over initial conditions. I need to confess (however) that I am much less confident than David seems to be that anything like the Simple Dynamical Conjecture is, in fact, going to turn out to be true. The conjecture (as David himself points out) straightforwardly fails in the classical case—and the case of quantum mechanics seems to me, at best, and as yet, undecided. But there is an enormously interesting suggestion here—something that surely merits much more careful and detailed investigation.

And the papers by Sean Carroll and Chris Meacham both, in one way or another, bring up questions about the character and the origin of the statistical-mechanical probabilities—and (more particularly) about the project of deriving those probabilities from something along the lines of a principle of indifference.

Sean's paper is mostly taken up with a clear and penetrating and learned and altogether magisterial critique of the sorts of reasons that are usually given in favor of hypotheses of cosmic inflation. All of those reasons fall under the general category of *fine-tuning*. All of them (that is) begin with an observation to the effect that this or that simple and general and robust and pervasive feature of the universe (its uniformity, say, or its flatness, or its smoothness, or the fact that it is far from equilibrium) is somehow, on the face of it, unlikely—and therefore in need of explanation. And Sean's paper is mostly taken up with the business of quibbling with the standard judgments about what sorts of universes are, on the face of it, unlikely, and in need of explanation, and what sorts aren't. And all I want to do here is to remind the reader that there is, hidden at the bottom of debates like that, a puzzle.

Whenever anybody suggests that the universe as a whole, or the entirety of what we can see of it, or the entirety of what we can safely infer about it, or (for that matter) any sufficiently simple and general and robust and pervasive feature of it, is *unlikely*—there is a puzzle, there is a worry, there needs to be a further question, about where they can possibly have gotten their ideas of what is likely and what isn't. The worry (in particular) is that anybody who suggests anything like that can apparently *not* have gotten their ideas about what's likely and what isn't from any empirical experience of what it is that tends, or tends not, to actually *happen*—because it is precisely those tendencies themselves that they seem somehow to have decided are unlikely! And the question (of course) is where, outside of or apart from experience like that, it could possibly be reasonable to get such ideas.

Sean is clearly aware of worries like that—and he certainly makes no claim to have somehow definitively put them to rest—but the way he passes over them in his contribution here is nonetheless (it seems to me) too quick. All he has to say on the matter is: "In the case of the initial state of the universe, one might reasonably suggest that we simply have no right to have any expectations at all, given that we have only observed one universe. But this is a bit defeatist. While we have not observed an ensemble of universes from which we might abstract ideas about what a natural one looks like, we know that the universe is a physical system, and we can ask whether there is a sensible *measure* on the relevant space of states for such a system, and then whether our universe seems generic or highly atypical in that measure." But the question, of course, is precisely about what sorts of grounds could there possibly be—other (again) than grounds of empirical success—for calling one rather than another of the available measures on the space of possible physical states a "sensible" one.

There is, of course, a famous and long-standing and widely-relied-upon attempt at answering questions like that—something that goes back at least as far as Laplace—which (as I mentioned above) goes under the name The Principle of Indifference. But the literature of the philosophy of science is positively swarming with very familiar and very compelling reasons to be skeptical (see, for example, pages 62–65 of *Time and Chance*) that any such attempt is ever going to work.

Chris Meacham is skeptical of attempts like that as well. But the arguments he offers in his essay here are new, and unfamiliar, and (at least for the moment) puzzling to me.

I'm used to thinking of principles of indifference—insofar (at any rate) as they're supposed to apply to the foundations of statistical mechanics—as a way of getting (or pretending to get) a very particular job done, as a way of filling in (or papering over) a very particular gap in our fundamental scientific account of the world: We start with the fundamental physical laws, and with whatever we may happen to know about the way the world presently, contingently, happens to be. And what the principle of indifference is for, and all the principle of indifference is for—insofar (again) as it is supposed to apply to the foundations of statistical mechanics—is to get us from there to a probability distribution over the possible present microstates of the world.

But Chris seems to think that an investigator who is committed to getting that particular job done in that particular way must also be committed—on pain of irrationality or dogmatism or silliness or something like that—to distributing his credences among different imaginable fundamental physical laws about the initial macroconditions of the world, to distributing his credences (that is) among different imaginable Past Hypotheses, by means of a principle of indifference as well. And the trouble (so Chris says) is that that latter procedure is going to end up, by the usual Boltzmannian reversibility arguments, assigning a very low probability to the truth of the correct Past Hypothesis—the low-entropy Past Hypothesis—that actually underwrites our empirical experience.

And one way of at least displaying my confusion about all of this might be to ask what's so special about the Past Hypothesis among the fundamental physical laws. If Chris is right—if an investigator who appeals to a principle of indifference in order to get the above-mentioned very narrow job done must also be committed to distributing his credences among different imaginable laws about the initial macroconditions of the world by means of a principle of indifference—shouldn't he also be committed to choosing among different imaginable dynamical laws in that way as well? And what about the space of possible microstates itself? Shouldn't that be chosen by maximizing ignorance too? But of course the reversibility arguments—the ones (that is) that Chris appeals to in order to argue that the investigator in question here will need to assign a very low probability to the truth of the low-entropy Past Hypothesis—depend crucially on the microdynamics and the space of possible microstates being what we normally take them to be. And at this point, I just sort of lose my grip on where we are and on how to proceed.

Dustin Lazarovici's contribution to this volume strikes me as the clearest and most sober and most balanced and most sensible defense of typicality approaches to the foundations of statistical mechanics that I know of—and I think it is likely to stand as a touchstone as the debate about the logical and metaphysical status of statistical-mechanical probabilities continues and develops into the future. But (for whatever it may be worth) I am still unconvinced. The crucial question here is whether any further substantive claims about the world need to be added to the deterministic classical laws of motion in order to get an explanation of the approach to equilibrium—and Dustin is quite right to say that "David Albert (in private communications) has been particularly forceful in rejecting *this* idea, that one could appeal to some a priori notion of 'nearly all,' that typicality facts could come more or less for free once the rest of the theory is fixed." And I think everything is going to hinge on how the equivocation implicit in "more or less for free" eventually gets unpacked. And it seems to me that (in a nutshell) things are either free or they aren't.

Eddy Chen's paper is a clear and careful and beautiful reflection on the nature of the Past Hypothesis—including a thoughtful and penetrating response to a host of different worries about whether that hypothesis can have the status of a natural law—but it is also a great deal more than that. Eddy has been at work, for a couple of years now, on a novel and original account of the foundations of quantum statistical mechanics—something he calls the Wentaculus—which (like David Wallace's Simple Dynamical Conjecture—but in an altogether different way) promises to do away with the need for any probability distribution at all over possible initial quantum states of the world. And Eddy offers us a brief and very intriguing and preliminary sketch of that project—more or less in passing—in his paper here. I have some questions about this project. I'm not sure (for example) if it can be carried through in a way that will allow us to hang on to the idea that the fundamental dynamical laws of the world are invariant under time-translation—and I'm not sure how much of a problem it might amount to if it can't. But my worries are pretty small compared to my interest and enthusiasm—and I am very eager to see all this explored a great deal further.

Tim Maudlin is exactly right, it seems to me, when he says: "The main point of Albert's discussion of Maxwell's demon is exactly to emphasize the gap between

> (P) No closed macropredictable system can exist whose entropy macropredictably goes down.

and

> (Q) No closed system can exist whose entropy macropredictably goes down."

And Tim does nothing at all by way of disputing that exactly that sort of a gap exists—and he does nothing at all by way of disputing that the existence of that sort of a gap completely demolishes the large and influential tradition of attempts to argue that violations of what Tim calls the "Modern Entropic Version" of the second law of thermodynamics are somehow ruled out, as a matter of fundamental principle, by the laws of classical mechanics. And Tim is perfectly clear about all this in his paper. The question with which Tim's paper is primarily concerned—and this is just the question that I myself bring up at the end of chapter 5 of *Time and Chance*—is (instead) whether or not the existence of a gap like that can actually be parlayed into a *practical* method of extracting significantly more energy from the world than is allowed by the second law of thermodynamics. And on this (it turns out) he has much, of considerable interest, to say. Tim (for example) proposes a natural measure of the "efficiency" with which an Albertian demon of this or that particular design manages to reduce the entropy of the thermodynamic system on which it operates—and he constructs a very ingenious and completely rigorous proof that the efficiencies of demons that lower the entropies of two-gas systems by raising and lowering a shutter, in the way that Maxwell originally suggested, cannot possibly be anywhere in the neighborhood of perfect. And Tim makes it very vivid, by means of an explicit calculation, that the number of macroscopically different possible final states that any Albertian demon whatsoever would need to have in order to lower the entropy of a gas enough to allow for the extraction of as much energy as would be required to lift an apple a few feet into the air—even (mind you) if the demon were operating with perfect Maudlinian efficiency—is astronomically large. And all of this represents a significant advance in our understanding of what it might even in principle be possible to do with a Maxwellian demon. And all of it is obviously and staggeringly discouraging—on the face of it—vis-à-vis the practical possibilities of making any money.

But it would still be nice (I think, if we can) to do better than that. It would still be nice to have a clean and simple and general demonstration—something that starts out with nothing over and above the fundamental microscopic Hamiltonian of the world, and the Statistical Postulate, and (maybe) the Past Hypothesis—such that (as I put it at the end of chapter 5 of *Time and Chance*) "the construction of a Maxwellian Demon system, or the operation of a Maxwellian Demon system, or the extraction of mechanical energy from a Maxwellian Demon system by means of heat engines (once its operations are done), or the exploitation of that energy (once it's been extracted) will necessarily—as a statistical matter—somehow prove prohibitive or self-defeating or otherwise uncircumventably pointless." I have no idea (of course) whether or not that can actually be done.

There are four papers about the asymmetries of knowledge and intervention—by Nick Huggett, Mathias Frisch, Alison Fernandes, and Sidney Felder.

Nick (insofar as I can see) is just making a mistake. But it's a particularly helpful and well-motivated and clearly described sort of a mistake—the sort of a mistake (indeed) that decisively advances the conversation. Nick is quite right to point out that nothing whatsoever can safely be inferred from the "memories" of an "information gathering and utilizing system" (IGUS) unless the IGUS in question can safely be assumed to have been in a very particular kind of computational state—its "ready" state—before those memories were formed. And Nick's worry is that what I call the complete contraption for making inferences is not going to be in a position to account for the fact that IGUSs manage to reliably know the sorts of things about the past that (in fact) they do—because that contraption is not going to be in a position to make any particular initial computational state of a system like that more or less probable than any other. Nick seems to think (in particular) that all that the complete contraption is ever going to be capable of telling us about whatever IGUSs there may happen to be in our world is that they are unlikely to have fluctuated out of equilibrium!

But that's just not right. We have a famous argument from Darwin (after all) about how it is that whatever IGUSs there may happen to be, in worlds like ours, are actually likely to have gotten here. We have a famous argument from Darwin (in particular) to the effect that whatever IGUSs there may happen to be in worlds like ours—worlds, that is, that have evolved to their present states, in accord with the microdynamical laws and the Statistical Postulate, over billions of years, from some much lower-entropy initial macrostate[6]—are likely to have gotten here by means of random mutation and natural selection. And IGUSs that get here like that are likely to open their eyes in the morning in some reasonable approximation of their "ready" states—because the ones that don't (so the argument goes) will have long since died out![7]

Mathias's position is more complicated. He agrees that the Past Hypothesis + the Statistical Postulate + the microdynamics (which is to say: the Mentaculus) does indeed amount, at the end of the day, to a successful scientific account of the asymmetries of knowledge and intervention. He says, near the end of his

---

[6] Or rather: Worlds that have evolved to their present states, in accord with the microdynamical laws and the Statistical Postulate, over billions of years, from some much lower-entropy initial macrostate, and in which (as in our own) there is no evidence of a massive conspiracy to flood the world with artificially manufactured IGUS's that come out of their boxes in something *other* than their "ready" states.

[7] The idea, then, is that Darwin has given us an argument to the effect that the past hypothesis and the Statistical Postulate and the microscopic equations of motion, all by themselves, are going to entail—in the absence of any evidence of a vast conspiracy aimed at making things otherwise—that whatever IGUSs there may happen to be are likely to open their eyes in the morning in some reasonable approximation of their "ready" states.

paper, that "Albert and Loewer's premises do after all seem to entail that traces or records of the past are overall reliable, and hence that there are certain counterfactuals—those associated with small macroscopic changes to the present—that are temporally asymmetric, and hence, perhaps, that there is a causal asymmetry." But he thinks that the Past Hypothesis ends up contributing a good deal less to that account than Barry and I think it does.

Mathias considers the question of how (for example) we manage to infer, from measurements of the electromagnetic field near the surface of the earth, that there are, or were, stars. And he suspects that there is nothing along the lines of a Past Hypothesis—he suspects (that is) that there is nothing along the lines of a simple postulate about the initial macrocondition of the universe—that could imaginably play a crucial role in underwriting an inference like that. But I'm not sure I see why. What we're in need of here, after all, is nothing more or less than an excuse to throw away advanced-potential solutions to Maxwell's equations. And the usual such excuse—the one (that is) that one finds in any of the standard textbooks on classical electrodynamics—has to do with imposing an initial boundary condition to the effect that, at sufficiently early times, the radiation fields vanish. And a condition like that is, of course, and for that very reason, exactly the sort of thing one expects to find in the Past Hypothesis.

And Alison's paper is (to my mind) brilliant. It isn't addressed directly to the text of *Time and Chance* itself—but (instead) to one of the discussions that have arisen in its wake. Alison is concerned (in particular) with a response of mine to one of Mathias's critiques of the account of the asymmetry of intervention in Chapter 7 of *Time and Chance*—and the long and the short of it is that I have studied her paper very carefully, and learned a great deal from it, and (as I mentioned above) have significantly changed my own approach to these matters as a result.[8]

And Sidney Felder's essay is, finally, in all sorts of ways, in a category by itself. The first thing to say is that I have known Sidney—and I have been learning from him—for something on the order of thirty years now. And notwithstanding the fact that I have been cajoling and admonishing and imploring him, for all that time, to publish his work—his contribution to the present volume is in fact the very first piece of his philosophical writing ever to appear in print. And I am very

---

[8] The degree to which my own thoughts on these matters have evolved under the influence of Alison's critique can be gauged by comparing my response to Frisch's example in "The Sharpness of the Distinction between the Past and the Future"—which was written before Alison began to talk to me about all this—with my response to that same example in Chapter 2 of *After Physics*.

pleased and proud that his introduction to the wider world should finally have been occasioned, however incidentally, by some work of my own.

Sidney's essay is much too rich to confront, in any substantive way, in the space I have available here—and I'm not sure I would know where to begin (to tell the truth) if I had more space. Let it suffice, for the time being, to say that it is a subtle and beautiful and difficult and prodigiously learned and immensely rewarding meditation on the nature of agency, and of knowledge, and of causation, and of probability, and of counterfactual dependence, and of a host of other things besides, and that it deserves to be read repeatedly, and with care, and that it represents only the extremest tip of the gigantic submerged iceberg of Sidney's thought, and that I hope that it will bring some small measure of the attention that (it seems to me) that thought has so long and so richly deserved.

Contributors

Index

Contributors

- DAVID Z ALBERT (*Columbia University*)
- SEAN M. CARROLL (*California Institute of Technology*)
- EDDY KEMING CHEN (*University of California, San Diego*)
- SIDNEY FELDER (*Rutgers University*)
- ALISON FERNANDES (*Trinity College Dublin*)
- MATHIAS FRISCH (*Leibniz University Hannover*)
- NICK HUGGETT (*University of Illinois Chicago*)
- DUSTIN LAZAROVICI (*Technion – Israel Institute of Technology*)
- BARRY LOEWER (*Rutgers University*)
- TIM MAUDLIN (*New York University*)
- CHRISTOPHER J. G. MEACHAM (*University of Massachusetts Amherst*)
- DAVID WALLACE (*University of Pittsburgh*)
- BRAD WESLAKE (*NYU Shanghai*)
- ERIC WINSBERG (*University of South Florida*)

Index

The letter *f* following a page number denotes a figure; the letter *t* following a page number denotes a table.

Adequacy, 151–154, 159, 163–165, 171–174
After Physics (Albert), 362
agency, 8, 250, 325, 364, 374
agency, deliberative, 327–331, 329f, 330n
Albert, David: *After Physics*, 362; Albertian demon, 253–260, 265–267; early universe, fine-tuning, 136; evidential undermining, 326; fly case, 324–325; Mentaculus, 13–14, 20–33; MEV, 253, 254; Past Hypothesis, the logic of, 103–108; PH and physical laws, 211–214, 219, 228–230, 236; reading the past in the present, 289; responses, 9, 351–374; statistical mechanics, 58–68, 72–79; Statistical Postulate (SP), 127; *Time and Chance*, overview of, 1. *See also* Albert's responses
Albert and Loewer's entropy account, 302–310
Albertian demon, 249–251, 251–258, 258–267
Albertian demon, how to make, 258–261, 259f, 261–263, 263–267
Albertian demon, the perfectly efficient, 261–263
Albert's responses: agency, 364; asymmetry of influence, 365, 365n; Sean Carroll, 367–368; Eddy Chen, 370; classical statistical mechanics, 351–352; Sidney Felder, 365, 373–374; Alison Fernandes, 365, 373, 373n; fiction of agency, 364; Mathias Frisch, 365, 365n, 372–373; Nick Huggett, 372, 372nn6–7; inference, 355; intervention, 364; Dustin Lazarovici, 369–370; Barry Loewer, 365; Tim Maudlin, 370–371; Chris Meacham, 367, 368–369; measurement, idea of, 360–363; measuring devices, 358–359; Mentaculus, 354–357; Wayne Myrvold, 357–358n; Newtonian mechanics, 352, 352n, 359, 363; past, present, future, 363–364, 364n; Past Hypothesis, 352–353, 356–357, 359–360; predictions and retrodictions, 359n; Predictive Accuracy of Macrophysics, 366–367; probability distribution, 353–354; time-asymmetry of intervention, 364; David Wallace, 366–367; Eric Winsberg, 365–366
antihistories, 84, 94
anti-thermodynamic initial conditions, 25
Arntzenius, Frank, 359
arrows of time, 3–4, 180, 228–230, 230–231, 232

379

asymmetry, causal / counterfactual: causes, randomness, and the Past Hypothesis, 294–295, 301–304; Mathias Frisch, 373; introduction, 8; Mentaculus, 32–37, 46; Past Hypothesis and physical laws, 231; statistical mechanics, 57; *Time and Chance*, 360–365

asymmetry, epistemic: causes, randomness, and the Past Hypothesis, 302; introduction, 7; Mentaculus, 29–30; Meta-Reversibility Objection, 158–159, 170; reading the past in the present, 273–292; statistical mechanics, 57

asymmetry, thermodynamic: causes, randomness, and the Past Hypothesis, 294–295, 301; introduction, 6; Mentaculus, 27, 33, 46; Past Hypothesis and physical laws, 226; statistical mechanics, 57

asymmetry of influence, 314–317, 314nn4–5, 315n, 316n

Atlantis, 32–33, 317–318, 363

atypicality, 194–195

backward dynamics, 92–93

Bayesian, 100, 145, 156–157, 171

Bayesianism, 146, 152–154, 156–157, 164–165, 171, 173–174

BBGKY hierarchy, 89

Bernoulli, Jacob, 47

Best System Account (BSA), 40, 41–44, 49, 180–181, 194, 232–233

Betting, 48–49, 182, 186, 366

Big Bang: Albert and Loewer's entropy account, 309; asymmetry of influence, 314; causal representations and initial randomness, 300; early universe, fine-tuning, 111–117, 124–126, 134–137; entropy status of, 3; evolutionary response—a reply, 289–290; Mentaculus, 14, 16, 27; metaphysical accounts of laws, 233; PH and physical laws, 212; statistical mechanics, 72

black holes, 115, 120, 124, 125

Bohmian mechanics (BM), 197, 199, 218–219, 235, 238, 239, 243

Bohmian Wentaculus, 223, 237

Bold Dynamical Conjecture, 96–97

Boltzmann, Ludwig, 3–5, 18–20, 114, 180, 208–212, 351–354

Boltzmann, variations on a theme from: classical Mentaculus, 209–214, 209n, 210f, 212nn11–12; development conjecture, 213; introduction, 208; quantum Mentaculus, 214–219, 215nn13–14, 217n15; Second Law for X, 213; Second Law for Ψ, 217; Wentaculus, 219–224, 223nn16–17

Boltzmann Brains, 4, 5, 73, 136, 227, 366

Boltzmann entropy, 118–119, 183, 211–213, 216, 221, 254

Boltzmannian account, 17, 18, 206, 208, 214–217, 219–224

Boltzmann's Brain (BB) paradox, 70–73

Borde-Guth-Vilenkin (BGV) theorem, 125n

Boring distribution, 92

boundary condition laws, 238–239

branching, 23, 79, 82, 89, 106

BSA. *See* Best System Account (BSA)

butterfly effect, 24n

canonical measure, 126–129, 127n, 135, 198

Carroll, Sean M., 5, 70, 72, 110–137, 233, 367

Cartesian product, 86

Cartwright, Nancy, 16, 24, 70

Cauchy problem, 296, 300

causal fork, 302, 303

causal handles, 314, 316, 361

causal inference, 8, 294, 296, 300, 302

causal reasoning, 294, 300, 301

causal representations and initial randomness, 295–302

Chance-Credence Principle, 145, 153, 154, 157, 165, 173

chancemaking patterns, 48, 183, 184

chances, definition of, 142–143

Chen, Eddy Keming, 6, 204–244, 370

classical mechanics: background, 2; canonical measure, 126; classical Mentaculus, 213; coarse-grained dynamics, 95, 98; initial conditions and histories, 124; macropredictions of microdynamics, 80, 82; Mentaculus, the, 15; microdynamical origins of irreversibility (quantum case), 101; nomic accounts, 148, 149; Past Hypothesis and physical laws, 208, 231, 244; typicality measures, 196, 198, 199; typicality versus Humean probabilities, 180, 193, 195

INDEX

classical Mentaculus, 209–214, 209n, 210f, 212nn11–12
closed system: Albertian demon, 253, 257, 264–265; causes, randomness, and the Past hypothesis, 304–305; fine-tuning, 118; intervention, concept of, 344; introduction, 2–3, 7; Maxwell's demon, 370; Mentaculus, 14, 28, 42; physical laws, 205; reading the past in the present, 284
CMB. *See* cosmic microwave background (CMB)
coarse-grained dynamics, 84–90, 86n, 88n
coarse-grained dynamics, microdynamical underpinnings of, 93–98
coarse-grained dynamics, time-reversibility in, 90–93
coarse-graining: causes, randomness, and the Past hypothesis, 309; Meta-Reversibility Objection, 155, 158; Past Hypothesis, the logic of, 81, 84–88, 90–101, 107; statistical mechanics, 59, 62
coarse-graining map, 85
Coen Brothers, 13, 127
complexity, 61, 115, 236–237
conditionalizing, 21, 28–29, 129, 145, 297, 303–304, 336
conditional probability, 30, 32, 40, 44–45, 49, 82, 336
Consequence Argument, 33–35, 365
constraint argument, 8, 306–308
cosmic expansion, 113, 116, 120, 122, 125–126
cosmic microwave background (CMB), 112–113, 116, 119, 121–122, 229
cosmology: causes, randomness, and the Past hypothesis, 300–302, 309; fine-tuning, 110–114, 120, 126–136; introduction, 4; Mentaculus, 14, 16; Wayne Myrvold, 357; Past Hypothesis, the logic of, 77, 100; Past Hypothesis and physical laws, 206; reading the past in the present, 278; statistical mechanics, 64
counterfactual arrow, 231
counterfactuals, 226, 231–232, 315–317
Cournot's principle (CP), 6, 47, 51, 186–188, 190–191, 192, 194, 200

Cournot's principle versus Principal Principle, 186–188
CP. *See* Cournot's principle (CP)
credence function, 93, 144–145, 146

dark matter particles, 112, 120
decelerating universe, 116
decision counterfactuals, 31, 35
decoherence theory, 96
deliberation, 313, 326–332
deliberative agency, 327–328, 327n, 328–330, 329f, 330n
density matrix, 219–224, 240, 243
Density Matrix Realism, 220
de Sitter phase, 115, 136
de Sitter space, 71, 73, 136–137
determinism, 2, 26, 31–32, 34, 43, 284
deterministic chances, 188, 239
disjoint, use of term, 81
distributional variant, 80
distributions, 28, 49, 105, 107–108, 117, 127. *See also* PROB
Doppler effect, 113
dynamical governing, 224
dynamical law primitivism, 234
dynamical laws, definition of, 65. *See also* fundamental dynamical laws (FDL)

early universe, fine-tuning: coarse-graining, 114; discussion, 136–137; features of the state, 112–115, 112n, 115n; introduction, 110–112, 110n, 111n; trajectories. *See also* horizon problem; trajectories, fine-tuning early universe
Earman, John, 16, 58, 77, 301, 364–365
Easy Part, 204–205, 206, 211, 212, 213
Eddington, Arthur, 17, 226
efficacy, 328–330
eigenstates, 125, 215
Einstein–Hilbert action, 128, 130
Einstein-Hilbert Lagrangian, 130
electromagnetic phenomena, 17, 296, 373
Elga, Adam, 32, 36, 43, 317–319, 323–326, 363
entropy, 2, 17n, 18, 114
entropy of a closed system never decreases, 253–254
epistemic arrow, 230–231

epistemic guides, 235
equilibration, 77, 117–119, 121, 122, 124, 262–263
equilibrium, 14–15
equilibrium state, 18–19, 19n15, 23nn22–23, 121n8, 307, 314
equivariance, 197–198
ergodicity, 92
eternal recurrence, 70–71
Everettian version, 218, 238, 239, 243
Everettian Wentaculus, 222–223, 237
evidential undermining, 313, 325–327, 331
exemplar rule, coarse-grained, 86, 88, 91
explanation, 27, 226, 229, 232, 234

Felder, Sidney, 8, 335–349, 365, 373–374
Fenton-Glyn, Luke, 42
Fernandes, Alison, 8, 33, 365, 373
fiction of agency, 315, 325, 364
fine-tuning, 111–115, 121–124, 126–129, 132, 134–137, 367
fine-tuning (the early universe). *See* early universe, fine-tuning
first principle of population dynamics, 38–39
fit: Mentaculus, 41–43, 48–49; Past Hypothesis, the logic of, 77; Past Hypothesis and physical laws, 232, 238, 241–242; reading the past in the present, 289; statistical mechanics, 59–61; typicality versus Humean probabilities, 183
flatness problem, 5, 111–112, 129–130, 132, 134
fly case, 319–323, 323n, 323–327, 326n13
Fodor, Jerry, 38
forward compatibility, 94–95, 96
forward dynamical trajectory, 85
forward dynamics, 85, 89, 90–94, 97, 101
forward predictable, 94, 96, 97, 98, 99, 101
Fourier mode, 133
Fourier space, 126, 133
Friedmann equation, 116, 118, 129, 130, 131
Friedmann-Robertson-Walker universe, 111, 120, 131
Frisch, Mathias, 8, 33, 63, 313, 317–320, 323–326, 330, 364–365
FTSM. *See* fundamental theorem of statistical mechanics (FTSM)

fundamental dynamical laws (FDL): asymmetry of influence, 314–315; classical Mentaculus, 213, 214; fly case, 319; intervention, 335, 341; Mentaculus, 15, 19, 24, 36, 40; Past Hypothesis and physical laws, 226, 228, 234; quantum Mentaculus, 218; Wentaculus, 222
fundamentalism, 59, 70, 73, 74
fundamental theorem of statistical mechanics (FTSM), 180, 181, 189
Future Hypothesis (FH), 169, 316

Galilean symmetries, 198–199
Gaussian, 87, 98, 136, 189, 195, 215
GHS. *See* Gibbons, Hawking, and Stewart (GHS)
Gibbons, Hawking, and Stewart (GHS), 128–133
Gibbs, Willard, 91–92, 94, 96–97, 351–354
Gibbs entropy, 91–92, 94–95, 96, 97
God, conversation with, 59–60
Goldstein, Sheldon, 77–78, 103, 118–119, 213
governing, 6, 40, 224
graceful-exit problem, 123
gravitation, 29, 63, 73, 116, 209
gravitational and nongravitational degrees of freedom, distinguish between, 121
gravity, 16–17, 117, 119–122, 125–126
gravo-thermal catastrophe, 120
GRW. *See* GRW theory (Ghirardi, Rimini, and Weber)
GRW theory (Ghirardi, Rimini, and Weber): Mentaculus, 26; metaphysical accounts of laws, 232; Meta-Reversibility Objection, 155; Past Hypothesis and physical laws, 223, 232, 235, 237–238, 242–243; quantum Mentaculus, 218–219; Wentaculus, 223, 237
GUT (Grand Unified Theory) scale, 135
Guth, Alan, 111, 123–125

Hajek, Alan, 37, 45
Hamiltonian: Albertian demon, 260; classical Mentaculus, 209, 213–215; demon, how to make, 259–262, 267; flatness, 131; initial conditions, role of, 64; laws of nature, 237–238, 243; macropredictions of micro-

INDEX 383

dynamics, 79–80; special sciences, underwriting the laws of, 70; trajectories, 124–134; typicality, 193–196; Wentaculus, 220
Hard Part, 204–205, 206, 212, 213
heat, background, 2–3
heat energy, 249–251, 255–256, 263–265, 267
Hilbert-space: horizon problem, 118; nomic vagueness, 240; Past Hypothesis, the logic of, 79, 81; quantum Mentaculus, 215–216, 218; Wentaculus, 220, 222
history operator, 83–84
history super-operator, 83, 88
Hoefer, Carl, 50
horizon problem, 5, 111–112, 116–119, 121, 124, 135–137
horizon problem (casual version), 117
horizon problem (equilibration version), 117
H-theorem, 89, 212
Hubble distance, 116–117
Hubble Parameter, 115, 116, 118, 120, 128, 133
Hubert, Mario, 48–49, 50
Huggett, Nick, 7, 271–292, 372
Humeanism, 6, 50–51, 232–239, 242
Humean mosaic (HM), 41–42
Humean picture, the basic, 59–60
Humean probability measure, 182
Humean supervenience, 41, 192

ignorance, 327–328, 329
indifference accounts, statistical mechanics, 6, 143–144, 148–150, 150–151, 152, 162–163, 173
indifference approach: Meta-Reversibility Objection, 143–144, 162, 164, 166–167, 172; rational agents, 146; Reversibility Objection, replies to, 154, 157, 159–160; statistical mechanics, 149
Indifference Principle: Meta-Reversibility Objection, 143, 162, 163, 166, 170–172; rational agents, 146; Reversibility Objection, 153; Reversibility Objection, replies to, 154, 157; statistical mechanics, 149
inflation, 111–112, 118, 122–126, 132, 137
influence and control, 30–32, 33–34, 207–208
Information Gathering and Utilizing System (IGUS), 272–273, 279–282, 282–286, 286–289, 289–291, 291–292

informativeness, 5, 41–42, 49, 59–68, 167, 232–239, 366
informativeness, definition of, 60–62
informativeness and simplicity, measuring, 60–68
initial conditions, role of, 60–68, 62n, 63–64n
initial micro-randomness, 8, 302, 310
Initial Projection Hypothesis (IPH), 221–223, 228, 237–239, 243
initial randomness, 298–302, 303, 309, 310
intervention, 335, 336, 337, 341, 346, 364
intervention, concept of, 336–337, 341, 344
intervention, concept of (*Time and Chance*), 335–349
intervention arrow, 232
IPH. *See* Initial Projection Hypothesis (IPH)
irreversibility, microdynamical origins (classical case), 98–101
irreversibility, microdynamical origins (quantum case), 101–103
Ismael, Jenann, 364

Jeans instability, 119–120
Jeans length, 119–120

kinetic theory of heat, 251
knowledge and control, 33
knowledge and intervention, underwriting the asymmetries of, 7–8
knowledge asymmetry in *Time and Chance*: knowledge asymmetry, the, 291–292; presently surveyable condition, the, 273–276, 274nn4–5, 275n6, 276n; records and the PH, 276–279, 277n, 278n. *See also* memories; mnemonic knowledge
Kutach, Douglas, 317–318

Laplacian Demon, 7, 179, 253–254, 255
law of large numbers (LLN), 185
laws of nature: Albertian demon, 254; intervention, 336; Mentaculus, 17, 34–36, 356–358; Past Hypothesis and physical laws, 207, 224–226, 231, 237, 244; statistical mechanics, 58–65; typicality versus Humean probabilities, 182, 194

laws of nature, conflicts with concept of: boundary condition laws, 238–239; introduction, 237; nomic vagueness, 210f, 240–244, 240nn26–27, 241f; non-dynamical chances, 239–240
laws of special sciences, 69–73, 70–71n
Lazarovici, Dustin, 6, 48, 178–201
Lebesgue measure, 25, 147, 199
LEPH. *See* Low-Entropy Past Hypothesis (LEPH)
Lewis, David: causal representations and initial randomness, 295; chancemaking patterns, 183; Fly case, 322; Humean Best System Account, 40; Mentaculus, laws and probabilities in, 41–48; metaphysical accounts of laws, arguments from, 232–233; Nixon example, 35–37; Past Hypothesis, the logic of, 103; Principal Principle, 188; PROB and PH, on the status of, 58–60; statistical mechanics, 64, 67. *See also* Principal Principle (PP)
Lewis-Loewer theory of objective chance, 188
linear scalar perturbations, 133
Liouville distribution, 25, 48
Liouville measure: Albertian demon, 256, 257; canonical measure, 126–127; early universe, fine-tuning, 114–115; flatness, 131; horizon problem, 118; Mentaculus, the, 19–20; Mentaculus and Typitaculus, 46–50; Meta-Reversibility Objection, 160–162, 166–169; probability measures versus typicality measures, 189, 190; rational agents, 147–150; Reversibility Objection, 150–152; stationarity, 197; typicality measures, justification of, 195; typicality versus Humean probabilities, 180–184, 199; uniformity, 198
Liouville-smooth probability distribution, 90
Liouville's theorem, 7, 132, 256–258, 261, 265–266
L-laws, 64–65
local beables, 237
Loewer, Barry: canonical measure, 127; causes, randomness, and the Past Hypothesis, 294–306; classical Mentaculus, 214; Adam Elga, responding to, 363; exceptional cases, dealing with, 318; laws of special sciences, underwriting, 69–70; metaphysical accounts of laws, 233; Meta-Reversibility Objection, 142–143; Principal Principle versus Cournot's principle, 188; Reversibility Objection, replies to, 155; statistical mechanics, 58–65; *Time and Chance,* overview, 4–5; typicality versus Humean probabilities, 180–183
Low-Entropy Past Hypothesis (LEPH), 103–106, 369

macrodeterministic, 87, 94
macro history, 22, 23–24, 23f, 43
macrohistory α, 81
macrohistory space, 81
macropredictable, 7, 256–258, 264
macropredictions of microdynamics, 79–84
macroproperties, 81–82, 85–86, 90, 94, 95–96
macrostates: entropy of, 16, 18, 168; initial, 103, 167, 180, 206, 240, 257–263, 372
make money, 255–256, 257, 260
Malthus's exponential law, 38–39
Maudlin, Tim, 6–7, 187, 189, 234, 249–267, 370–371
Maxwell, James, 6, 18, 89, 249–255, 257, 264, 267, 371
Maxwell's demon: Albertian demon, 249–251, 253, 257, 264; introduction, 1, 4, 6; Tim Maudlin, 370; Mentaculus, 37
Maxwell's equations, 17, 226, 373
Meacham, Christopher J. G., 6, 142–175, 367, 368–369
measurements, definition of, 29
measure of simplicity, 59, 194, 240
measuring devices, 322, 358–359
melting, 18, 204, 257
memories, 279–291
Mentaculus, 13–51; Albert's responses, 354–357, 364, 367, 372; introduction, 8; laws and probabilities in, 40–46, 43n45, 45nn47–49; nomic vagueness, 244; overview, 13–14, 13n, 14n, 15nn3–5, 16nn6–8, 17n; Past Hypothesis and physical laws, 214, 226–244; probability map of the world, 17–26, 18nn10–14, 19nn15–16, 20nn17–18, 23f, 23nn22–23; probability versus typicality, 46–51, 47n, 50n; reading the past in the present, 275–291; three components of, 4; three ingredients, 14; time's arrows and, 26–40, 33nn31–32, 36n37, 37n, 38nn39–41; typicality versus Humean probabilities, 181–200; Typitaculus, difference between, 51

INDEX

Mentaculus, classical, 209–214
Mentaculus, quantum, 214–219
Meta-Reversibility Objection: μ-constraint, 168; against I2, 164–165; appendix, 172–175; background, 144–150; conclusion, 172; formalizing, 163–165, 163n; introduction, 142–144, 142n, 161–162, 162n; replies to, 165–172, 166nn31–32, 168n, 169n, 170n, 172n; reversibility objection, 156–160
Meta-Reversibility Objection, rational agents, 144–146, 145–146n6, 145nn3–4, 146n7
Meta-Reversibility Objection, reversibility objection: adequacy, 152; against I1, 153–154; against N1, 152–153; formalizing, 151–154, 151n, 152nn16–17; introduction, 150–151; replies to, 154–160, 154n19, 156n22, 157nn23–25, 160–161n28
Meta-Reversibility Objection, statistical mechanics, 146–150, 146n9, 147nn10–13, 148n
MEV. *See* Modern Entropic Version (MEV)
MEV, prospects for, 254–256
microdynamical origins of irreversibility: classical case, 98–101
microdynamical origins of irreversibility: quantum case, 101–103
microdynamics, macropredictions of, 79–84
micro history, 22–24, 26, 32, 34–36, 49, 223
microstates: actual, 99, 210, 215, 315; probability distribution over, 25, 50, 99, 205, 302–304, 351–353
Mill, John S., 58
minimal primitivism, 234
mnemonic knowledge: evolutionary response—a reply, 289–291, 290nn17–18; fine-tuning, 116; IGUS, 279–291, 279n, 280n, 281t; IGUS remembers, 281t, 282–286, 283nn12–13, 284n, 285n; Mentaculus, 286–289
Modern Entropic Version (MEV), 253–254, 254–256, 264, 267, 371
mother of all ready conditions, 29, 277, 285, 303, 316
mother of all ready states, 229, 308
Myrvold, Wayne, 357–358n

narratability, 236
Near Past Hypothesis (NPH), 227
Near Past postulate (NP), 73–74
Newcomb's Problem, 346–347
Newtonian gravitation, 29n, 63, 133, 209
Newtonian gravitational theory, 29n
Newtonian gravity, inverse-square law of, 120
Newtonian laws of motion, 314
Newtonian mechanics, 26, 60, 64, 262, 352, 363
Newtonian theory, 193, 240
Nixon example, 35–37
nomic approach, 143, 154, 157, 159–160
nomic vagueness, 240–244, 241f
non-dynamical chances, 239–240
non-Humean, 6, 40–41, 50, 207, 224, 233–235, 237, 239
notion of evidence, 158

objective probability, 44–45, 47–48, 101, 214, 223, 231
Oppenheim, Paul, 14

Package Deal Account (PDA), 43
PAμ. *See* Predictive Accuracy of Microphysics (PAμ)
Papineau, David, 38
Past Hypothesis (PH): Albert's responses, 352–373; asymmetry of influence, 314–315; asymmetry of records, 316; causes, randomness, 295, 302–310; definition of, 3; fine-tuning, 111–113, 127, 135–136; initial conditions, role of, 63–64; intervention, 336, 346; intervention, concept of, 336; Mentaculus, 14–16, 23–36; Meta-Reversibility Objection, 143–149, 160–172, 161; physical laws and, 205–227; reading the past in the present, 271; records and, 276–279; Reversibility Objection, 144, 160; typicality versus Humean probabilities, 180–181, 193, 197; Wentaculus, 221
Past Hypothesis, the logic of: coarse-grained dynamics, 84–90; coarse-grained dynamics, microdynamical underpinnings, 93–98; coarse-grained dynamics, time-reversibility in, 90–93; conclusion, 106–108; introduction, 76–79; irreversibility, classical case, 98–101; irreversibility, quantum case, 101–103; low-entropy past, 103–106; microdynamics, macropredictions of, 79–84

Past Hypothesis and physical laws: asymmetries, arguments from other, 228–232; conclusion, 244; counterfactual arrow, 231; dynamical governing, 224; dynamical law primitivism, 234; epistemic arrow, 230–231; epistemic guides, 235; fundamental law of nature, as a, 224–237; intervention arrow, 232; introduction, 204–208, 205nn1–2, 207nn4–5, 208nn7–8; laws of nature, conflicts with concept of, 237–244; metaphysical accounts of laws, 232–235, 234n, 235n; minimal primitivism, 234; Near Past Hypothesis, 227; quantum entanglement, 235–237; records arrow, 228–230, 229nn20–21, 230n; Second Law, 225–228, 225n, 226n. *See also* Boltzmann, variations on a theme from
Past law hypothesis, 102–103
Pearl, Judea, 38
Penrose, Roger, 16, 114–115, 123, 137, 206
perfectly natural properties, 41–43, 233
perturbation mode, measure on, 134
PH. *See* Past Hypothesis (PH)
physicalism, 13, 15–16, 227
physical laws. *See* Past Hypothesis and physical laws
Planck density, 137
Planck length, 137
P-laws, 64–68, 366
Poincaré's recurrence theorem (PR), 70
PP. *See* Principal Principle (PP)
prediction: causes, randomness, and the Past hypothesis, 296, 303; fine-tuning, 124; intervention, 337; introduction, 7; Mentaculus, 62; Past Hypothesis, the logic of, 95, 104; reading the past in the present, 277–278; time, flies, and why we can't control the past, 312–320, 328; typicality versus Humean probabilities, 183
Predictive Accuracy of Macrophysics (PAM), 105, 106, 366–367
Predictive Accuracy of Microphysics (PAμ), 105, 106, 367
presently surveyable condition, 273–276, 279, 282–284, 286–287
Prigogine, Ilya, 95
Principal Principle (PP): introduction, 6; Mentaculus, 42, 44–46, 49; Meta-Reversibility Objection, 145, 175; rational agents, 145; typicality versus Humean probabilities, 182–186, 186–188, 200
Principal Principle and the meaning of Humean probabilities, 182–186, 184n
Principal Principle versus Cournot's principle, 186–188
principle of indifference, 45, 186, 196, 200, 357, 367–369
PROB, 4, 57–74
probabilism, 145
probabilistic dependence, 337, 342–343
probability, deterministic, 188, 233, 239
probability, epistemic: causes, randomness, and the Past hypothesis, 298; introduction, 6; Mentaculus, 30, 44–45, 51; Meta-Reversibility Objection, 144–146, 152–158, 161–165, 173; Past Hypothesis, the logic of, 93; reading the past in the present, 287–288, 290; statistical mechanics, 69; typicality versus Humean probabilities, 187
probability, notion of, 181
probability, objective: intervention, 336; Mentaculus, 40; Meta-Reversibility Objection, 142–148, 160–166, 173; Past Hypothesis, the logic of, 101–105; Past Hypothesis and physical laws, 207, 232, 238–244; statistical mechanics, 58–59, 69; typicality versus Humean probabilities, 180–188, 200
probability, use of term in Mentaculus, 25–26
probability density, 14, 22, 26, 189
probability measure, 82, 181, 184, 192, 197, 212–213, 217
probability measures versus typicality measures, 189–191
probability of a history, defining, 88
PROB and PH, 4, 57–74. *See also* statistical mechanics, metaphysical foundations of
putative, use of term, 275
Putnam, Hilary, 14

quantum entanglement, new light on, 235–237
quantum field theory (QFT), 64
quantum gravity, 16–17, 118, 125, 206
quantum mechanics: Bohmian, 196–197, 199, 218–219, 223, 235–239, 243; Everettian, 79,

218, 223, 230, 235–239, 243; GRW, 26, 155, 218–219, 223, 232, 235, 237–238, 242–243; Past Hypothesis, the logic of, 79
quantum Mentaculus, 214–219, 215nn13–14, 217n15
quantum statistical mechanics, 210, 214–215, 219–220, 370

Ramsey, Frank, 58
rational agents, 144–146
reading the past in the present, 271–273, 271–272n2, 271n1, 272n3, 273–279, 279–291, 291–292
ready conditions, 29, 277, 285, 303, 306, 316
ready-state argument, 308–309
records and the PH, 276–279
records arrow, 228–230
redshift, 113, 116, 122
reduction, 8, 233, 255
regularity theory of probabilities, 184–186
retrodiction, 7, 69, 104, 213, 277–278, 296, 304, 359
Reversibility Objection, 6, 71, 143–144, 148, 150–154, 157–167, 172–174
reversibility paradox, 20, 21
reversibility problem, 3, 21
reversibility worries, 143, 157, 166, 169
Robertson-Walker universe, 129, 130, 132, 136
rolling dice, 90

Schrödinger equation, 125, 197, 215, 218–220, 244, 357
Schwarzschild black hole, 124
scientific explanation, 27, 207, 224, 226, 229, 232, 313
SDC. *See* Simple Dynamical Conjecture (SDC)
second law of thermodynamics: Albertian demon, 251–254; Albert's responses, 371; arguments from, 225–226; causes, randomness, and the Past hypothesis, 294, 306; introduction, 2, 14–17; limitation of, 251–258, 257n; Past Hypothesis and physical laws, 213; probabilistic version, 21–22; reading the past in the present, 277; time, flies, and why we can't control the past, 313; typicality versus Humean probabilities, 178–181
self-gravity, 115, 119, 121
seriality, 8, 327–330

set Mentaculus, 25
SH. *See* Skeptical Hypothesis (SH)
Shafer, Glen, 186
Shannon Entropy, 167, 168–169
Simple distribution, 97–98, 99, 101, 105, 107
Simple Dynamical Conjecture (SDC), 97–98, 100–101, 103–106, 367
Simple Past Hypothesis (classical version), 101
Simple Past Hypothesis (quantum version), 102
Simple Past Hypothesis (SPH), 101–106, 367
Simplicity: Albert's responses, 367; fine-tuning, 115; introduction, 5; Mentaculus, 41–43, 48–49; Past Hypothesis, the logic of, 81, 97–105; Past Hypothesis and physical laws, 213–215, 228–240; statistical mechanics, 59–65; time, flies, and why we can't control the past, 314; typicality versus Humean probabilities, 181–184, 188, 192–197
Simplicity, as a condition, 99–101
Simplicity criterion, 99–100
Skeptical Hypothesis (SH), 150, 166
Sklar, Lawrence, 78, 89
slow-roll inflation, 123
Sober, Elliot, 299
SP. *See* statistical postulate (SP)
spacelike surfaces, defining, 113
space-phase volume, 120, 308
special sciences, underwriting the laws of, 69–73
SPH. *See* Simple Past Hypothesis (SPH)
states, as records of the past, 316
state space: fine-tuning, 124; Mentaculus, 39; Past Hypothesis, the logic of, 79–82, 86, 94; Past Hypothesis and physical laws, 242; statistical mechanics, 66; typicality versus Humean probabilities, 193, 197–198
stationarity, 196–197
stationary distributions, 87, 92
statistical mechanics, 1–4, 78–79, 146–150. *See also* typicality versus Humean probabilities
statistical mechanics, metaphysical foundations of: account of, 4, 148, 152, 159; conclusion, 73–74; Humean picture, 59–60; informativeness and simplicity, 60–68, 62n, 63–64n; initial conditions, role of, 60–68, 62n, 63–64n; introduction, 57–59, 58n; laws of the special sciences, 69–73, 70–71n, 72n10; PROB and PH, 68–69

statistical mechanics, nomic accounts of, 143–144, 148, 160–162
statistical mechanics, philosophical foundations of, 4–7
statistical postulate (SP): asymmetry of influence, 315; canonical measure, 127; classical Mentaculus, 214; fly case, 320; intervention, concept of, 336–337, 341; introduction, 3; Mentaculus, 16, 40; metaphysical accounts of laws, 232–233; Meta-Reversibility Objection, 165; non-dynamical chances, 239–240; Past Hypothesis and physical laws, 205–206; probability versus typicality, 46–48, 51; quantum Mentaculus, 218–219; second law, 228
superposition, 98, 216, 220
surface of last scattering, 112, 113, 116, 122, 134, 136
symmetry, definition of, 198–199
synchronous gauge, 113

temporal asymmetries, 15–16, 17, 33, 170. *See also* time's arrows
temporal invariance, 2, 3
temporally asymmetric processes, 3
tensor product, 86
thermal equilibrium, 89, 90, 210–211, 216, 221, 250
thermodynamic arrow of time, 180, 214, 228
thermodynamic entropy, 91, 103, 106–107, 204, 305
thermodynamic equilibrium, 18, 341
thermodynamics, definition of, 2–3, 18
Time and Chance (Albert), 1, 4, 57–58, 272–273, 335–349
time-asymmetric, 79, 90–91, 205, 291, 298, 301
time-asymmetry of intervention, Albert's responses, 364
time-reversal invariance: Albertian demon, 266; Albert's responses, 352, 359, 362–363; causes, randomness, and the Past hypothesis, 294, 302; Mentaculus, 15–17; Meta-Reversibility Objection, 158, 170; Past Hypothesis, the logic of, 77–85, 91–107; Past Hypothesis and physical laws, 204, 213, 232; time, flies, and why we can't control the past, 313–314; typicality versus Humean probabilities, 179

time-reversal operator, 79–80, 91
time-reversibility, 79, 90–93, 91n, 92n, 99, 304
time's arrows, 26–40, 33nn31–32, 36n37, 37n, 38nn39–41
trajectories, fine-tuning early universe, 124–135, 125n, 127n, 136–137, 136n
transition probability, 82
Turing machine, 280, 283, 285, 292
typical, defining, 190–191
typical frequencies, 185, 188, 190
typicality: Albert's responses, 369–370; introduction, 6; Mentaculus, 25, 46–47; Past Hypothesis, the logic of, 107; Past Hypothesis and physical laws, 205–222, 239; versus Humean probabilities, 178, 181–182, 187, 189–201
typicality, concept of, 46
typicality, epistemic and metaphysical status of, 191–195
typicality measure, 181, 191–195, 195–200
typicality measures, justification of, 195–200, 197f, 197n
typicality measures versus probability measures, 189–191
typicality postulate (TP), 46–47, 50–51, 206, 239–240
typicality versus Humean probabilities: conclusion, 200–201; epistemic and metaphysical status of typicality, 191–195, 192n; microscopic laws to macroscopic regularities, 178–182, 179n, 180n; Principal Principle and the meaning of Humean probabilities, 182–186; Principal Principle versus Cournot's principle, 186–188; probability measures versus typicality measures, 189–191, 189n, 190f; regularity theory of probabilities, 184–186; typicality measures, 195–200
Typitaculus, 46–47, 50–51

underwriting: Albert's responses, 367, 373; causes, randomness, and the Past hypothesis, 295; introduction, 4, 7; Mentaculus, 25; statistical mechanics, 58, 69, 73
uniformity, 107, 113, 121, 134, 198–199
Uniform Past Hypothesis (UPH), 105–106
untraceable arbitrariness, 240–241, 243, 244
UPH. *See* Uniform Past Hypothesis (UPH)

veridicality, 71–73, 151, 275–276, 280, 282, 287–288, 290

Wallace, David, 5, 76–108, 364, 366–367
wave function, 95–96, 98, 125–126, 136, 196–199, 215–221, 236, 244

Wave Function Realism, 220
Wentaculus, 222–223, 223nn16–17, 243
Weslake, Brad, 1–9, 186, 364
Weyl Curvature Hypothesis, 206
Wheeler-DeWitt equation, 125, 236
Winsberg, Eric, 1–9, 57–74, 227, 364, 365–366